D0218402

Environmental Change: Key Issues and Alternative Perspectives

Environmental Change: Key Issues and Alternative Approaches describes and explains the current and future significance of past and contemporary environmental, and especially climatic, change. It outlines the conceptual framework for studies of environmental change by posing a series of key questions and presenting the results of the most recent and relevant research. It thereby establishes a basis for evaluating projections of future environmental change.

The crucial role of modelling and the basic concepts and techniques used to research past environmental change are introduced for non-specialists in the field. The main time frame is the last 400 000 years, with special emphasis on past periods of rapid warming, on the nature of climatic variability over the last 1000 years and on the dramatic and accelerating changes in the Earth system heralded by the industrial revolution. The book is extensively referenced and illustrated.

This book provides a balanced basis for understanding and exploring the scientific issues underlying global change for advanced undergraduates in environmental, earth, biological and ecological sciences and geography. By providing a non-specialist introduction to models in environmental change research and to the study of past environmental changes, it seeks to draw a wider group of students into the field.

FRANK OLDFIELD is an Emeritus Professor and Senior Research Fellow at the Department of Geography, University of Liverpool. He obtained his first degree at the University of Liverpool in 1956 and his Ph.D. at the University of Leicester in 1962. Between 1958 and 1975, he held academic positions at various universities in the UK, and following a brief period as Acting Vice-Chancellor at the University of Papua New Guinea, was Director of the School of Independent Studies with a Personal Chair in Geography at the University of Lancaster. From 1973 to 1996, he was John Rankin Professor of Geography at the University of Liverpool, holding visiting positions at institutions across the world. Until 2001, he was Executive Director of the International Geosphere-Biosphere Programme (IGBP) Past Global Changes (PAGES) Core Project in Bern, Switzerland, where he was responsible for promoting and coordinating environmental research worldwide. He was also on the editorial team responsible for the writing, coordinating and finalising the IGBP Synthesis.

During this time, Professor Oldfield's research has covered many areas of environmental change, including palynology, palaeolimnology, environmental magnetism and the study of man-made radioisotopes in sediment and peat chronologies. More recently, his research interests have included coastal contamination from nuclear fuel reprocessing, the mass-balance in ombrotrophic peat bogs in response to climate changes and the human impact on global environmental change. The over-riding theme uniting these diverse research methodologies has been to document, quantify and understand past and current environmental changes at ecosystem level resulting from the interaction between natural variability and human activities.

In recognition of his research, he received the Linton Award from the British Geomorphology Research Group in 1992 and the Murchison Prize from the Royal Geographical Society and Institute of British Geographers in 1995. He was elected President of the Quaternary Research Association between 1994 and 1997, and was awarded an honorary Doctor of Science Degree by the University of Plymouth in 2000. He has also served on numerous national and international committees linked to research, as well as on the editorial boards of leading journals.

Environmental Change

Key Issues and Alternative Perspectives

Frank Oldfield
University of Liverpool

CAMBRIDGE
UNIVERSITY PRESS

CAMBRIDGE UNIVERSITY PRESS
Cambridge, New York, Melbourne, Madrid, Cape Town, Singapore, São Paulo

CAMBRIDGE UNIVERSITY PRESS
The Edinburgh Building, Cambridge CB2 2RU, UK
Published in the United States of America by Cambridge University Press, New York

www.cambridge.org
Information on this title: www.cambridge.org/9780521829364

© F. Oldfield 2005

This book is in copyright. Subject to statutory exception
and to the provisions of relevant collective licensing agreements,
no reproduction of any part may take place without
the written permission of Cambridge University Press.

First published 2005

Printed in the United Kingdom at the University Press, Cambridge

A catalogue record for this book is available from the British Library

Library of Congress Cataloguing in Publication data

ISBN-13 978-0-521-82936-4 hardback
ISBN-10 0-521-82936-4 hardback

Cambridge University Press has no responsibility for
the presistence or accuracy of URLs for external or
third-party internet websites referred to in this book,
and does not guarantee that any content on such
websites is, or will remain, accurate or appropriate.

For my wife Mary

Contents

Preface

The nature of the science we undertake hinges crucially on the kinds of question we ask. Over the last four decades, the questions that have dominated the concerns of a substantial part of the environmental sciences research community have changed dramatically, first in response to the realisation that human activities were creating contemporary problems that cried out for deeper understanding and urgent solutions. More recently, and especially since the late 1980s, the research agenda has become increasingly dominated by the over-arching issues of global change and its implications for the future. These issues are the starting point for, and underlying motif throughout this book. I begin by defining some key questions, then proceed to examine the different approaches taken by the whole range of research communities that consider global change from a predominantly environmental science-based perspective. The book then considers past, current and possible future changes before returning to a reconsideration of the questions posed at the outset.

I hope that this book will serve as both stimulus and orientation. It rests on a personal perspective developed over the past 50 years spent in the fascinating business of trying to understand better the course, nature, causes and consequences of environmental change. I have had the good fortune to be inspired and guided along the way by a great many colleagues whose friendship, insight and generosity deserve more thanks than I have space to record. Several members of my personal pantheon require special acknowledgement. Donald Walker, who supervised my first attempts at postgraduate research in the Cambridge Botany School, set standards of originality, caution and intellectual honesty that I have never forgotten, though rarely, if ever, attained. At a time when I was being drawn into research fields for which I was all too poorly prepared, John Mackereth inspired me with an unrivalled combination of patience and shear genius. Without his help and that of Roy Thompson I would have floundered even more than I actually did. Although my meetings and conversations with Ed Deevey were all too few, each one was a revelation and a new beginning. I came to place incalculable value on his wisdom and guidance. Gordon Manley's combination of painstaking empirical research and marvellous flights of fancy spun magic for all who knew him. More recently, Ray Bradley gave me more of his time, insight and critical acumen than I had any right to expect at the stage when I was desperately trying to get to grips with the wider, and to me, still unfamiliar issues of past global change. To these and many others I record my heartfelt thanks.

My main aims in writing the book are to:

- Set concerns about current trends and future changes into longer-term perspective.
- Provide, within the limitations of individual insight, a comprehensive and balanced view of the key issues surrounding present and future global change.
- Encourage students and younger scientists to develop a greater awareness of the influence that particular methodologies and frames of reference have on the nature of environmental change science.
- Heighten awareness of the strengths, limitations and complementarity of different approaches.
- Highlight the need for research that bridges methodologies and not merely disciplines.

Acknowledgements

I am especially grateful to Keith Alverson, John Dearing and Jan Bloemendal for finding time to read parts of the manuscript, thereby sparing me more blunders than have survived, as well as for making many positive suggestions.

Sandra Mather and Suzanne Yee have patiently and skilfully prepared all the diagrams and I am eternally grateful to them for this, and for bearing with my frustrating indecision and impossible requests.

List of Permissions

Figure 2.3   Modified from Wasson, R.J. and Claussen, M. Earth system models: a test using the mid-Holocene in the southern hemisphere. *Quaternary Science Reviews* **21** (2002), 819–24, with permission from Elsevier.

Figure 3.4   Based on Blaauw, 2003, with kind permission of the author.

Figure 3.6   Modified from Zillen, 2003, with kind permission of the author.

Figure 4.1   Modified from Jones, P.D., Briffa, K.R., Osborn, T.J., Moberg, A. and Bergström, H. Relationships between circulation strength and the variability of growing season and cold-season climate in northern and central Europe. *The Holocene*, **12** (2002), 643–656, with permission from Edward Arnold.

Figure 4.2   (a) Modified from Holzhauser, H. and Zumbuehl, H.J. Reconstruction of minimum glacier extensions in the Swiss Alps. *PAGES Newsletter*, **10**:3 (2002), 23–25.
(b) Modified from Mock, C.J. Documentary records of past climate and tropical cyclones from the southeastern United States. *PAGES Newsletter*, **10**:3 (2002), 20–21.
(c) Modified from Brazdil, R., Glaser, R., Pfister, C. and Stangl, H. (2002). Floods in Europe – a look into the past. *PAGES Newsletter*, **10**:3 (2002), 21–23.

Figure 4.3   (a) Adapted from Grudd, H., Briffa, K., Karlen, W., Bartholin, T.S., Jones, P.D. and Kromer, B. A 7400-year tree-ring chronology in northern Swedish Lapland: natural climatic variability expressed on annual to millennial timescales. *The Holocene*, **12** (2002), 657–667, with permission from Edward Arnold.

Figure 4.4   (a) From Heikkila, M. and Seppä, H. A 11 , 000 yr palaeotemperature reconstruction from the southern boreal zone in Finland. *Quaternary Science Reviews*, **22** (2003), 541–554, with permission from Elsevier.

Figure 4.5   Modified from Bigler, C., Larocque, I. Peglar, S.M., Birks, H.J.B. and Hall, R.I.. Quantitative multiproxy assessment of long-term patterns of Holocene environmental change from a small lake near Åbisko, northern Sweden. *The Holocene*, **12** (2002), 481–496, with permission from Edward Arnold.

Figure 5.5   Modified from Ruddiman, W. F. Orbital insolation, ice volume and greenhouse gases. *Quaternary Science Reviews* **22** (2003), 1597–1629, with permission from Elsevier.

Figure 5.6   Adapted from Labeyrie, L., Cole, J., Alverson, K. and Stocker, T. The history of climate dynamics in the late Quaternary. In: Alverson, K., Bradley, R. S. and Pedersen, T. F. (eds). *Paleoclimate, Global Change and the Future*. Berlin, Springer Verlag, 2003.

Figure 5.7   Seki, O., Ishiwatari, R. and Matsumoto, K. Millennial scale oscillations in NE Pacific surface waters over the last 82 kyr: new evidence from alkenones. *Geophysical Research Letters* **29**:23 (2002), 2144. Copyright [2002] American Geophysical Union; modified by permission of American Geophysical Union.

Figure 6.1   Adapted from Labeyrie, L., Cole, J., Alverson, K. and Stocker, T. The history of climate dynamics in the late Quarternary. In: Alverson, K., Bradley, R. S. and Pederson, T. F. (eds). *Paleoclimate, Global Change and the Future*. Berlin, Springer Verlag, 2003.

Figure 6.5   Reprinted with permission from Monnin, E., Indermühle, A., Dällenbach, A. *et al*. Atmospheric CO_2 concentrations over the last glacial termination. *Science* **291** (2001), 112–114. Copyright [2001] AAS.

Figure 6.6   The GRIP, Huascaran and Vostok graphs are from: Raynaud, D., Blunier, T., Ono, Y., and Delmas, R. J. The late Quaternary history of atmospheric trace gases and aerosols: interactions between climate and biogeochemical cycles. In Alverson, K. D., Bradley, R. S. and Pedersen, T. F. (eds.). *Paleoclimate, Global Change and the Future*. Berlin, Springler-Verlag, 2003. The Dome C graphs are from: Delmonte, B., Petit, J. R. and Maggi, V. Glacial to Holocene implications of the new 27 000-year dust record from the EPICA Dome C (east Antartica) ice core. *Climate Dynamics* **18** (2002), 647–660.

Figure 6.7   Modified from Stocker, T. F. Global change: south dials north. *Nature* **424** (2003), 496–499.

Figure 6.8   Modified from Ridgwell, A. J. Implication of the glacial CO_2 'iron hypothesis' for Quarternary climate change. *Geochemistry, Geophysics, Geosystems – Research Letter* **4**:9 (2003), 1–10.

Figure 7.1   (b) From Heikkila, M. and Seppä, H. A 11,000 yr palaeotemperature reconstruction from the southern boreal zone in Finland. *Quarternary Science Reviews* **22** (2003), 541–554, with permission from Elsevier.

(c) From Anderson, C., Koc, N., Jennings, A. and Andrews, J. T. Non-uniform response of the major surface currents in the Nordic

Seas to insolation forcing: implications for the Holocene climate variability. *Paleoceanography* **19** (2004), PA2003, 1–16. Copyright [2004] American Geophysical Union; modified by permission of American Geophysical Union.

(d), (e), (f) From Fisher D. A. and Koerner, R. M., Holocene ice core climate history: a multi-variable approach. In: Mackay, A. W., Battarbee, R. W., Birks, H. J. B. and Oldfield F. (eds.). *Global Change in the Holocene*. London, Edward Arnold, 2003.

Figure 7.2   **(a)** Modified from Gasse, F. Hydrological changes in the African tropics since the last glacial maximum. *Quaternary Science Reviews* **19** (2000), 189–211.

(b) Modified from Overpeck, J. T., Anderson, D., Trumbore, S. and Prell, W. The southwest Indian monsoon over the last 18 000 years. *Climate Dynamics* **12** (1996), 213–225.

(c) Modified from Nuñez, L., Grosjean, M. and Cartajena, I. Human Occupations and climate change in the Puna de Atacama, Chile. *Science* **298** (2002), 821–824. Copyright [2002] AAS.

Figure 7.3   Modified from Overpeck, J., Whitlock, C. and Huntley, B. Terrestrial biosphere dynamics in the climate system: past and future. In Alverson, K. D., Bradley, R. S. and Pederson, T. F. (eds.). *Paleoclimate, Global Change and the Future*. Berlin, Springer Verlag, 2003.

Figure 7.5   Modified from deMenocal, P. B., Ortiz, J., Guilderson, T. *et al.* Abrupt onset and termination of the African humid period: rapid climate responses to gradual insolation forcing. *Quarternary Science Reviews* **19** (2000), 347–361. Copyright (2000), with permission from Elsevier.

Figure 7.7   (a) Adapted from Moy, C. M., Seltzer, G. O., Rodbell, D. T. and Anderson, D. M. Variability of El Niño /Southern oscillation activity at millennial timescales during the Holocene epoch. *Nature* **420** (2002), 162–165.

(b) and (c) Adapted from, Gagan, M. K., Hendy, E. J., Haberle, S. G. and Hantaro, W. S. Post-glacial evolution of the Indo-Pacific warm pool and El Niñ-Southern oscillation. *Quarternary International* **118–119** (2004), 127–143. Copyright (2004), with permission from Elsevier.

Figure 7.8   Modified from Labeyrie, L., Cole, J., Alverson, K. and Stocker, T. The history of climate dynamics in the late Quarternary. In: Alverson, K., Bradley, R. S. and Pedersen, T. F. (eds). *Paleoclimate, Global Change and the Future*. Berlin, Springer Verlag, 2003.

(b) Adapted from Moore, G. W. K., Holdsworth, G. and Alverson, K. Extra-tropical responses to ENSO 1736–1985 as expressed in an

Figure 12.5   Modified from Cazenave, A. and Nerem, R. S. Present-day sea-level change: observations and causes. *Review of Geophysics* **42** (2004), RG3001. Copyright [2004] American Geophysical Union; modified by permission of American Geophysical Union.

Figure 12.6   (a) Modified from Siddall, M., Rohling, E. J., Almogi-Labin, A. *et al*. Sea-level fluctuation during the last glacial cycle. *Nature* **423** (2003), 853–858.
(**b**) Modified from Goodwin, I. D. Unravelling climatic influences on late Holocene sea-level variability. In Mackay, A., Battarbee, R. W., Birks, H. J. B. and Oldfield, F. (eds.). *Global Change in the Holocene*. London, Edward Arnold, 2003, pp. 406–421.

Figure 12.8   (a) Modified from Kunkel, K. E., Easterling, D. R., Hubbard, K. and Redmond, K. Temporal variations in frost-free season in the United States:1895–2000. *Geophysical Research Letters* **31** (2004), L0321. Copyright [2004] American Geophysical Union; modified by permission of American Geophysical Union.
(b) Modified from Root, T. L., Price, J. T., Hall, K. R., Schneider, S. H., Rosenzweig, C. and Pounds, J. A. Fingerprints of global warming on wild animals and plants. *Nature* **421** (2003), 57–60.

Figure 12.10   (b) Modified from Broccoli, A. J., Dixon, K. W., Delworth, T. L., Knutson, T. R. and Stouffer, R. J.. Twentieth-century temperature and precipitation trends in ensemble climate simulations including natural and anthropogenic forcing. *Journal of Geophysical Research* **108** (2003), ACL 16–1–16–13. Copyright [2004] American Geophysical Union; modified by permission of American Geophysical Union.

Figure 12.11   Modified from Hegerl, G. C., Crowley, T. J., Baum, S. K., Kim, K-Y. and Hyde, W. T. Detection of volcanic and greenhouse gas signals in paleo-reconstructions of northern hemisphere tempera-ture. *Geophysical Research Letters* **30**:5 (2003), 46–1–46–4. Copyright [2004] American Geophysical Union; modified by permission of American Geophysical Union.

Figure 12.12   Modified from Jones, P. D. and Mann, M. E. Climate over past millennia. *Review of Geophysics* **42** (2004), 2003RG000143. Copyright [2004] American Geophysical Union; modified by permission of American Geophysical Union.

Figure 13.1   Modified from Schneider, S. H. Can we estimate the likelihood of climatic changes at 2100? *Climatic Changes* **52** (2002), 441–451.

Figure 13.5   Modified from Stott, P. A. and Kettleborough, J. A. Origins and estimates of uncertainty in predictions of twenty-first century temperature rise. *Nature* **416** (2002), 723–726.

Chapter 1
Defining and exploring the key questions

1.1 Global changes present and past

Any of our ancestors living a full three score years and ten in Western Europe some 11 600 years ago would have experienced, during their life time, truly remarkable changes in climate. Evidence from that time shows that the main changes took place over a period of 50 years at most. Although different lines of evidence give different figures for the degree of warming, it would be difficult to argue for an increase of less than 4 °C in mean annual temperature over much of Western Europe. In many areas, the shift would have been substantially greater. Parallel changes varying in nature and amplitude, but often synchronous in timing, took place over much of the Earth. The stratigraphic signal of these changes in the records from sediments and ice cores marks the transition from glacial times to the opening of the Holocene, the interglacial in which we live. Here then, was a period of rapid 'global change'. We may infer from this that there is nothing so very special about what we now think of as global change, that is, the current and impending changes in the Earth system driven by human activities. We would be quite wrong. The changes under way at the present day are of a different kind. At this stage, we need consider only three key differences:

- The rate of change in atmospheric CO_2 concentrations exceeds the mean rate during glacial–interglacial transitions by one to two orders of magnitude (Raynaud *et al.*, 2003).
- The human population is now many orders of magnitude greater than it was at the opening of the Holocene.
- The degree to which the full range of human activities has transformed the world and the way it functions, especially over the last 50 years, has created a biosphere with no past analogues. Humans have become agents of change with diverse and increasing impacts on almost every aspect of the Earth system.

Another way of highlighting the unique nature of the global changes currently under way is to consider present-day and projected greenhouse-gas concentrations in relation to the values typical of the Earth system over the last four glacial cycles. Figure 1.1 shows how far outside this envelope of recent natural

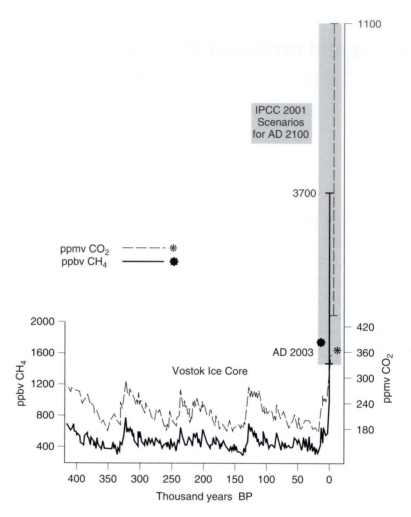

Figure 1.1 Past, present and the range of projected future atmospheric concentrations of carbon dioxide and methane. (Modified from Alverson *et al.*, 2001.)

variability current and projected future atmospheric concentrations of carbon dioxide and methane lie. Trying to tease out the climatic and broader environmental implications, as these values surge beyond what were their natural limits over the last four glacial cycles, will be a recurrent theme throughout the book. Contemporary and future global changes serve as substantive themes in their own right. In addition, by sharpening the focus on those lines of inquiry most likely to shed light on the *processes* involved in present and future changes, they also provide criteria of relevance in relation to evidence from the past.

1.2   Earth-system science

Central to any exploration of environmental change is the concept of an integrated Earth system of which climate is a part and in which human activities now play a major, integral role (Figure 1.2). As we shall see from numerous examples throughout the book, climate interacts with all the other components of the Earth system in complex ways involving multiple feedbacks between ocean, terrestrial biosphere, cryosphere and atmosphere. To view climate as a self-contained atmospheric component of global change, separated from all the other processes that interact with it, is unrealistic. Equally, to view human activities as separated from, or running counter to the 'natural' world, rather than as key agents in the processes that are transforming the Earth system, would be to revive an earlier and now totally inappropriate conceptual framework.

It follows from the above that any realistic scientific appraisal of present and future global change, indeed the whole field of study that we are addressing under the heading 'global change' rests on the emerging field of Earth system science. Using the definition given in Steffen *et al.* (2004): 'The Earth System has come to mean, broadly, the suite of interacting physical, chemical, and biological global-scale cycles (often called biogeochemical cycles) and energy fluxes which provide the conditions necessary for life on the planet.'

A comprehensive account of Earth-system science lies outside the scope of this book. Nevertheless, it forms the framework within which the key questions, research themes and case studies will be considered. Some of the main processes involved and some generic characteristics of the Earth system and its major, interacting components, including human populations, are set out below.

1.2.1 Key Earth-system processes

Central to any understanding of how the Earth system operates and how it changes on timescales of greatest relevance to future human welfare, are the processes that control changes in the fluxes of energy within the atmosphere and between atmosphere, hydrosphere, cryosphere and biosphere. These in turn largely control climate. Figure 1.3 is a schematic representation of the main processes driving change in the climate system. Even at this level of general-isation, many processes and transformations are involved. Several questions arise from this schematic diagram. What modulates changes through time in the external input of energy through solar radiation? What are the main ways in which the various processes shown in Figure 1.3 can be changed? How do the most important mechanisms redistributing incoming energy around the globe vary through time?

Solar energy received at the top of the atmosphere averages approximately 1365 W m^{-2}. Only since 1980 has it been possible to make direct measurements of the way in which this solar 'constant' actually varies through time as a result of quite a diverse range of processes (see e.g. Beer *et al.*, 2000). For periods further back in time and on longer timescales, changes in incoming solar radiation have to be inferred from proxy measurements (see Section 4.3.2) and from calculations based on the behaviour of the Solar System (see 4.3.1). Figure 4.6 shows aspects of

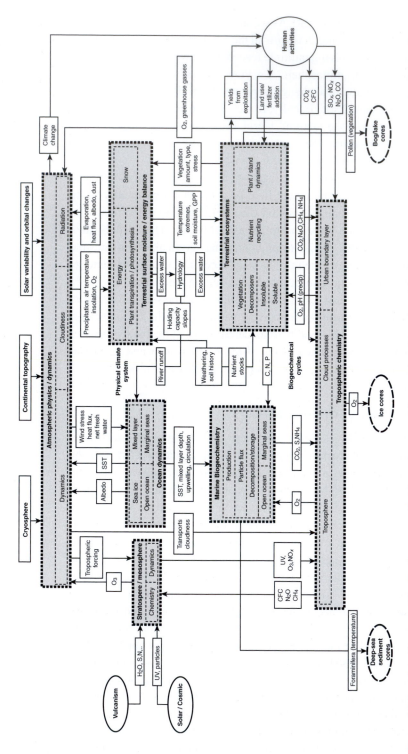

Figure 1.2 A wiring diagram of the Earth system. Key boundary conditions and major external forcings are at the top and left-hand side of the diagram. Human activities are indicated down the right-hand side. Some of the main environmental archives recording past changes are shown along the bottom. (Based on Schellnhuber, 1999.)

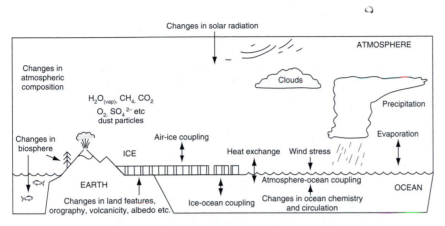

Changes in solar radiation

ATMOSPHERE

Changes in
atmospheric
composition

Clouds

$H_2O_{(vap)}$, CH_4, CO_2
O_2, SO_4^{2-} etc
dust particles

Precipitation

Changes in
biosphere

Air-ice coupling

Evaporation

ICE

Heat exchange   Wind stress

EARTH

Atmosphere-ocean coupling

OCEAN

Changes in land features,
orography, volcanicity, albedo etc.

Ice-ocean coupling

Changes in ocean chemistry
and circulation

Figure 1.3 Schematic diagram of the main components of the climate system, showing many of the linkages and exchanges that play an important role in climate variability. (Based on Bradley, 1999.)

the reconstructed variability on a range of timescales from sub-annual to multi-millennial. As the timescale lengthens, the importance of low-frequency changes in the Earth's orbit increases. These changes are now considered the likely 'pace-makers' of the major switches between glacial and interglacial conditions that dominate the recent climate history of the Earth (see 3.4.1 and 5.2.1).

Although the existence of significant links between solar variability and climate change on Earth is generally accepted, the extent to which it dominates other factors is controversial. Moreover, the variations in energy received at the top of the atmosphere are so small in relative terms that much uncertainty surrounds the mechanisms responsible for the links on all timescales.

The above account deals only with the incoming arrow at the top of Figure 1.3. We must next consider briefly the major forces responsible for redistributing the energy received around the globe – the large-scale systems of atmospheric and ocean circulation. These are shown schematically in Figures 1.4 and 1.5. They serve here mainly as points of reference for subsequent sections of the text, though several initial observations can be made. The response times of the two types of system vary across many orders of magnitude. The radiative balance of the atmosphere equilibrates in a matter of hours or days, whereas in the deep ocean it may take hundreds of years. The ocean provides both storage and transport of heat on a massive scale. Coupling through time between these two systems is complex. Both systems are strongly influenced by the actual config-uration of land and sea, leading to major contrasts in circulation between, for example, the atmosphere in the northern and southern hemispheres and between the Atlantic and Pacific oceans. In the case of the atmosphere, one of the main contrasts arises from the different thermal capacity of the land masses and oceans, leading to the dominance of monsoon-type climate systems in lower latitudes in many parts of the northern hemisphere. In the case of the oceans, the main contrasts arise from the dominance of the Atlantic by the system of meridional overturning circulation (see, e.g., Ganachaud and Wunsch, 2000.) and from the uniquely powerful westerly circum-polar flow in the southern

Figure 1.4 A schematic diagram of the main components of atmospheric circulation. H – high pressure; L – low pressure; ICTZ – inter-tropical convergence zone. (From www.planetearthsci.com.)

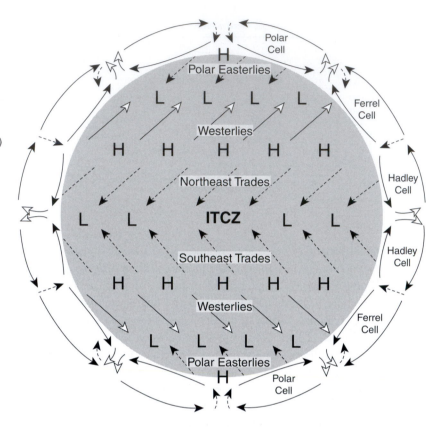

hemisphere. The major systems of persistent energy flux (in the atmosphere: the westerlies, the tropical anticyclone cells, the equatorial Hadley cell; in the oceans: the thermohaline circulation for example) all include complex subsystems that vary continuously in terms of their spatial domains and relative strength. All aspects of each system are subject to modulation, not only by the strength and distribution of the external energy flux from the Sun, but also by feedback processes linking the atmosphere and oceans and operating between them and the other components of the Earth system.

The last of the above points leads to a preliminary review of some of the additional feedback processes involved in regulating the fluxes of energy in the Earth system and the all-pervasive climate component. At the interface between land and atmosphere, one of the most important factors is surface albedo, which controls the amount of incoming radiation that is retained or reflected back into the atmosphere. The reflected fraction can vary from around 80% where snow cover is continuous, to around 40% for un-vegetated hot deserts and to well below 20% in densely forested regions. Changes in climate that affect snow and ice cover and vegetation thus lead to changes in albedo that may reinforce or counter the initial change. Where ice extent varies through time over the ocean, there are at least equally significant consequences. Not only does this change the

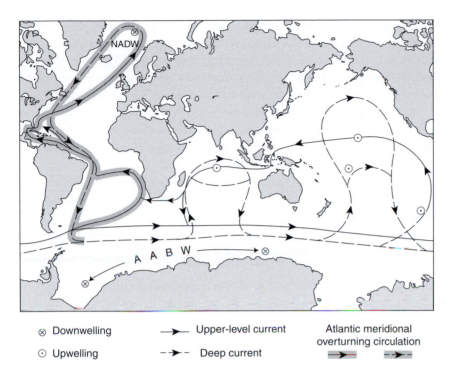

Figure 1.5 The main
features of ocean
circulation. NADW – North
Atlantic deep water;
AABW – Antarctic bottom
water. The circulation
within the Atlantic is
marked by southerly
flowing deep water and
northerly flowing upper-
level water. Together
these comprise the
meridional overturning
circulation system, which,
in turn exerts a dominant
influence on the
thermohaline circulation
that links the Atlantic
circulation to that in the
Indian and Pacific oceans
(Based on WMO/WCRP
sources.)

albedo and thermal capacity of the surface, it also modifies salinity, hence the
density gradients within the water column. This can affect heat exchange
between surface and deep water and the pattern of ocean circulation. These, in
turn, have important implications for climate.

Feedbacks arise from processes within the atmosphere and the oceans as well
as in the boundary layer between the two. Clouds have the effect of both
screening the Earth from incoming radiation and reflecting back to Earth outgoing long-wave radiation. The balance between the two varies for any given
cloud type and the altitude at which it develops. Indeed, the role of clouds in the
climate system continues to be notoriously difficult both to measure and to
model. Dust and aerosols also serve to absorb and scatter incoming radiation.
Over the last two decades especially, the role of greenhouse gases and more
recently aerosols has attracted great attention. Water vapour, carbon dioxide and
methane are the most important and commonly considered greenhouse gases.
Recent, rapid increases in the atmospheric concentration of the latter two as a
result of human activities are beyond doubt. The rate of any future increases
remains uncertain. The precise implications of any given increase in terms of
surface warming remain controversial, partly because quantifying the past and
likely future role of water vapour as a greenhouse gas has proved so elusive. The
whole question of greenhouse feedback will thread through later chapters.

Simply to calculate the implications of increased CO_2 emissions for atmospheric concentrations, without any regard for wider consequences, requires an

understanding of the carbon cycle. This involves every major component of the Earth system. There are major carbon reservoirs in the terrestrial biosphere and soils as well as in the oceans and atmosphere (see Figure 5.3). Fluxes between these reservoirs are extremely difficult to calculate on a global scale even for the present day, let alone the past or future. In consequence, current best estimates of the main components of the present-day global carbon budget are subject to large uncertainties. The main controls on the system include complex biological and geological processes that interact on timescales ranging from seasonal through inter-annual and decadal to multi-millennial.

Despite the complexity of the Earth system, there are some generic characteristics of environmental systems that can be introduced at the outset and these are considered next.

1.2.2   Some generic characteristics of environmental systems

Although a great deal of research in environmental science still uses concepts of simple cause and effect, as well as linear approximations to forcing–response relationships, we are increasingly obliged to realise that the Earth system does not behave in this way. Most of the changes that we seek to understand involve non-linear responses to whatever set of processes influences them; that is to say, there is no simple proportionality between the driving force and the response. Nor is there necessarily any immediately detectable response to forcing. What we often see is a delayed response reflecting the transgression of some kind of threshold in the system. Below the threshold, the response may be within the range of variability typical of the unforced state. Once the threshold is crossed, the response may then be out of all proportion to the forcing. This is due to the power of internal feedback mechanisms that may interact to amplify the effects of the original forcing, once the threshold beyond which they can maintain the unforced state is crossed. Thresholds thus combine with the intrinsic (i.e. internal), non-linear system dynamics to create lags, hysteresis (whereby the trajectories that systems follow as they are perturbed from an initial state differ from those followed as they recover from the perturbation), and disproportionate responses. This is a dangerous combination, for it confronts us with the possibility that there may be critical thresholds in the future of which we have, as yet, no warning, but which, once crossed, may herald major, non-linear responses. Part of the challenge is therefore to identify crucial thresholds and characterise key feedback mechanisms in the Earth system so that those posing the greatest dangers to human populations may be avoided.

This task is made all the more difficult by other features of the Earth system characterised as emergent behaviour and contingency. Many components of the Earth system at all spatial scales, when unperturbed, tend to evolve towards increasing internal order, complexity and differentiation. For example, within a developing ecosystem, the minor environmental variations that may exist on a newly exposed land surface evolve divergently to become expressed as spatial variations in soil profiles, vegetation and food webs. One implication of emergent behaviour is that many features of the Earth system that we see at the present day are contingent upon their histories. This adds greatly to the difficulties involved in developing quantitative predictive statements from a limited number of studies.

1.3   The key issues: a preliminary analysis

What are the first-order issues of greatest importance to environmental scientists addressing the challenge of global change? The following paragraphs attempt to define and briefly explore these in light of the foregoing introduction to Earth system science and in terms of the major research issues that each one raises. They introduce the main themes addressed in the remainder of the book.

1.3.1   Global climate change

What will be the amplitude and rate of global climate change over the next century and beyond?

To address this question, we need to consider further the biophysical forcing and feedback mechanisms that control the energy budget of the Earth system. These include natural, external forcing mechanisms such as solar variability and volcanic activity, as well as forcing and feedback processes strongly impacted by human activities: for example, atmospheric greenhouse gases, dusts and aerosols. The need to characterise and quantify human impacts on these latter, and also on other chemical species that strongly influence the Earth system, leads to a consideration of industrialisation and land cover changes since both are key drivers of changes in atmospheric composition. The way in which this complex combination of natural and anthropogenic influences affects climate at any point on the Earth's surface is partly a function of the way in which the atmosphere and the oceans redistribute energy around the globe. The processes involved in this redistribution are not only strongly interactive, they are also influenced by feedback mechanisms that link them to the terrestrial biosphere. Thus the whole Earth system is intimately involved. Moreover, all the human activities that come into play reflect cultural, economic, technological and political processes operating on a range of scales from personal to global. For example, future rates of increase in atmospheric greenhouse gases will depend not only on the biophysical processes by which CO_2, methane and the other gases are processed and ultimately transformed or sequestered in components of the Earth system; they will reflect economic and technologically based decisions about, for example, energy production and the reduction of emissions. Human priorities and aspirations are just as crucial as trace-gas atmospheric residence times or the rate at which sediments in the deep ocean can provide long-term carbon storage (see 13.1.2).

We can study the way most of these processes operate and interact in the present day, though many methodological difficulties remain and major uncertainties persist when we attempt a fully quantitative analysis, or a level of explanation that captures all the interactions. We can also reconstruct the operation of many of the key processes and interactions for the past. What is much more difficult is to *predict* with confidence, or even with statistically quantifiable uncertainty, what will happen in the future. Even developing conditional projections or attaching degrees of probability to alternative future scenarios has proved to be extremely challenging. Poorly understood thresholds and inadequately quantified non-linearities in the biophysical realm, coupled with unpredictable discontinuities in human affairs, make all attempts to look into the future to a greater or lesser degree speculative. Yet, paradoxically, it is the imperative nature of that forward look that now drives much of the science upon which this book draws. Increasingly, we have come to rely on future projections, scenarios or 'story lines', each based on a plausible set of trajectories for the main processes responsible for future climate change.

1.3.2   Sea-level change

What will be the likely response of sea-level to global climate change?

Our understanding of the climate system and its future behaviour is vital to any estimate of the way in which sea-level may change in the future. A recent, at least century-long rise in global mean sea-level is well documented and can be ascribed in part to the warming, hence expansion, of the surface layers of the ocean (the so-called steric effect) and partly to the melting of glaciers. In order to assess the likely future rate and amplitude of any further sea-level rise, we need to be able to quantify better the recent trends and ascribe them more precisely to the main processes involved. Complementary research on the influence of climate change on sea-level in the past is also necessary. The links between climate and sea-level are complex and even the simplest cause–effect relationships are subject to time-lags, hence the value of a longer time perspective. Beyond this, that essential look into the future will, as in the case of climate change, remain a matter of alternative scenarios, each dependent on a trajectory of climate change and a series of calculations designed to estimate the effects of the climate change on the main controls of sea-level – ocean surface temperature and the mass balance of both polar and temperate ice.

When we move on one step further and attempt to estimate the likely effect of a given rise in global sea-level on a specific site or region, we need to know how to translate a mean rise world-wide into impacts on a particular coastline, with a specific morphology, tidal regime and sediment budget. We also need to be able to develop realistic relationships between *mean* sea-level change and the changing magnitude–frequency relationships of extreme events such as storm surges.

1.3.3   Ecological and human implications of climate change

How will the changes in global mean climate be expressed in terms of extremes (droughts and floods, for example) at continental, national and regional level?

Addressing these kinds of question involves bringing the implications of global processes down to the regional and local scale, just as developing global insights from observations or experiments at a specific location involves aggregating these in ways that draw out their significance for global processes. These translations, respectively downscaling and upscaling, pose major conceptual and practical problems. Additionally, projected future climatic variability, often expressed as changes in mean values, needs to be reframed as changes in the likely future incidence of extreme events, whether these be, for example, droughts, killing frosts, periods of prolonged heat stress, or storm damage. Future climate scenarios that incorporate both downscaling and articulation in terms of extremes, or other biologically relevant variables, can then be linked to models based on, for example, ecosystem responses, crop physiology and performance, or pest and disease prevalence and transmission, under changing conditions. Ultimately, the effects of human influences ranging from individual decision making to trans-national or global influences, must be integrated in the scenarios in a fully interactive way. In this example, the three components – climatic, eco-physiological and human – do not interact in simple cause–effect chains, in which climate controls eco-physiological function, which in turn determines what human populations can and cannot do. The three types of process interact to generate multiple feedbacks. Thus, in any attempt to develop realistic future scenarios, they must be coupled in ways that allow for these feedbacks.

1.3.4   Wider implications of human activities and global change

What are the future implications of other ongoing processes resulting from human activities as they proceed alongside climate change, e.g., loss of biodiversity, changes in ecosystem functioning, soil degradation, deforestation?

One of the most important aspects of global change has often been sidelined by the current preoccupation with climate and 'global warming'. In reality, the cumulative impact of human activities on the functioning of the biosphere over the last 200 years has been far greater than any resulting from climate change, despite the fact that the range of global temperature changes since the mid nineteenth century has been large in relation to the amplitude of variability recorded during the second half of the Holocene. The human-induced changes in question are those characterised by Turner *et al*. (1990) as 'cumulative' rather than 'systemic'. These changes – deforestation, the expansion of agricultural and range lands, soil degradation through erosion and salinisation, loss of biodiversity through habitat fragmentation and over-exploitation, pollution of water resources, dramatic changes in atmospheric composition in addition to those affecting the major greenhouse gases, as well as growing domination of some of the key biogeochemical cycles – are considered more fully in a historical context in Chapters 8 to 11. It is important to note here that we cannot simply regard these changes, brought about by the combination of population growth, exploding technology and the urge for economic development at any cost as somehow separate and detached from climate change. Nor is the distinction between 'cumulative' and 'systemic' changes entirely straightforward, for some of the former now appear to be having systemic effects. There are, as ever, multiple interactions and feedbacks between the two types of process.

1.3.5   Implications for human vulnerability and sustainability

How will the interacting complex changes encompassed in all the above affect key issues of vulnerability and sustainability for human populations and the resources upon which they depend?

Clearly in order to create any plausible scenarios for the future, we need to integrate the processes and impacts outlined above into a fully holistic evaluation that embraces human as well as biophysical dimensions. One approach involves viewing the ecosystems of the world not as functioning entities in their own right, but rather from an anthropocentric standpoint, as providers of 'goods' and 'services' (see, e.g., Costanza *et al*., 1997). This is consistent with a realistic recognition that the perceived threat from global change is not the *survival* of the planet as a functioning system, but rather its capacity to deliver the security and the resources, both renewable and non-renewable, upon which human life depends. At the same time, it is important to recognise that any scheme of ecosystem goods and services is inevitably limited to the cultural system within which it is defined. Moreover, ecosystem qualities that defy simple economic valuation are not necessarily of no significance.

Any realistic appraisal of the future sustainability of the Earth system from a human perspective must rest not only on the responses to all the foregoing questions but on a parallel and interwoven consideration of the growing needs and impacts of future human populations. The issues of resilience, vulnerability, risk, sustainability and adaptive capacity arise at a variety of spatial and organisational scales, from local to global. Increasingly, at local, regional and national levels, they are being evaluated through the use of impact assessment models that seek to integrate all the processes and interactions, both biophysical and human, that are considered likely to affect significantly the future success of human populations and their endeavours. Chapter 15 approaches these themes mainly from an environmental science perspective.

1.3.6   Future surprises

What is the likelihood that future changes will include major perturbations of the Earth system such as a reorganisation of ocean circulation patterns or a collapse of the west Antarctic ice sheet?

Among the many great unknowns about the future, there are what have often been termed potential 'surprises'. The two most commonly debated are the close-down of North Atlantic deep water (NADW) formation and the collapse of a major part of the west Antarctic ice sheet. The former would dramatically alter ocean circulation, which in turn would strongly modify continental climate regimes, especially around the Atlantic Ocean in mid and high northern latitudes. The latter would lead to a rise in sea-level over an order of magnitude greater than that projected in the current range of scenarios. It would be unrealistic to see these as the only examples of potential surprises. The notion of surprise is inherent in complex, non-linear systems with unknown thresholds and strong hysteresis.

1.3.7   The IPCC approach

The closest the scientific community has come to addressing the kinds of question posed above is in the successive publications of the Intergovernmental Panel on Climate Change (IPCC). The IPCC Third Assessment Report (IPCC TAR, 2001) attempts to integrate the findings of scientists working across a wide range of disciplines and methodologies and present them in the form of a broadly consensus-based assessment of the state of knowledge on present-day and near-future climate change, its causes and likely consequences, especially in human terms. What kinds of science contribute to this sort of synthesis? How do they differ in their approaches, what they study, the assumptions they make, the methods they use, the goals they seek to attain, their strengths and their limitations? What more has been learned since the IPCC TAR? These are some of the questions this book seeks to address.

1.4   Scientific methodologies

The nature and practice of any branch of science can be viewed in a variety of ways. At one extreme are its philosophical underpinnings. At the opposite, practical end of the spectrum, are the techniques used to obtain scientific results – radioactive tracing, remote sensing, controlled laboratory experiments, the deployment of instrumental arrays or pollen analysis, for example. Each of these tends to be used as part of the repertoire of techniques employed by scientists working within particular methodological frameworks. Methodologies thus sit somewhere between the philosophy underpinning a field of study and the techniques its practitioners employ.

1.4.1   Methodological frameworks

Below are listed and briefly defined what I see as the six main methodological approaches that have been used to address the global change questions posed above:

1. *Monitoring*. By this are implied routine observations, usually within a coordinated programme. Two key elements in an effective monitoring programme are consistency of method and continuity of observations, preferably over a long period. The meteorological services of most countries, coordinated by the World Meteorological Organization (see www.wmo.ch), fulfil this function in respect of the weather. The Permanent Service for Mean Sea-Level (see www.pol.ac.uk/psmsl) coordinates world-wide observations. In recent years there has been a move to coordinate and harmonise a wide range of monitoring programmes, most using a combination of satellite remote-sensing and surface observations, into a framework that can allow data assimilation and integration on a global basis. These tend to group into hierarchies of monitoring systems. For example, the Global Sea Level Observing System (GLOSS), forms a major part of the Global Ocean Observing System (GOOS), which in turn is linked to a wider range of initiatives concerned with the atmosphere, the land surface and the cryosphere. The need for an over-arching organisation has led to the establishment of the Integrated Global Observing Strategy (IGOS) partnership, which aims to build an institutional structure that will be able to provide early warning of major changes in Earth-system function (see www.igospartners.org).

2. *Coordinated observational campaigns*. One of the key characteristics of this type of initiative is a clear focus on a specific range of environmental processes. Whereas some types of monitoring programme had their beginnings many decades ago, these are much more recent in origin and have mainly developed rapidly over the last 30 years. They do not, therefore, have the same long continuity of observation. Moreover, although the observations may include routine ones, many techniques and, sometimes, new types of instrumentation are specially developed for the purpose of the campaign. In some cases, the campaigns are referred to as 'experiments', but I have chosen to restrict the use of that word to a narrower and more classical definition (see 3, below). The recently completed World Ocean circulation Experiment (see www.soc.soton.ac.uk; wmo.ch/web/wcrp; www.ecco-group.org) is one of the most ambitious and successful programmes of coordinated observations.

3. *Field experiments*. Although the types of campaign noted above may have some of the qualities of experimental design and may also have 'true' experiments embedded within them, the idea of perturbing the system in a more or less controlled way and

attempting to isolate the effects of the perturbation is not central to their research
design. In that sense, they are less close to the classic notion of an experiment than are
the research initiatives described in this section. In the case of the Free Air Carbon
Enrichment (FACE) experiments (Norby, 2004), the aim is to measure the direct and
indirect effects of subjecting particular species (crop plants, for example), or types of
vegetation, to increased concentrations of CO_2. Large-scale experiments have also
been designed to simulate warmer air and soil conditions at a whole-ecosystem scale,
or to explore the links between biodiversity and ecosystem function by manipulating
species composition (see, e.g., Diaz *et al.*, 2002). In the case of the IRONEX, SOIREE,
SEEDS and SOFeX experiments (see 6.5.3), the main aim has been to test the
hypothesis that iron is a crucially limiting nutrient in some open ocean environments,
by actively seeding an area of ocean with biologically available iron and observing any
consequent changes in primary productivity.

4. *Palaeo environmental studies*. Here, the over-riding aim is to deepen our
   understanding of the longer-term processes and interactions that operate within the
   Earth system and of the ways in which they change through time, 'concentrating on
   those aspects of past environmental change that most affect our ability to understand,
   predict and respond to future environmental change' (Alverson and Oldfield, 2000).

5. *Modelling*. The only scientific tools available for exploring the future (beyond
   'informed opinion', the use of analogues, or some kind of simple extrapolation) are
   simulations using models. By now, some form of modelling is an integral component
   of almost all the other types of methodological approach discussed here. At the same
   time, there is often something of a division between modellers and researchers who use
   predominantly empirical methods. For this reason, it is realistic to consider modelling
   as a separate and distinctive methodology or, perhaps more properly, range of
   methodologies, for there are many different types of model, many approaches to
   modelling and a wide range of goals in modelling exercises. Of particular interest are the
   ways in which model simulations interact with empirical evidence through the kind of
   two-way trade that can provide critical 'reality checks' for model output in one direction,
   and a coherent framework for incomplete and disparate empirical observation in the other.
   So important and all pervasive are models, that Chapter 2 is devoted entirely to them.

6. *Global 'base-line' surveys*. Gaining a better perspective on future changes will always
   be limited by the extent to which our understanding of contemporary conditions is
   flawed. This realisation has provoked at least one major, global initiative, the
   Millennium Ecosystem Assessment Project (see www.millenniumassessment.org)
   aimed at documenting the contemporary state of the world's ecosystems, using all the
   information currently available.

1.4.2 Integrating programmes

The various methodological approaches outlined above are complementary. For
a fuller understanding of the Earth system there is a need to combine them and
build insights on the whole range of contributions that each can make. At a

global scale, there have been earlier attempts to do this, of which the Man and the Biosphere Programme (see www.unesco.org/mab) is a notable example. From the late 1990s onwards, this has also been a major concern of the International Geosphere-Biosphere Programme (IGBP) (see www.igbp.kva.se). This programme embraces globally oriented international projects covering most of the research that forms the scientific basis for this book. In recognition of the close coupling between physical, ecological and human systems, there is growing liaison and cooperation between IGBP and WCRP, the World Climate Research Programme (see www.wmo.ch/web/wcrp), DIVERSITAS (see www.diversitas-international.org), the international group dealing with biodiversity, and IHDP, the International Human Dimension Programme (see www.ihdp.org).The goal is to create the Earth System Science Partnership.

Integration is also essential at the regional level. (The System for Analysis, Research and Training) START, (see www.start.org), has responsibility for coordinating a range of regional projects mainly in less economically developed parts of the world. One notable example of highly focused scientific coordination at a regional scale is the Large-Scale Biosphere Experiment in Amazonia (LBA) (Nobre *et al*., 2001; Nobre, 2004). This project has been designed to study the whole complex ecosystem of Amazonia and its role in the Earth system.

Finally, a whole new methodology of impact assessment has developed, aimed at using data and models from both the biophysical sciences and socio-economic research to develop future scenarios at national or regional level, designed to assist directly in policy making, (see 15.3).

1.5   Linking methodologies

To some extent, the *methodologies* outlined above are emerging as coherent, self-defining entities, each with powerful protagonists, but there is a growing need to unite them (see, e.g., Szeicz *et al*., 2003). This implies working across two major divides that are as yet inadequately bridged: that between many empirical and modelling studies, and that between the combination of methodologies focused on contemporary changes with a shallow, recent-time perspective, and research designed to bring to light longer-term environmental changes in the past. Building the latter bridge and illustrating the need for two-way traffic across it is a recurrent theme in the rest of the book.

The reasons for a strong focus on past changes and longer-time perspectives in the present account are as follows:

- Reconstructions of past conditions from empirical data provide the only test of the skill with which models, calibrated to present-day or recent observations, perform under different boundary conditions and forcings.
- Only by studying the past is it possible to gain direct insight into the long-term processes that condition what we observe at the present day, and that drive all the

interactions taking place on timescales longer than the short span of direct observations.

- By the same token, only in the evidence from the past are we likely to encounter fully evolved manifestations of some of the crucial generic properties of the Earth system, including non-linearities, threshold-linked behaviour, hysteresis and contingency.
- In many areas of global-change research, models designed to capture system dynamics are developed and tested using only short-term observations and/or spatially distributed elements that have been linked serially into a time sequence. Before being used to generate putative future time series, formulations of this kind should be tested with reference to evidence from the past.
- The past is rich in evidence for the long-term implications of different types of human–environment interactions, for both social and environmental systems.
- Projected future changes, of climate, for example, need to be compared with those documented for the past in order to gain any realistic impression of their potential implications, and of the extent to which they pose new challenges that lie beyond those already met by previous societies.
- For every aspect of the Earth system, the present day is a dynamic, not a static, baseline. Monitoring needs to be set in this dynamic context, rather than seen simply as an activity starting from a fixed point in time. The insights provided by longer-term historical studies provide information that is essential in developing strategies for future management or conservation. This can be illustrated from ecosystems as diverse as temperate forests (Foster, 2002: Foster *et al.*, 1998; 2002a, b; Hall *et al.*, 2000), heathland (Walker *et al.*, 2003) and coral reefs (Bellwood *et al.*, 2004).

It follows from the above that the approach to palaeo-science, both by researchers in the field and by those looking from other methodological frameworks at what it has to offer, must go beyond the idea of palaeo-science as 'history'. Above all there is a need to see it as making a vital contribution to understanding processes and interactions. This theme is taken up once more in Section 3.2. Finally, it is important to recognise that any full appraisal of the Earth system demands a deeper understanding of the role of human decision and actions. This then defines a third major bridge for us, that between the biophysical and the social sciences.

1.6   The thematic sequence

The chapters that follow describe in more detail and illustrate the power of each of the main methodologies and research modes outlined above. Chapter 2 considers models. These must be the first substantive theme, since their use pervades all the other sections. Chapters 3 to 7 look at different aspects of palaeo-science as it contributes to our understanding of the changing function of the Earth system through time. Chapters 8 to 12 consider the evidence for human impacts and climate change over the last 300 years. Chapters 13 to 15 outline the ways in which projections of future environmental change are developed and evaluated and Chapter 16 revisits the essential questions already posed in this first chapter. The last chapter in the book also considers briefly some of the underlying characteristics of 'global-change science' that make it qualitatively so different, even strange, to someone brought up in earlier traditions.

Chapter 2
An introduction to models and modelling

2.1 The role and rationale for modelling

Models increasingly permeate every aspect of Earth-system science. In part, this reflects the need to acquire a better understanding of likely changes in the future; but even without this imperative, models would have an essential role to play. The complex, interactive and non-linear nature of the Earth system limits the extent to which Earth-system science can progress by traditional methods in which variables are isolated experimentally, or simple cause–effect linkages defined. Linking together observations to the point where they provide a coherent view of the functioning of the complex system frequently requires the development of models that satisfy at least the following criteria:

- Internal consistency.
- Compatibility, within uncertainties, with all applicable biophysical laws.
- Compatibility with the constraints imposed by secure and relevant data, bearing in mind the uncertainties attached both to the model and to the data.
- Robust performance under a range of credible boundary conditions.

Models that satisfy these requirements can play a role in many ways similar to that played by hypotheses in the classic Popperian formulation of deductive science. In Popper's scheme of reasoning (Popper, 1963), hypotheses are developed such that they are potentially refutable through crucial tests in the form of observations or experiments. Refutation leads to the rejection or modification of the hypothesis, after which the process of hypothesis building and testing can enter a further stage using the newly gained insights. In the case of model development and testing, refutation is, in practice, usually a less rigorous and narrowly defined concept. Nevertheless, the last two requirements noted above open the way to establishing the extent to which the performance of a model is consistent with all the relevant data available, both for the present day and for times in the past when boundary conditions and external forcings were different. Significant inconsistencies point to the need for model improvement. Weak, or contradictory data constraints highlight the need for better-validated data. Thus models, in their development and testing, interact with present-day observations and the reconstructions of past conditions, to create an inferential framework with increasing power and reliability over time, as improved models are developed and more reliable data become available. Models may not only serve as 'hypotheses' in the way outlined above. They can also serve as tools for interpolation, as well as tests of the extent to which a set of processes, or a scheme of interactions that have been inferred from incomplete empirical data, generate credible and self-consistent outcomes when they are incorporated into a model.

As well as becoming increasingly integral to the reasoning required to explore the complexities of the Earth system, models have become ever more powerful as a result of a steady, exponential growth in computing power since the early 1950s. This has led to an increase of seven orders of magnitude over that time interval (McGuffie and Henderson-Sellers, 1999). One outcome of this increase in computing power, alongside the growing realisation that all the components of the Earth system need ultimately to be linked in model simulations, is the development shown in Figure 2.1.

Models make an increasingly crucial contribution to the science of environmental change in every research area and at every scale, from the exchange of moisture and energy at the contact between the individual leaf and the atmosphere, to the whole Earth system. Ideally, each of the relevant areas of environmental science where models are used – atmospheric science, oceanography, hydrology, ecology and so on – merits a chapter or more, with full explanations and examples. This is clearly not realistic. The text that follows first outlines some of the generic concepts common to most of the *climate*-oriented modelling activities that contribute to Earth-system science. We then consider first atmospheric models, then the Earth-system models that link the atmosphere to the other major subsystems – the oceans, cryosphere and terrestrial biosphere. There then follows a brief note on ocean modelling, as an example of one of these subsystems. The main aim of the chapter is to provide for students of environmental science and related subjects, the vast majority of whom are unfamiliar with modelling at a practical level, enough of the basic concepts and vocabulary to allow them to appreciate the importance of modelling in all aspects of Earth-system science, and to make sense of references to models and model simulations in the later parts of the book.

2.2   Some essential generic concepts in climate and Earth-system modelling

The present section attempts to identify and explain some of the widely applicable concepts and definitions that underlie the business of climate and Earth-system modelling.

2.2.1   Boundary conditions and forcing variables

Numerical models are discrete simplifications of reality in which the initial conditions, boundary conditions and forcing variables have to be prescribed in

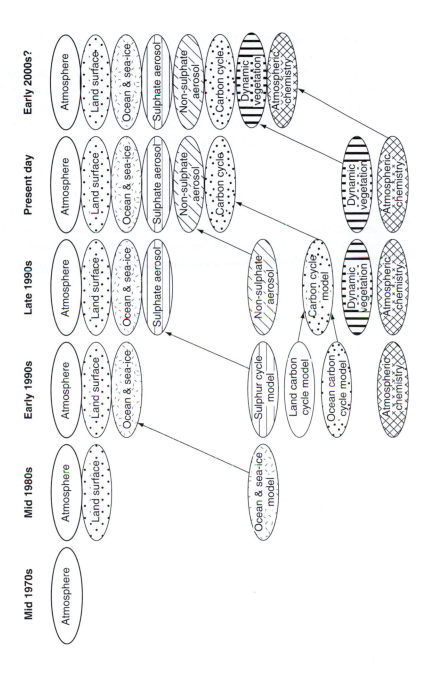

Figure 2.1 The development of climate models since the mid 1970s. The diagram shows the development of successive subsystem models and their eventual coupling into increasingly complex and realistic climate models. (From IPCC TAR, 2001.)

some mathematical form. The boundary conditions chosen will depend on the purpose of the model. The boundary conditions and purpose of the model will, in turn, influence the choice of processes that drive, or 'force' the model. Thus the simplest global model of present-day energy balance at the Earth's surface uses terms to represent the flux of energy reaching the Earth from external sources and the flux of energy re-radiated from the Earth. Each term must adequately approximate these fluxes as they occur under contemporary conditions of solar radiative input, the albedo of the present day Earth's surface and atmosphere (since this determines the proportion of incoming heat absorbed or re-radiated) and the 'greenhouse' properties of the atmosphere, (since these determine the extent to which re-radiated energy is captured and retained rather than 'lost'). If we seek to simulate realistically the energy balance either in the future or the past, the boundary conditions, hence the forcing variables, would have to be changed. For the future, at the very least we would need to find some way of estimating the likely effect of increased greenhouse-gas concentrations in the atmosphere. If we chose the boundary conditions at last glacial maximum as our example of past conditions, there would be a need to reduce the value for incoming solar radiation and for greenhouse-gas concentrations, as well as to modify the surface albedo of the Earth to take into account, at the very least, much more extensive ice and snow cover.

2.2.2   Equilibrium (quasi-steady state), versus transient simulations

In the above example, we envisaged the simulation of energy balance at the Earth's surface for a given period of time, a 'snap-shot', (Figures 2.2 and 2.3) with no regard to changes that may have led up to or followed the conditions characteristic of that time. Models that generate such simulations are generally referred to as equilibrium models. If, instead, we are interested in modelling changes as they occur through time we need to perform transient simulations. This can be achieved by allowing the variables forcing the model to vary in prescribed ways during the course of the simulation. Transient behaviour can also be simulated when an equilibrium condition is perturbed by altered forcing, and the simulation captures the response of the model to the perturbation. This type of perturbation 'experiment' is generally used to explore the effects of varying individual parameters. Relatively simple transient models have been used to simulate, for example, changing ice extent during the last two glacial cycles. Transient simulations have also been used to explore the likely causes of dramatic shifts in climate such as the onset and termination of the Younger Dryas period (see 6.4 and 6.5) or the mid-Holocene desiccation of the Sahara (see 7.4).

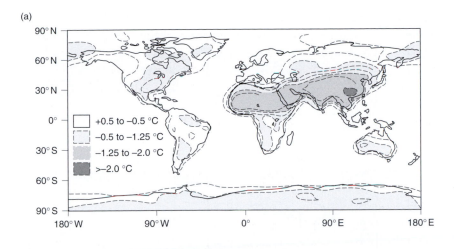

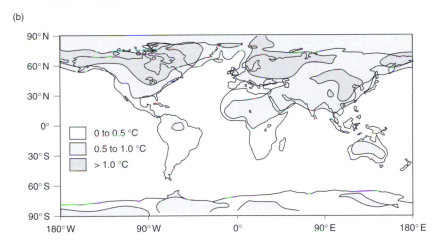

Figure 2.2 (a) Global climate model simulation of the difference between present-day and mid-Holocene winter (December, January, February; DJF) temperatures. The contours plot the mean values from the PMIP (Paleomodel Intercomparison Project) simulations. Positive values indicate that the modelled mid-Holocene winter climate was warmer than the present day and vice versa. (Modified from Valdes, 2003.) (b) The standard deviation between the model simulations in 2.2 (a). (Modified from Valdes, 2003.)

2.2.3   Initialisation and spin-up

Complex models that incorporate a range of time-dependent variables and capture their interactions not only require that their initial state be prescribed; it is also necessary for such models to run for a while before they reach equilibrium conditions. The need for a period of 'spin-up' reflects the fact that the initial conditions do not satisfy the model equations. This is especially the case with models designed to simulate components of the Earth system that have a long response/equilibrium time. Such times vary by six to seven orders of magnitude, with atmospheric processes at one extreme and polar ice sheets, with response times of several thousand years, at the other (Saltzman, 1985).

Figure 2.3 Differences in
near-surface winter
temperature (DJF)
between present-day and
mid-Holocene winter (DJF)
temperatures. Each plot
shows the differences as
computed from
simulations using
increasingly complex and
complete representation
of major Earth-system
components. Note the
differences in resolution
when compared with full
GCM simulations (Figure
2.2). The ATM + OCE
simulation compares most
closely with the PMIP
pattern of changes.
Addition of vegetation
feedbacks introduces
significant changes. The
area south of 60° S has
been omitted. (Modified
from Wasson and
Claussen, 2002.)

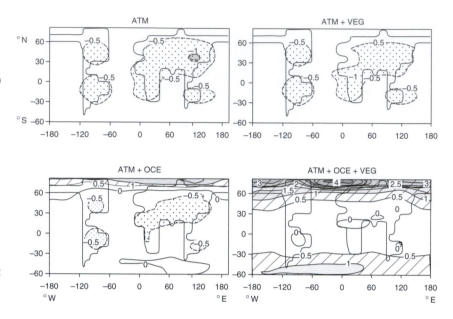

Figure 2.3 Differences in near-surface winter temperature (DJF) between present-day and mid-Holocene winter (DJF) temperatures. Each plot shows the differences as computed from simulations using increasingly complex and complete representation of major Earth-system components. Note the differences in resolution when compared with full GCM simulations (Figure 2.2). The ATM + OCE simulation compares most closely with the PMIP pattern of changes. Addition of vegetation feedbacks introduces significant changes. The area south of 60° S has been omitted. (Modified from Wasson and Claussen, 2002.)

2.2.4 Parameterisation

Such is the complexity of the Earth system and its major components that models cannot realistically portray all the known processes and variables in a complete and accurate way. Moreover, many variables are too poorly understood to allow accurate representation. The solution to these problems is often to represent variables by simplified approximations. This process is known as parameterisation. In some cases, this involves actually ignoring processes believed to be of negligible effect in relation to the spatial or temporal scale under consideration. Other approaches to parameterisation include the use of empirical approximations in the form of mathematical equations, or time-averaged representations. In practice, the structure of a model will often include some variables that are explicitly represented in terms of physical laws, or fully representative equations, and others that have been more crudely parameterised.

2.2.5 Flux adjustments

Until quite recently global climate models (GCMs) tended to be unstable unless the energy fluxes between the ocean and atmosphere were arbitrarily adjusted to allow the model simulations to stabilise. Increasingly, the most recent models coupling atmosphere and ocean (AOGCMs) can be run without flux adjustments.

2.2.6 Process-based and empirical/statistical models

The somewhat idealised outline of model development presented so far relies on the skill with which the component processes of a given environmental system can be expressed mathematically, either through dynamic equations or adequate parameterisation. Some models more or less by-pass representation of the processes involved and focus on matching to the pattern or sequence of effects to which the processes give rise – in terms of the last paragraph, this may be seen as the ultimate extension of one kind or paramaterisation. Thus, modelling the link between climate and vegetation may be done either through process models incorporating such features as plant physiology and soil respiration, or it may treat these as a black box and rely instead on empirical relationships, ideally expressed in statistical terms, between the present-day distribution of species or groups of species and spatial variations in some properties of the climate – for example mean annual rainfall, growing season or July temperature. There are circumstances under which statistical models are more appropriate and effective than process-based, dynamic models.

2.2.7 Sensitivity

Where many variables contribute to the output of a model, it is essential to be able to distinguish between those variables to which the model outcomes are most sensitive and those that are least likely to affect the results. Exploring the sensitivity of a model involves carrying out multiple runs using credible combinations of variables. Models may prove to be especially sensitive to changes in initial conditions, boundary conditions, forcing variables or internal dynamics. Separating out the sensitivity of the models to these different types of variable, then exploring the effects of alternative parameterisations within each, may generate vital information both on the strengths and weaknesses of each model, and on the real world processes the model is designed to simulate. It is in this respect that modelling comes closest to conducting experiments and it is only in relation to sensitivity experiments that it may be justifiable to use the word experiment in preference to simulations or scenarios. Sensitivity is often used to describe the quantitative relationship between a forcing variable and a response. For example, estimates of the sensitivity of the climate system to radiative forcing (often expressed, in the case of CO_2, as the temperature increase forced by a doubling in atmospheric concentrations) are essential for developing projections of future climate change. Similarly, estimates of the sensitivity of North Atlantic deep water formation to particular rates of warming and freshwater input are needed in order to assess the probability of future climate change leading to a shutdown of the thermohaline circulation (see 13.3).

2.2.8   Ensembles

In complex climate models, for example, each individual simulation using the same initialisation will give somewhat different results for the equilibrium state once it has been reached. The outcome values will usually be calculated from the range of values generated over a given period of time during an extended run of the model, and/or through multiple runs of the model. The values quoted will therefore be ensemble averages. In some studies, ensemble averages generated by several models are combined to provide multi-model ensembles. These will embrace the variability arising from the different model configurations, as well as the variability encompassed by each of the individual model ensembles.

2.2.9   Time steps and coupling

Models designed to reach a target equilibrium state or to simulate dynamic changes require that calculations for each variable be made at a given time interval appropriate to the processes under consideration, the demands of the model, and the computing power available. The GCMs considered later may have time steps of an hour or less, with major implications for the computational power required. Less complex models or models dealing with long-term processes will often have much longer time steps. Where models are coupled in order to explore the linkages between subsystems, for example, or to give a more complete representation of all the interactions and feedbacks involved in climate change, the choice must be made between linking the models in a totally synchronised way, or coupling them asynchronously. In the latter case, processes within one subsystem may be modelled on an appropriate timescale, then the results used as input to a model simulating another subsystem. Take, for example, the case of coupling atmospheric and ice-sheet models. Since the latter need to simulate processes on much longer timescales than are currently feasible for GCMs, one solution is to generate initial-climate conditions using a GCM, then drive the ice-sheet model using the GCM output.

2.2.10   Temporal and spatial scale

The time steps used in a simulation determine the temporal resolution that can be achieved. They should therefore be matched to the kind of outcomes required, whether, for example, these are estimates of inter-annual variability or multi-millennial-scale changes. The spatial scale resolved by a model is also of critical importance, for there is an inevitable trade-off between spatial resolution and computing requirements. The spatial scale needed to gain a realistic impression of global energy balance is orders of magnitude less than that required to provide simulations of future climate for the whole world that are directly applicable at river-catchment or ecosystem level.

2.2.11   Upscaling and downscaling

Because of the limitations in spatial resolution imposed by computing constraints, a range of techniques has been developed to 'downscale' model results by nesting areas of higher resolution within simulations of lower resolution. In the case of climate models, for example, this involves taking the output from GCMs and deriving patterns and processes at finer spatial scales than those resolved by the model. There are several approaches to downscaling (for example, using the climate simulated for the adjacent surrounding grid squares as a boundary condition for a spatially high-resolution model within the grid square of interest) and each produces different results.

In other cases, the reverse problem may arise, for example, where there is a need to use widely scattered observations, discontinuous in time and space, as part of the basis for developing or testing a global model. The usual goal of upscaling is to develop, from discrete, unevenly distributed measurements, a scheme of processes and interactions that apply at the scale of the grid square used in the model. A range of upscaling techniques is available using statistical procedures to minimise the uncertainties introduced as a result of the inevitable interpolations and extrapolations involved.

2.2.12   Uncertainty

One of the most important, complex and controversial aspects of modelling is that of uncertainty, and there are many ways of dealing with it. Provided the uncertainties associated with the inputs to a model are sufficiently well known, statistically derived estimates of model uncertainty can be calculated. Where models are linked together serially, uncertainty can cascade through the models with consequences that can seriously reduce their reliability. Increasingly, probability distributions are being attached even to scenarios where the eventual tests of performance lie well into the future. This is not always the case, though, and many model simulations come with only subjective assessments of uncertainty. This is especially the case where a future model scenario is dependent on inputs that cannot be subject to rigorous statistical evaluation (see 13.1.1). Such is the case with future climate scenarios that are partly contingent on the rate of increase in fossil fuel combustion, as well as on the poorly quantified effects of changing cloud cover. In such cases the concept of *conditional* uncertainty may help to deal with the problem. In all cases any attached uncertainties, whether subjective or objective, conditional or not, apply to the model and not to the realities that the model is designed to simulate. These uncertainties will often be substantially greater.

2.2.13   Inverse models and data assimilation

'Forward' models are based on physical laws and processes, either accurately represented or parameterised. Where they are used to predict a particular outcome, or response, and data are available for comparison, their performance can be evaluated in terms of the extent to which the simulations generated match the data. Inverse models use all the available and relevant data, and the dynamics, sometimes in simplified form, to estimate, to the best of our knowledge, the state and/or evolution of a system. Ideally, uncertainties should be attached to both the dynamics and the data. In order to use data in model building, a range of methods for assimilating data into models has been developed.

2.3   Climate and Earth-system models of varying complexity

All climate and Earth-system models are necessarily simplifications of reality, but the degrees and types of simplification vary enormously. As we have already seen, the greater the degree of simplification, the less is the demand of simulations on computing power. Other things being equal, simple models can therefore be run more frequently, or over longer periods, than complex ones and this gives them significant advantages with regard to some aspects of Earth-system function.

One of the computationally and conceptually simplest types of model is the box model in which the box serves as a reservoir, or major component of the Earth system, and the linkages between provide quantitative estimates of fluxes. Such models are often used to simulate the behaviour of chemical species – carbon or nutrients for example – in the environment, though they have also been used successfully to generate long-term simulations of the global-climate system.

Simplification can also be achieved through the dimensionality of the model. The simplest ones are what are often termed 'zero-dimensional' energy balance models (EBMs), developed from the late 1960s onwards, which consider the energy balance of the whole Earth as a single unit. The energy fluxes taken into account are the incoming and outgoing radiative fluxes at the Earth's surface and the results of the model simulations are expressed as surface temperature. The simplest one-dimensional EBMs differentiate the Earth as a set of latitudinal bands between which heat transfer occurs horizontally by diffusion. Two-dimensional models add more detail and complexity by differentiating the Earth's surface into more categories on the basis of, for example, longitude as well as latitude, or land versus ocean. Such models are much less demanding of computing time than more complex ones and this allows them to be used many times over with slight variations in the way in which key processes are approximated. Such repeat simulations constitute tests of model sensitivity, as defined above, and they have been widely used from the 1980s onwards to examine the effects of, for example, changing albedo, solar input and atmospheric green-house-gas concentrations.

Statistical dynamic models (SDMs) are models, generally two-dimensional, that bring us closer to a realistic portrayal of long-term changes in the climate system through time. These models use equations expressing statistical approximations to represent what are considered to be the main processes under given boundary conditions, thereby significantly reducing computing time. By coupling models representing different components of the Earth system and their contrasted response times, it has been possible to develop reasonable simulations of the main changes in ice volume and climate over the last 200 000 years.

In all the above cases, computational efficiency has been gained at the expense of detail; resolution, both temporal and spatial; and the explicit representation of complex processes and interactions. As computing power has grown, increasingly detailed and complex models have become possible. Most of these are variants or extensions of GCMs (see Figure 2.2). The aim of GCMs is to calculate the full three-dimensional character of the climate. The global atmosphere is the main realm modelled and the ocean is linked to it in a variety of ways, depending on the goals of the research, hence the complexity of the model. The atmosphere is modelled through a set of equations that represent the relevant physical laws and parameterised processes. At its simplest, the ocean component is a set of prescribed sea-surface temperatures (SSTs). Slightly more complex and realistic is an ocean that can change heat capacity, but within which convection is not allowed, the so-called 'swamp' ocean. In more complex models, SSTs are calculated and horizontal advection (but no diffusion) is allowed – the 'slab' ocean. More complex representations of the ocean allow for both advection and diffusion into the deep ocean and even, in the most sophisticated cases, an ocean with internal dynamics in three dimensions with exchanges in energy, moisture and momentum at the ocean–atmosphere interface.

Dynamic ocean models as such have become increasingly detailed and realistic, largely as a result of the research carried out as part of the World Ocean circulation Experiment (WOCE: see especially the posters at www.ecco-group.org). This has made available huge amounts of data on ocean temperatures, salinity and circulation, the assimilation of which into inverse models (see 2.2.13) has provided the basis for capturing the three-dimensional dynamics of the ocean and its inter-annual variability.

Models in which GCMs are coupled to a fully dynamic ocean in this way are known as atmosphere–ocean GCMs (AOGCMs), but problems of coupling arise, as noted above, since the response time of the deep ocean is many orders of magnitude longer than that of the atmosphere. The problems are compounded when consideration is extended to major ice sheets and, if GCMs are used and long-term processes simulated, asynchronous coupling may be the only solution.

In recent years, the crucial role in the climate system of feedback from the terrestrial biosphere and from both land and sea ice has been recognised. In response to this, models are being developed in which all these components are included, in addition to biogeochemical cycling in the atmosphere and the ocean. Such models approach complete simulation of the Earth system as considered here, but their computing requirements are enormous. One response to this has been the development of the Earth Simulator in Japan – a computing system capable of generating what is, in effect, a 'virtual planet', using some 5120 processors and capable of performing 40 trillion calculations per second. The possible benefits of this initiative lie in the future.

Valuable insights into the dynamic operation of the Earth system have come from the use of EMICs – Earth-system models of intermediate (or reduced) complexity (Figure 2.3). By reducing the spatial resolution, skilfully parameterising many of the key processes and using longer time-steps than those used in GCMs, it has become possible to include the full range of forcings and feedbacks over thousands of years, thus allowing realistic simulations of processes for which increasingly strong data constraints exist in the palaeo-record. This allows for a degree of synergy between data and model development that comes close to realising the rather idealised role for models envisaged at the opening of this chapter.

2.4 Subsystem models, with special reference to the oceans

2.4.1 Introduction

Linking atmospheric GCMs to ocean, terrestrial biosphere and cryosphere models naturally depends on the skill with which these components of the Earth system can themselves be modelled, as well as on the representation of linkages between the different subsystems through processes such as gas and energy exchange, or the hydrological cycle. When we consider all these aspects the range and diversity of relevant models is overwhelming. Rather than attempt any kind of comprehensive listing, the aim of the following sections is to summarise the underlying rationale for some of the main types of model developed. In many cases, a fuller exploration of particular models is left for later chapters where the interaction between models and data can be considered in more depth.

Three kinds of goal in purely biophysical subsystem models can be identified at the outset and they lead to different types of model. The first ultimately relates to integrating subsystem models into more comprehensive Earth-system models. The focus is therefore on the processes that control the variability and between-system fluxes of key Earth-system properties: energy in all its forms; biologically reactive chemical elements such as carbon and nitrogen; and water, whether as liquid, vapour or ice. The second kind of model is oriented more towards simulating subsystem *responses* (for example the changed distribution of vegetation types in response to climate change) and may be less concerned with the feedbacks that may arise from these responses. The third kind of model arises from the need to base reconstructions of past conditions such as air temperature, surface-moisture balance, sea-surface temperature or salinity on what are termed 'proxy' records – variations in the past distribution and abundance of organisms preserved in the fossil record or physico-chemical properties preserved in ice or sediments. Making quantitative reconstructions from proxy records involves developing process-based or empirical/statistical models that link variability in the proxy data to variability in the environmental property to be reconstructed. Consideration of this third kind of model is deferred until 4.2.

2.4.2 Ocean models

Of most direct interest to us are those models concerned with ocean circulation, with the role of the oceans in the carbon and nitrogen cycles and with energy and material fluxes across the ocean–atmosphere boundary layer.

Our knowledge of ocean circulation has been greatly improved over the last 50 years. One result has been a much more detailed representation of the thermohaline circulation (THC) that links the world's oceans (see Figure 1.5). The development of process-based models of the THC in its present form have proved essential in understanding the role of the ocean in past environmental change as well as in predicting the possible consequences of future changes. Note, havever, that any attempt to generate simulations of the THC for the past or the future depends on the accuracy with which other linked subsystems can be modelled. For example, changes in ocean salinity in key areas of the ocean are highly dependent on variations in the hydrological cycle, since these, in turn, generate freshwater fluxes that can critically change ocean salinity.

The ocean is the main reservoir of carbon in the Earth system on timescales of millennia or less (see Figure 5.3); therefore quantifying its role in modulating variations in atmospheric CO_2 is of vital importance. Many attempts have been made to model, with varying degrees of sophistication, the carbon cycle in the oceans. One major problem is the lack of quantitative data on some of the key fluxes. This still limits the success of even the most recent attempts to establish the role that ocean variability currently plays in controlling variations in atmospheric CO_2 content from year to year, as well as during past periods of major climate change. In consequence, the extent to which, and ways in which, the ocean modulates atmospheric CO_2 concentrations remain debatable, even for the present day, let alone the past or future.

Over the last two decades, enormous scientific effort has gone into characterising and quantifying the fluxes of moisture, energy and many chemical species from atmosphere to ocean and vice versa, and many of the key processes have been successfully modelled. Nevertheless, the full range of fluxes across the ocean–atmosphere boundary layer remains poorly quantified for the world as a whole and this remains one of the crucial scientific challenges for the future, in response to which the IGBP (see 1.4.2) has recently launched a new project, SOLAS – the Surface Ocean–Lower Atmosphere Study. Any improvement in attempts to model these fluxes is likely to depend on the acquisition of a great deal more data than are presently available.

2.4.3   Other subsystems

Changes in the *cryosphere* affect the Earth system on a wide range of timescales. Ice and snow, whether covering land or sea, have a strong feedback on the climate system and it has become increasingly common to incorporate them in climate models. Changes in glaciers and polar ice sheets affect sea-level and atmospheric circulation over much longer timescales; the latter have the longest response times of all the major components of the Earth system of direct interest to us. This means that in practice they are often modelled separately or else coupled asynchronously to models designed to simulate processes with shorter response times.

The interactions between climate and the *terrestrial biosphere* are many and complex. Global modelling of the terrestrial biosphere is a relatively recent development, but one that is essential if we are to understand the full range of feedbacks that affect climate on all spatial scales from local to global. Characterising and quantifying the actual processes governing exchanges of energy and gases across the boundary layer between the terrestrial biosphere and the atmosphere have required detailed studies at the scale of small stands of vegetation. Capturing the way these exchanges affect climate as major biomes change requires massive upscaling of the integrated effects of the processes involved, alongside the development of models that can represent the response of ecosystems and of the world's major biomes to changing climate and biogeochemistry (see, e.g., Foley *et al.*, 1996; Cramer *et al.*, 2001; Sykes *et al.*, 2001).

There is also a need to model the responses of ecosystems to perturbations, whether biophysically driven or largely anthropogenic. Models have been developed to capture the interactions involved over a wide range of spatial scales. One type of model that has proved powerful in this regard is the forest gap model (see, e.g., Bugmann, 1997; 2001; Bugmann and Pfister, 2000).

By now, models have been developed for virtually every process of interest to the student of environmental change. The range and diversity are just too great to encompass in a single chapter. In all the subsequent chapters, many of these approaches to modelling will be considered alongside other lines of evidence for the nature of past, present and future environmental change. For example, models of land-use/land-cover change are considered in Section 9.2. Land cover affects climate at the very least at regional and, in some cases, at continental scale. Moreover, land-cover change is one of the main threats to

biodiversity at species, landscape and global levels. It is therefore hardly surprising that the modelling of processes driving land-cover change has been given
high priority in recent years.

Increasingly, integrated assessment models that link biophysical processes
with both human drivers of, and responses to, environmental change are being
used to assess the likely consequences of, and responses to, future changes at
regional or national scale. These are considered briefly in Section 15.3. Finally,
brief mention must be made of the modelling approach using the concept of
cellular automata (Wolfram, 2002). Cellular automaton-type models build on
the demonstration that complex behaviour can be generated by models to which
a set of relatively simple rules are applied at the outset. Used realistically, they
may have several important advantages. For example, they can mimic emergent
behaviour in many kinds of system; they can generate non-linear responses
through the combined interaction of linear processes; the simulations they
provide can be tested and the rules refined through comparison with historical
time series, and they can be applied to human, biophysical and combined
human–biophysical systems. They have been tested with some success against,
for example, the historical growth of a major conurbation (Clarke *et al.*, 1997),
land-use change on a small Caribbean island (White and Engelen, 1997) and
Holocene sediment delivery in response to changing climate and land use
(Coulthard and Macklin, 2001).

Chapter 3
The palaeo-record: approaches, timeframes and chronology

3.1 Introduction

We now turn to the past for what can be learned about those aspects of Earth-system function and environmental change that bear most directly on the present and future. This involves critical choices about the geological timespan and the types of palaeo-evidence of greatest relevance. The geological record contains evidence from periods of even higher atmospheric CO_2 concentrations and higher global temperatures than are anticipated for the coming century. At first sight, it may seem logical to turn to these periods as possible analogues for the future, but there are good reasons for rejecting any simple search for past analogues of this kind. Many Earth-system components with major functional implications were crucially different – the disposition of the continents, land-mass topography, ocean circulation and plant cover at the very least. These changed characteristics imply completely different boundary conditions, forcings and feedbacks. Rather than turn to the past for analogues of state with respect to a limited number of Earth-system characteristics, we need to use the evidence from the past to address questions about amplitudes and modes of variability, process interactions and rates of change. The emphasis here is therefore on the Quaternary period, though one aspect of the pre-Quaternary geological record may prove to be of some relevance. Recent evidence (Svensen *et al.*, 2004; Dickens, 2004) provides strong support for the hypothesis that the initial Eocene thermal maximum, some 55 million years ago, was triggered by a massive release of methane from below the Norwegian Sea. If confirmed by further research, this would strengthen the case for the powerful warming effects of greenhouse gases, as well as lend credibility to the 'clathrate gun' hypothesis put forward by Kennett *et al.* (2000; 2003) as an important factor in global warming at the transition from glacial to interglacial conditions (see 6.5.4).

Any future changes in the Earth system will reflect the interplay between natural processes and the effects of anthropogenic perturbations. It is therefore necessary to reach as full an understanding as possible both of the unperturbed Earth system as it has functioned prior to significant human impact, and the increasingly perturbed system within which we now live. For the former, we must choose a timeframe within which what we learn about these questions is most directly applicable to the present and near future and for which the best range of evidence is available.

In addition to selecting the most appropriate timeframe, it is important to consider the kind of evidence required to reconstruct past states, processes and rates of change. This calls for archives of all kinds, instrumental, documentary and environmental. From environmental archives such as sediments, or tree rings, the researcher must retrieve the evidence that best records the environmental properties under study, whatever these may be, for example air temperature, ocean salinity or vegetation types. This requires the use and interpretation of what are usually termed proxies. Deriving consistent and, where possible, quantifiable palaeo-reconstructions from proxy records requires that they be calibrated with respect to the environmental properties they are required to capture. Finally, the evidence from the past must be dated. In the sections that follow, we consider dating methods and chronologies. Archives, proxies and calibration are considered in Chapter 4.

One of the main aims in writing Chapters 3 and 4 is to give readers whose background and experience have been largely focused on contemporary systems, whether physiographic, ecological or human, some sense of the nature of palaeo-research, its requirements, different approaches, strengths and limitations. My hope is that this may help to bridge one of the methodological gaps referred to in Section 1.5, that between contemporary and 'historical' studies.

3.2 Alternative approaches to the palaeo-record

Using relevance to current environmental changes as the touchstone, the main modes of research adopted in the palaeo-research community may be classified by means of a rather loose taxonomy based on the dominant purpose underlying each:

1. Narrative reconstruction, both quantitative and qualitative, of the past sequence of events.
2. Provision of empirically based quantitative constraints on scenarios for past periods derived from numerical simulations or conceptual models. In some cases, such reconstructions provide time-slice 'realities' on a global scale.
3. Elucidation of processes and process interactions.
4. *Post hoc* hypothesis testing.
5. Determining the recent antecedents to present-day environmental systems and current trends.

In order to draw out the crucial insights that improve understanding of the ways in which the Earth system functions and changes, it is essential to look beyond narrative. Even the most temporally or spatially discontinuous evidence from the past can, by establishing the presence of diagnostic features for a specific time and place, provide essential constraints on reconstructions that have been made using other lines of evidence, or on scenarios developed through *post hoc* modelling. Drawing together as much evidence as possible for a particular time slice, the last glacial maximum (LGM) or the mid Holocene, for example, has become one of the key ways in which model performance, under changed boundary conditions and forcings, can be tested against empirical data. This is especially so in the case of global climate models (GCMs) that place such heavy demands on computing power that they cannot be run for long periods in transient mode (see 2.3).

The most important insights from palaeo-research come from attempts to elucidate processes and process interactions, often by combining both empirical and modelling approaches. Where the evidence available brings the history of climate or ecosystem change right up to the present day, this type of approach can be linked to contemporary studies in ways that provide not only the ante-cedents to contemporary conditions, but also an improved basis for understand-ing long-term dynamics and for setting any notion of a base-line state into a realistically dynamic framework.

In all of the above modes of research, there is often scope for hypothesis testing (Oldfield, 1993). Palaeo-research has proved adept at testing hypotheses relating to contemporary problems, their origins and consequences, as in the case of the disputed causes of surface water acidification (see 10.2.2). Equally, we may turn to the past to evaluate hypotheses that arise from model simulations, or use simulations themselves to test the coherence and credibility of hypotheses based on empirical evidence.

3.3   The Vostok timeframe

The Vostok record, spanning the last four glacial cycles (Petit *et al.*, 1999), has become the main timeframe to study for clues from the past about current and impending climate change. It provides an unparalleled range and quality of information. This is largely because continuous, long-term reconstructions of atmospheric CO_2 concentrations depend entirely on the analysis of the air bubbles trapped in Antarctic ice, and for the moment, Vostok provides the longest record of these measurements. The concentrations measured in northern-hemisphere polar ice cores cannot be used in the same way, for they have been modified by chemical interactions with the higher levels of impurities contained within the ice (Haan and Raynaud, 1998). Moreover, although atmospheric CO_2 concentrations can be reconstructed from an index derived from fossil-leaf stomata (Rundgren and Beerling, 1999; Rundgren and Björck, 2003), this approach does not provide anything approaching a continuous record for the period before the end of the LGM. The special value of Antarctic ice-core records does not lie only in reconstructions of CO_2 concentrations. Analysis of the same air bubbles trapped in the ice also allows reconstruction of changes in the ratio $^{18}O/^{16}O$ in the atmosphere ($\delta^{18}O_{atm}$) on glacial–interglacial timescales. Variations in this ratio largely reflect changes in ice volume and global mean sea-level.

We can place the Vostok record in longer-term context by considering the records of multi-millennial scale changes in climate shown in Figure 3.1. This shows the variability in climate-related proxies preserved in a marine core and a long section of Chinese loess. In both sequences there are shifts in the amplitude and periodicity of the main cycles of variability. These take place between c. 900 000 and 430 000 years ago. Before this, most of the variability recorded is of lower amplitude and matches the 41 000-year obliquity cycle (see 3.4.1 and 4.3.1). By the time the Vostok record begins, the amplitude has increased and the periodicity of the main pattern of variability has lengthened to correspond more closely with the 100 000 year eccentricity cycle (see Figure 4.6). These shifts are superimposed on a gentle overall decline in mean ocean temperature and roughly coincide with the onset of successive major glaciations in the northern hemisphere.

The latest data from the Dome C core located some 560 km from Vostok Station sheds more light on the period of transition leading up to the beginning of the Vostok record (EPICA community members, 2004). Here, at an altitude of over 3200 m above mean sea-level, 1000 km from the nearest coast and with a mean annual temperature of –50° C, annual snowfall is only some 3 cm of water equivalent. As a result, ice accumulation is very slow and the timespan of the total thickness of ice to be cored is believed to represent the last 800 000 years. The pre-Vostok, isotopically inferred temperature record from 740 000 to 430 000 years (see Figure 5.2b) includes three periods of minimum temperature comparable to those typical of the peak of subsequent glaciations, but maximum temperatures never reach those typical of the last four interglacials. The evidence so far reinforces the view that the Vostok timeframe of just over 400 000 years, supplemented by overlapping evidence from Dome C for the glacial/interglacial transition immediately before the Vostok record begins, is an appropriate one within which to study environmental variability if the prime purpose is to shed light on the way in which the Earth system has operated in the absence of major anthropogenic perturbations.

3.4 Chronology – subdivisions, tuning and dating

3.4.1 The orbital pacemaker and marine isotope stages

The commonest way of dividing up our chosen timeframe, for the purposes of description and analysis, is into marine isotope stages (MIS). The chronological framework that they provide evolved during the 1970s and 1980s. By correlating the sequence of magnetic reversals (see 4.3.1) recorded in their sediment cores with dated ones in lava flows, Shackleton and Opdyke (1973) provided an initial framework for the major changes in global ice volume as revealed by stable isotope signatures in the foraminifera recovered from dated marine sediments (see 4.4.1). Hays et al. (1976) confirmed the role of the 100 000-year eccentricity

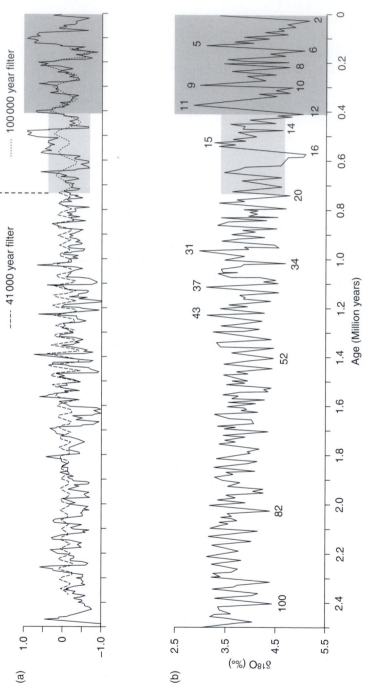

Figure 3.1 Setting the Vostok record into longer-term context. The upper graph plots grain-size variation in the loess sequence from Baoji (Ding *et al.*, 1994). The dotted lines show the 41 000-year filter prior to 740 000 years and the 100 000-year filter subsequently. The lower graph shows part of the benthic oxygen-isotope record from equatorial Atlantic core ODP-607 (Raymo, 1992). Numbers refer to marine isotope stages (see 3.2). Both records show the greater influence of variability paced by the 100 000-year eccentricity cycle over the last *c.* 430 000 years. The broad shaded area represents the timespan covered by the Vostok record. The narrow shaded area shows the timespan over which data from Dome C have, so far, extended the Vostok record. (Modified from Bradley, 1999.)

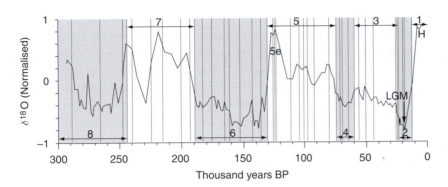

Figure 3.2 The SPECMAP composite sequence of marine isotope stages (MIS) over the last 300 000 years dated by an orbitally tuned chronology, as defined by Martinson *et al.* (1987). (Modified from Bradley, 1999.)

cycle as pacemaker of the glacial/interglacial sequence over the last *c*. 600 000 years and Martinson *et al.* (1987) published a chronology for marine sediments tuned to the orbital periodicities, that still serves as a template for changes in ice volume and climate over the last 300 000 years. Figure 3.2 shows the chronology for the last eight stages . The MIS nomenclature provides a convenient way of referencing major time intervals during the period with which we are concerned. Most of the main features of the MIS sequence, notably the interglacial stages 11, 9, 7, 5e and 1 (the Holocene) can be identified in the Vostok record (Figure 3.3) and it is convenient to make the link between the marine and Vostok ice core sequences in broad outline, though it would be misleading to imply at the outset that the parallels always denote exact synchroneity, or that the chronologies developed for each are either finalised, or perfectly interchange-able (see especially 5.2 to 5.4). Nevertheless, such are the parallels between the pattern of orbital changes and many long sequences of climate-linked changes in a wide range of environmental archives, that it has become common to date many of these sequences by tuning them to the orbital 'timescale'.

Despite any remaining uncertainties in chronologies and correlations, this common framework is sufficiently robust to allow its use as a basis for defining major episodes for which evidence can be drawn together from a wide range of

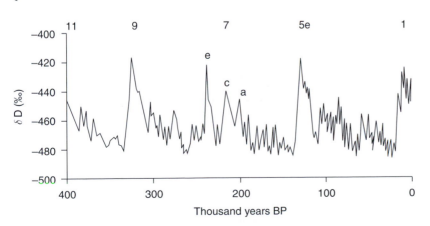

Figure 3.3 Isotopically inferred temperature variations over the last 400 000 years at Vostok. Interglacial MIS stages 1, 5e, 7 and 11 are indicated. (Based on Petit *et al.*, 1999.)

environments, marine; terrestrial and polar. Relying on tuning alone, however, precludes determination of the precise phasing between the records that have been synchronised to achieve the tuning. Thus, any evidence for leads and lags between signatures of environmental change in different archives requires some kind of independent chronological control. This issue becomes of outstanding importance in addressing the role played by greenhouse gases in the dramatic changes recorded during the transitions from glacial to interglacial conditions (see especially 6.4.7).

3.4.2   Seasonal signatures

The development of chronologies with finer temporal resolution than the comparison with the known pattern of orbital changes can provide depends upon the use of a great variety of methods. Ideally, accurate, absolute ages are required, which have been precisely and independently determined for each archive and proxy. Only in this way is it possible to establish the exact phasing of processes and events detectable at different sites and to calculate rates of change. In practice, the ideal is seldom realised, save in those cases where annual resolution can be achieved by counting growth increments, or varves – sediments characterised by consistent, undisturbed, seasonal patterns of accumulation that permit the recognition of annual layers (see, e.g., Overpeck, 1996; Zolitschka, 2003). Tree rings (see, e.g, Briffa et al., 2002a; 2002b; Baillie and Brown, 2003), corals (see, e.g., Cole, 2003; Gagan et al., 2000; 2004) and, in some instances, speleothem seasonal growth increments (Lauritzen, 2003), and varved sediments are thus of special value for the potential they offer in establishing rates of change, as well as for the proxy evidence for environmental change contained within them.

A variety of seasonal signatures have been used successfully to date ice cores (for a summary, see Bradley, 1999, p. 142) These include detecting the seasonality of the $\delta^{18}O$ signature (see 4.2.2), identifying seasonally deposited dust layers, visual stratigraphy, electrical conductivity measurements, trace-element and microparticle analysis. It is usual to develop chronologies of ice accumulation by combining different dating methods and also by using reference horizons such as tephra layers or acid peaks associated with volcanic eruptions, and spikes in the concentration of cosmogenic radionuclides such as ^{10}Be. These may serve to correlate cores, so that emerging chronologies can be used in combination to constrain each other. The precise age of many major volcanic eruptions, especially during the last thousand years or so, has become well established through a combination of chemical fingerprinting, links to documentary sources and dendrochronological dating of the transient impact of the eruption (see 4.3.3. and Briffa et al., 1998).

3.4.3   Dating older ice

Beyond a certain depth that varies from core to core, the capacity to resolve annual layers is lost as the ice becomes increasingly compressed and subject to flow. In this type of material, dating is mainly dependent on the development of theoretical flow models constrained by any correlations with other dated sequences that can be established. As a result, dates on the older layers of ice cores are subject to a significant degree of uncertainty that generally increases with age. In some cases, as for example in ice older than $c.100\,000$ years in the GRIP and GISP cores from central Greenland, the stratigraphic sequence can become distorted and unreliable (Taylor *et al.*, 1993).

Until recently it has been difficult to correlate and compare in detail the sequence of changes in Greenland and Antarctica. The problem is now largely resolved through inter-hemispheric correlations based on the changing record of methane trapped in the air bubbles (Blunier *et al.*, 1997; 2001; Caillon *et al.*, 2003a). Such methods achieve synchronisation, but not absolute chronology. Providing precise dating control for older ice is an important goal, especially for the Vostok record. Petit *et al.* (1999) adopted a chronology based on an ice-flow model pinned to the SPECMAP marine isotope chronology at $110\,000$ and $390\,000$ years before present (BP). More recently, Shackleton *et al.* (2000), Bender (2002), and Ruddiman and Raymo (2003) have proposed modifications to this chronology by tuning variations in trace-gas concentrations to orbital frequencies. Modifications in both the Vostok and MIS chronologies have important implications for resolving the phasing of events during periods of glacial initiation and termination. These are considered further in Chapters 5 and 6.

3.4.4 Radiocarbon dating and ^{238}U/^{230}Th ratios

Of the various radiometric methods available, radiocarbon dating has been by far the most useful for organic material formed within the last $\sim$40 000 years. Radioactive ^{14}C is produced in the atmosphere by cosmic bombardment. It is incorporated in living organisms and, on the death of the organism, provided it is excluded from exchange with the atmosphere, the ^{14}C decays in accordance with its half-life of 5730 years. Its decay thus constitutes a radiometric clock whereby determination of the radioactivity remaining in the sample allows calculation of the sample's age. Carbon-14 dates are normally quoted as dates BP, where the present is taken as AD 1950. They are normally given with one standard error derived from the counting statistics. Several methods are available for determining the ^{14}C activity in a sample. Increasingly, the more conventional methods using several grams of carbon extracted, then measured in gas or liquid form, are giving way to measurements using accelerator mass spectrometry (AMS), by means of which milligram-size samples can be dated. While the basic principles of radiocarbon dating are beguilingly simple, in practice establishing dates is often far from straightforward. Here it is possible to do no more than outline briefly, by way of illustration, some of the complications:

1. The atmospheric concentration of ^{14}C has varied through time, partly as a result of changes in the rate at which the radioisotope is received at the Earth's surface, partly because of changes in the residence time of ^{14}C in the ocean, which in turn affects ^{14}C concentrations in the atmosphere. It follows that organisms incorporating ^{14}C at different times will have different starting values from which decay begins. This means that ^{14}C 'dates' are not true age determinations. Ideally, they need to be calibrated to an absolute timescale established by methods that are not subject to any similar error. For the last 9500 years, this has been done by comparing the ^{14}C 'dates' and calendar ages of tree rings (Stuiver and Reimer, 1993).

2. Beyond the period for which direct tree-ring calibration of ^{14}C dates is possible, there is less certainty involved in converting them to calendar ages. Both varved marine sediments (Hughen et al., 1998) and Uranium/Thorium (U/Th) dating (see below) have been used for calibration. Using a combination of these, Stuiver et al. (1998) proposed calibrations back to 24 000 BP. More recently, Hughen et al. (2004) and Bard et al. (2004) have proposed calibrations back to 50 000 years.

3. One of the consequences of the changes through time in atmospheric ^{14}C concentrations is that, for certain periods, radiocarbon activity does not decline monotonically with age. It may plateau or even briefly increase. Quite a wide range of calibrated ages can thus be ascribed to individual ^{14}C dates within such time intervals. In order to establish the age of materials within these periods, it is often necessary to determine a close sequence of radiocarbon 'dates' and match their pattern to the calibration wiggle. This method of 'wiggle matching' has provided a basis for some rather precise age determinations (e.g., Blaauw et al., 2003, and Figure 3.4).

4. In every environment sampled, it is essential to establish that the dated material is contemporary with the event or stratigraphic horizon to which it is intended to apply. This is not always straightforward. Sedimentation processes often include the re-deposition of older materials, thus introducing older carbon into the sample to be dated. They may also re-deposit older organisms bearing palaeoclimatic signals that are not applicable to the period of sediment accumulation (Okhouchi et al., 2002). In hard-water environments, aquatic organisms may incorporate old carbon during photosynthesis. In the ocean, the age of the carbon used by marine organisms varies greatly with water depth and location, depending on the degree to which the water column in which the organisms live has been isolated from exchange with the atmosphere. There is therefore a 'reservoir' effect that must be estimated correctly if true age is to be calculated from ^{14}C activity. In most studies, it has been necessary to assume that in a given location, this effect remains constant over long periods. Recent work using tephra layers (see 3.4.6) deposited over both land and sea, or synchronous rapid climate shifts to correlate marine sequences with precisely dated terrestrial material (Waelbroeck et al., 2001; Björck et al., 2003) has shown that this assumption is not always valid.

One of the most useful dating methods for carbonate materials relies on the Uranium-238 decay series. During the course of the radioactive decay of ^{238}U, it is replaced by ^{230}Th. The rate of decay/replacement is known and, provided the

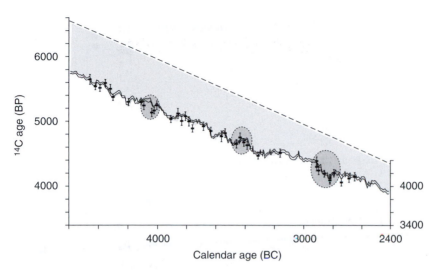

Figure 3.4 Carbon-14 dating – calibration and 'wiggle matching'. The dashed straight line at the top of the graph shows what the slope relating ^{14}C age to calendar age (BC) would be if no process other than radioactive decay controlled the relationship. The wavy, continuous lines below (plotted as the bandwidth of the standard deviation values), show the actual relationship between ^{14}C age and calendar age (BC), as determined by ^{14}C measurements of tree rings (Stuiver *et al.*, 1998). There is a significant offset that varies through time between the lines representing the 'ideal' and the actual relationship. Note the three clusters of individual radiocarbon determinations from a peat core, that allow the chronology to be pinned by 'wiggle matching' to some of the main high-frequency deviations between ^{14}C age and calendar age. (Based on Blaauw, 2003.)

carbonate is sufficiently pure and has remained a chemically closed system, the [238]U/[230]Th ratio constitutes a remarkably precise chronometer over an age range from a few years to *c.* 350 000 years BP. The method has been successfully applied to corals (Bard *et al.*, 1998), speleothems and lacustrine carbonates (Edwards *et al.*, 1987). From the mid 1980s onwards, [230]Th/[238]U dating has been carried out increasingly by thermal ionisation mass spectrometry (TIMS). This has allowed a significant reduction in sample size along with significant improvements in the precision achievable. Some sense of its importance in pinning down late Quaternary chronologies and correlations may be obtained from the articles by Bard *et al.* (2004) and Shackleton *et al.* (2004).

3.4.5   Luminescence dating

Neither of the above methods is applicable to sediments lacking in organic matter or pure carbonates. Over recent years, much progress has been made in the development of luminescence as a dating method. This relies on the fact that some of the electrons emitted during radioactive decay become trapped within the fabric of any sufficiently dense material, for example quartz or feldspar crystals. They can then be detected by means of the luminescence that they emit. All other things being equal, the luminescence resulting from the retention of trapped electrons will increase through time from the date on which the electron traps were last emptied. In dating sediments, use is made of the ability of light to empty electron traps. If, before deposition, material is exposed to sufficient light to empty the trapped electrons, then, provided subsequent deposition excludes the sediment from light, the luminescence clock has been reset and electrons build up through time. Increasingly sophisticated methods have been developed for isolating the time-dependent luminescence signal and using it to establish date of deposition. Problems can arise from, for example, incomplete zeroing of

Figure 3.5 A loess/
palaeosol profile dated by
optically stimulated
luminescence (OSL). Note
the good agreement
between the OSL age
determinations and all but
the most recent of the ^{14}C
dates, as well as the age
ascribed to the 'Eemian'
soil. This roughly
corresponds to MIS 5e,
currently dated
independently to
c. 125 000–118 000 years
BP. (Modified from Lang
et al., 2003.)

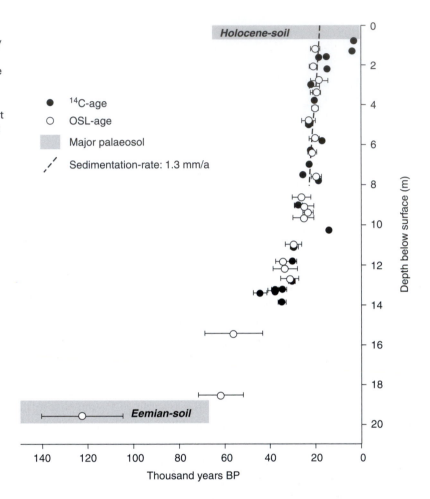

the luminescence signal before deposition, or changing water content after deposition. Nevertheless, optically stimulated and (its variant) infrared stimulated luminescence (OSL and IRSL respectively) are being applied with increasing success to alluvial and colluvial sediments (Lang, 2003), loess (Roberts *et al.*, 2001; Figure 3.5) and dune sands (Wintle, 1993; Wintle *et al.*, 1998).

3.4.6   Correlation methods – tephra and palaeomagnetism

In many sediment sequences, especially where opportunities for direct dating are lacking, methods are used that permit correlation with other sequences that have been directly dated. One of the most useful of these uses volcanic ash (tephra) layers, chemically and mineralogically 'finger-printed' so that they can be ascribed to volcanic eruptions that have been either identified in documentary records, or dated using one of the methods described above. For example, Zillen *et al.* (2002) use lake-sediment varves for this purpose. Improvements in the techniques for extracting and finger-printing tiny quantities of fine distal tephra, remote from their point of origin, have allowed the development in northern Europe of tephrochronologies linked to the history of volcanism in Iceland (Hall and Pilcher, 2002). Similar chronologies are emerging for many parts of the world and some tephra, for example the Mount Mazama ash that was produced as Crater Lake, Oregon was created around 7000 years ago, provides a dated marked horizon over many thousands of square kilometres.

Another technique relying on correlation with dated sequences elsewhere is that of palaeomagnetism. The long-term chronological framework for environmental change during the Quaternary was largely established by combining the record of reversals of the geomagnetic pole with dates established radiometrically, using the potassium/argon (^{40}K/^{40}Ar) method (Dalrymple and Lanphere, 1969). The Vostok timeframe lies entirely within the latest Bruhnes Normal Polarity Chron, or Epoch, that began some 780 000 years ago. Thus, the contribution of palaeomagmetism to establishing chronologies during the Vostok timeframe depends on detecting for shorter-term, less complete changes in polarity (polarity 'excursions'), as well as characterising the continuous pattern of palaeomagnetic secular variation (PSV). Both types of variation can be reconstructed from sequences of lavas and from sediments. Although several proposed excursions remain controversial, those like the Blake Excursion (108 000–112 000 years BP), that are well dated and authenticated, can provide a basis for wide-scale correlations. Secular variation sequences that have been dated may serve as 'master curves' for other undated sequences with which they can be correlated. Since secular variation at any point on the Earth reflects the changing behaviour of non-dipole (i.e., regionally differing) as well as dipole components, each region needs its own set of 'master curves'. Reconstructions of PSV curves from varve-dated lake sediments in Scandinavia have been able to provide a secure chronology for PSV variations in that region (Saarinen, 1999; Snowball and Sandgren, 2002; Zillen, 2003; Figure 3.6).

The quality of sedimentary PSV records varies enormously and is highly dependent on particle size, sediment structure, lack of severe bioturbation (sediment mixing by bottom-living organisms), the concentration and grain size of the magnetic minerals present and the extent to which post-deposition conditions have favoured their survival. Changes in palaeointensity records usually reflect variations in sedimentary properties as well as changes in the past strength of the Earth's magnetic field. Reconstructing true palaeointensity requires near ideal and either constant conditions of sedimentation, or only variations that can be fully accommodated by normalisation procedures

3.4.7   Dating recent sediments

Many human impacts on environmental systems have accelerated and become more widespread over the last two centuries and especially the last 50 years. In the case of climate, instrumental records are available for this period, but the

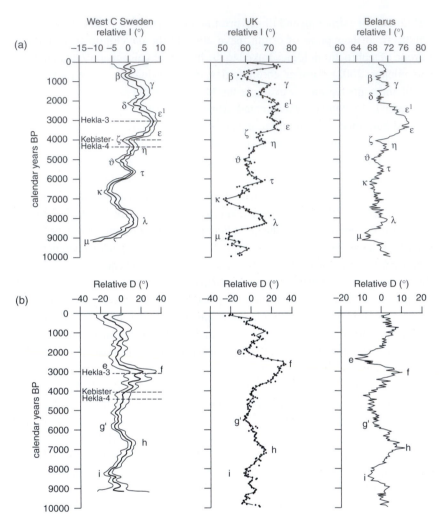

Figure 3.6 Correlations based on changes in palaeomagnetic secular variation (PSV) as recorded in lake sediments. The upper row of graphs shows changes in inclination (I), the lower in declination (D), at sites in Scandinavia, the UK and Belarus. The main inflections used for correlation between the sequences are denoted by Greek letters in (a) and by Roman letters in (b). The timescale is derived from varve-dated lake sediment records in Sweden and, in the case of the UK and Belarus data, from radiocarbon dates. Changes in the detailed expression of each feature from site to site reflect both sedimentological 'noise' and spatial variability related to non-dipole components of the magnetic field at each location. Differences in the position of the features relative to the timescale shown are mainly a function of the difference between the varve-based and radiocarbon chronologies. The dates of three Icelandic tephra identifiable in the western Swedish lake sediments have been added. (Modified from Zillen, 2003.)

situation is very different when we consider many other aspects of environmental change. Lack of any monitoring or direct and consistent observations
over that period has placed a premium on developing ways of establishing dates
for quite recent events and trends recorded in archives such as lake and near-
shore marine sediments, and peats.

Where a continuous sequence of sufficiently accurate and precise radiocarbon
dates is available, the very fluctuations in atmospheric ^{14}C that make isolated
dates meaningless for the last $c.$ 300 years can be used to advantage to provide a
'wiggle-matched' chronology (Oldfield $et\ al.$, 1995; 1997).

The history of radioactive fall-out from weapons testing and major nuclear
accidents has provided chronological markers for the period from the 1950s
onwards. The radioisotope most often used is one of the radioactive isotopes of
caesium (^{137}Cs) from which, in sufficiently well-resolved and unperturbed
sediment sequences, the onset of measurable atmosphere concentrations in
1954, the peak fall-out in 1963 and, in some cases, a subsidiary peak in 1958,
can be identified. Americium (^{241}Am) can also be used in the same way and,
though more difficult to measure, it has the advantage of being less chemically
mobile. Since their atmospheric concentrations changed on a global basis (with a
delay of a year between northern and southern hemispheres) the signatures of
both radioisotopes can be used globally. By contrast, fall-out from the Chernobyl
accident in 1986 provides a dating 'spike' only in the areas over which the cloud
of radioactivity emitted during the accident passed, and only at those sites where
deposition, usually through rain-out, took place.

In addition to artificial radionuclides, one natural decay series, that of ^{210}Pb,
has proved invaluable for dating sediments spanning the last 100 to 150 years –
the period of major human impact on most ecosystems. Radium-226, the parent
radioisotope of ^{210}Pb, occurs in the Earth's crust and decays to form the gas
radon which diffuses from rocks into the atmosphere, where further decay via
short-lived isotopes leads to the production of ^{210}Pb. Atmospheric ^{210}Pb is then
deposited onto land and water surfaces whence it finds its way into accumulating
sediments. In order to make use of ^{210}Pb as a dating tool, the atmospherically
derived ('unsupported', or 'excess') activity has to be separated from the activity
arising from the in-$situ$ decay of ^{226}Ra present in the sediments. Provided the
flux of unsupported ^{210}Pb to the system has remained roughly constant through
time, its changing activity versus depth can be used to develop a chronology of
sedimentation using one or more of the dating models available (Oldfield and
Appleby, 1984; Appleby and Oldfield, 1992). The accuracy and precision of the
method varies with location, the nature of the sediment delivery system, the rate
of sedimentation and the degree to which the fluxes and pathways of ^{210}Pb
incorporation in the sediments have been perturbed. Nevertheless, it has been
used with great success in many contexts, both lacustrine (Appleby $et\ al.$, 1979;
1990; Figure 3.7) and marine, as well as in dating peat profiles under favourable
circumstances (Oldfield $et\ al.$ 1995; Shotyk $et\ al.$ 1996; 1998).

(a)

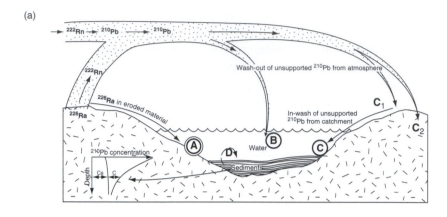

(b)

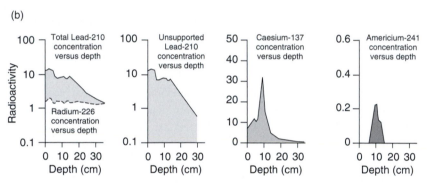

Figure 3.7 Recent chronologies using short-lived and artificial radioisotopes. (a) Schematic model of the incorporation of unsupported (excess) ^{210}Pb into lake sediments. A refers to the eroded input of ^{210}Pb supported by its parent isotope ^{226}Ra. B refers to direct deposition of unsupported ^{210}Pb onto the lake surface. C refers to unsupported ^{210}Pb either washed directly into the lake (C_1) or initially incorporated into surface soils (C_2). D refers to re-suspended and re-deposited surface sediment. The sketch, bottom left, shows a simplified graph of the concentrations of supported (C_s) and unsupported ^{210}Pb with depth. The dating parameter is the declining concentration of the latter. (Based on Oldfield and Appleby, 1984.) (b) The graphs in (b) show the depth distribution of the radioisotope measurements from which the chronology and sediment-accumulation rate shown in (c) are calculated. The graphs show, from left to right: total ^{210}Pb and ^{226}Ra concentrations; unsupported (excess) ^{210}Pb concentrations once the activity supported by ^{226}Ra has been subtracted; ^{137}Cs concentrations; ^{241}Am concentrations. (c) Plots of age and dry mass sedimentation vs. depth for a core from Windermere in the English Lake District. The upper graphs show the basis for the ^{210}Pb chronology and the depths at which datable features in the weapons-testing isotope record (^{137}Cs and ^{241}Am) occur. Chronologies of this kind are especially important for recording recent human impacts on terrestrial and aquatic ecosystems where the impacts have not been systematically recorded by direct observations or measurements (see Chapters 9 and 10).

(c)

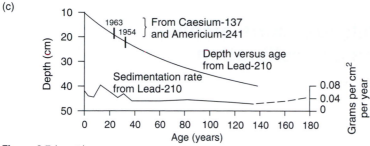

Figure 3.7 (cont.)

3.5   Temporal resolution

The notion of temporal resolution can be applied both to the time interval covered by a single sample and to the minimum time interval resolved by a given sampling interval. Both will depend on sampling strategy as well as on the nature of the material to be sampled. High temporal resolution requires closely spaced samples, each spanning a short time interval. In some types of depositional environment, temporal resolution is limited by slow sedimentation rates or by the effects of mixing and/or chemical diffusion, both of which smooth out the record and reduce the degree of temporal resolution that can be realistically achieved. High temporal resolution is required in order to resolve many issues of phasing and to establish rates of change where these are rapid relative to the rate of accumulation of the archive studied. In most of the environmental archives where seasonal, signatures provide an absolute chronology, annual, in some cases seasonal, resolution is possible. At the other extreme, bioturbation and slow sedimentation in some marine environments, or the combined effects of slow ice accumulation, and gas diffusion and mixing, can limit temporal resolution, for some archives and proxies, to centuries or millennia.

Chapter 4
The Palaeo-record: archives, proxies and calibration

4.1 Instrumental and documentary archives of past environmental change

4.1.1 Instrumental climate records

Instrumental data serve a variety of purposes in palaeo-environmental study:

1. They make it possible to quantify objectively how the climate has varied in recent times and provide the most statistically robust data from which to establish the short-term spatial and temporal coherence of those modes of variability (see 7.6) recognisable in the climate system at the present-day.
2. They provide the basis on which diverse proxies (see below) can be calibrated to measured climatic properties, using statistical expressions that either capture the links between their current distributions and present-day climate, or express their responses to climate variability over some recent time interval for which instrumental records are available.
3. They form a link between the longer, proxy records available from environmental archives and the period of present-day monitoring.

The longest series of instrumental observations of the weather is for the English Midlands and goes back into the seventeenth century. In this instance, careful evaluation of each instrumented site and measurement has made possible a detailed record of climate change from AD 1659 onwards (Manley, 1974) (see Figure 4.1). In most parts of the world, there are very few reliable instrumental series that go back more than 100 years. For part of the span of time over which reliable, instrumental records are available, one of the most powerful tools for reconstructing past atmospheric conditions is what is termed reanalysis. In reanalysis studies, daily observations have been assimilated into a comprehensive global representation of the daily state of the atmosphere, including vertical temperature and pressure gradients, clouds and precipitation. The longest reanalysis series available is the NCEP/NCAR (National Center for Environmental Prediction/National Center for Atmospheric Research) series from 1948 onwards. One of the key features of reanalysis is the use of consistent procedures for harmonising data from a vast range of initially inhomogeneous observations.

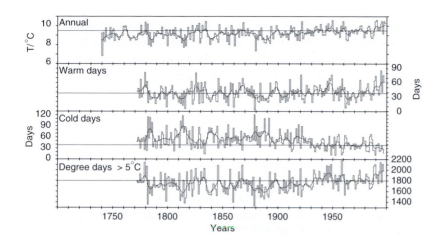

Figure 4.1 Climate reconstructions based on instrumental records from central England. The 'days' in the middle two graphs indicate the number of days above/below the 90th/10th percentile values for that particular day of the year. (Modified from Jones *et al.*, 2002.)

In order to recreate the atmospheric properties provided by the reanalysis, these data have to be assimilated into a global circulation model that generates an output for each day that best matches the full range of data for that day. Reanalysis provides a way of recreating many of the climatically significant properties of the atmosphere on a daily basis for the whole of the period covered. This in turn makes possible much greater insight into large-scale atmospheric processes, patterns of variability and longer-term trends. It also provides a much better basis for extrapolation from data-rich to data-poor regions. Reanalysis studies have been of great importance in global climate model (GCM) inter-comparison, as well as in the detection and ascription of climate change in the recent past (Chapter 12).

4.1.2   Direct measurements of other environmental change processes

Environmental change is about more than climate alone. If we now turn to other components of the Earth system, quantitative knowledge based on measurements and direct observations is much more diverse and less integrated. In most cases, coordinated observations on a global scale and with some continuity through time hardly existed before the start of satellite-remote sensing in the 1970s. For the most part, effective coordination has only come through the various global observation systems set up within the last two decades. Even individual case studies of environmental processes on a small spatial scale have only rarely stretched beyond a decade or two. Hydrological and tide-gauge records come closest to achieving potential coherence and continuity beyond the last few decades, but even in these cases, harmonising, cross-validating and integrating multi-decadal data on a global or even national scale pose severe problems.

Two types of recent observational records deserve special mention, for both served to alert the scientific community and the world at large to the potential dangers resulting from human activities – the monitoring of atmospheric CO_2 concentrations and of stratospheric ozone. Direct measurements of atmospheric CO_2 concentrations began at the Mona Loa Observatory in Hawaii in 1958, since when they have been of vital importance in linking up with Antarctic ice-core measurements to document the post-industrial increase, tracing the continued rising trend, and providing vital information on seasonal and inter-annual variability (see 8.2.1). Concerns about the effects of chlorofluorohydrocarbons (CFCs) on stratospheric ozone (see 8.4) were unheard of before 1974, but during the short period since then, demonstration of their effects has led to major changes in technology.

Instrumental climate records and other direct measurements cover, at best, less than 0.1% of the Vostok sequence. In order to lengthen the time perspective further, we turn next to documentary records.

4.1.3   Documentary records

The primary goal of historical climatology is to extend 'continuous monthly and seasonal temperature and precipitation data back into the pre-industrial period on the basis of documentary data' (Pfister et al., 2002). These data may be derived from both direct or indirect evidence of weather conditions. Documentary evidence is often highly specific with regard to time and place but its translation into more or less quantifiable descriptors of climate is a specialized task. The procedure outlined by Pfister et al. (2002) involves the construction of a range of standardised indices, calibration of these using a period when the documentary sources overlap with the period of instrumental measurement, and verification of the calibration by applying it to a separate period of overlap not used in the calibration exercise. Provided the response function that emerges from the above procedure is sufficiently precise and constant over the whole period of reconstruction, the index values can be used to reconstruct a record of, for example, temperature and precipitation variability. (Pfister et al., 2002). Pauling et al. (2003) have recently tested the reliability of a range of climate proxies against instrumental records from central Europe and conclude that documentary evidence can provide the most reliable proxy record available for changes in European winter temperatures. Documentary sources of the kind used for climate reconstruction are available for many parts of the world, although the timespan they cover varies from a couple of centuries in the case of Australia, for example, to five millennia in the case of Egypt. Not only land-based observations, but maritime logs and accounts of voyages can also be used.

If they are sufficiently detailed and reliable, document-based climate reconstructions can be interpreted in terms of synoptic patterns and even provide the basis for attempting a partial palaeo-reanalysis (Wanner and Luterbacher, 2002; Luterbacher et al., 2002).

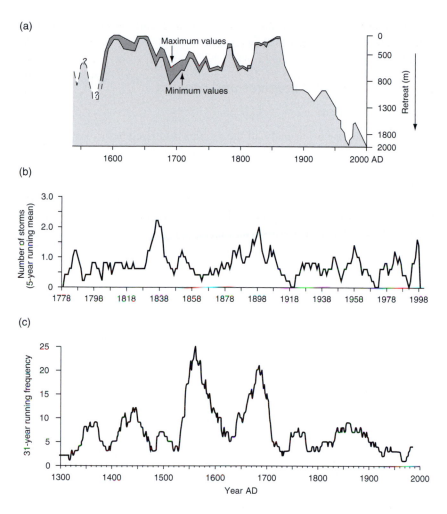

Figure 4.2 Examples of the use of documentary evidence in climate reconstruction. (a) The advance and retreat of the Lower Grindelwald Glacier (AD1535–2000). (Modified from Holzhauser and Zumbuhl, 2002.) (b) Time series of tropical cyclone frequencies for South Carolina (AD 1778–2000). (Modified from Mock, 2002.) (c) River Pegnitz flood intensities as recorded in Nuremberg (AD 1300–2000). Note that the peak values during parts of the sixteenth and seventeenth centuries greatly exceed any during the period since AD 1700. (Modified from Brazdil *et al.*, 2002.)

As well as providing the basis for the kinds of continuous reconstructions noted above, documentary evidence has also been used to reconstruct the history of El Niño (e.g., Ortlieb, 2000) and the advance and retreat of continental glaciers (Figure 4.2a). Documents can also contain vital evidence of extreme events such as cyclones (Figure 4.2b), hurricanes, floods (Figure 4.2c) and droughts. The record of Nile floods, for example, is one of the longest hydrological series in the world (Hassan, 1981). Proxy evidence often includes accounts of the effects of extreme events on phenology, crop yields and social organisation. Documentary evidence linking changing climatic conditions to contemporary indicators of harvests, human welfare, demographic changes and socio-cultural responses can deepen understanding of the nature of the interaction between environment and society during periods of stress.

Even the longest historical records from Egypt span little more than one percent of the Vostok timeframe, and to extend the record further back into the past we must now consider the great range and variety of environmental archives, as well as the proxies they contain.

4.2   Environmental archives and proxies

4.2.1   Calibration

In order to make use of virtually all the proxies discussed below some form of calibration is needed. This usually takes one of three forms:

1. Direct comparison, over the same time interval, of a time series of the proxy with a time series for the measured process or property which it is designed to reconstruct. This requires that an adequate period of overlap exists between the timespan covered by the proxy and the period of direct measurement. Figure 4.3 shows two examples of this type of calibration, relating to tree rings and corals, respectively.
2. A proxy time series may lack effective overlap in time with the property or process it serves to represent, often because the period of direct measurement is too short. In such cases, calibration is usually based on the assumption that variability in space at the present day can be used to represent variability in time. Thus, a range of measurements over an area spanning the kind of variability encountered in the palaeo-record can be

Figure 4.3 Examples of the calibration of proxy climate records by means of overlapping time series. (a) Tree rings: mean summer temperature (June–August) for the period 1869–1997 used to calibrate the tree-ring data from northern Sweden. In the graph shown, actual temperature variability is used and decadal-to-century scale variability is represented; RSC = regional curve standardisation. (Adapted from Grudd *et al.*, 2002.) (b) Corals: time series of element ratios and $\delta^{18}O$ versus SSTs for a site in Japan (Mitsuguchi *et al.*, 1996). (Adapted from Bradley, 1999.)

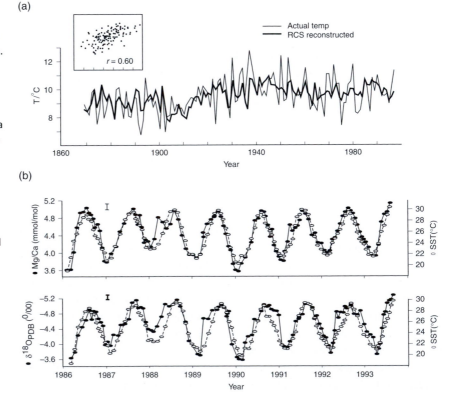

linked to values for the proxy derived from the same spread of localities. There are several alternative approaches to this type of calibration, depending on the nature of the proxy being used, but each allows some kind of empirical-statistical model to be developed to capture the relationship between the proxy and the direct measurements. Once validated, such relationships are often termed transfer functions (Birks, 2003). Figure 4.4 shows examples of the type of relationship between present-day distributions, or measurements, and climate, which can be used to develop transfer functions.

3. In neither of the above two approaches to calibration is it necessary to understand fully the processes that are responsible for the empirical relationships upon which the calibration depends, though some sense of the likely mechanisms involved can reinforce confidence. In other cases, calibration may lean more heavily on an understanding of the processes linking the proxy to the property it represents. For example, this is often the case when measurements of stable isotopes ($\delta^{18}O$, $\delta^{13}C$ and δD – see 4.2.2) are used to reconstruct variations in past climate from archives where the variations in stable-isotope ratios depend on several biotic and abiotic processes.

One of the key assumptions in all the above approaches to calibration is that the relationships established through the calibration process hold good over the

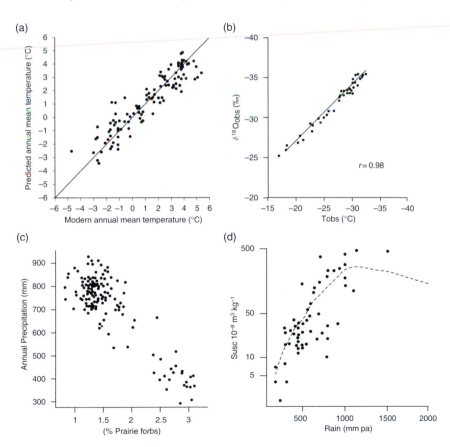

Figure 4.4 Examples of the calibration of proxy climate records by means of transfer functions derived from present-day distribution. (a) Pollen vs. temperature. Scatter plot showing the reliability of the pollen calibration to annual mean temperature for modern climatic conditions in northern Finland. (From Heikkila and Seppa, 2003.) This calibration function is used in the reconstruction of Holocene temperature variability in Figure 7.1b. (b) Stable isotopes vs. temperature. The $\delta^{18}O$ variations versus temperature over the Greenland ice sheet as recorded by Dansgaard (1964). (Adapted from Jouzel et al., (2000).) (c) Pollen vs. precipitation. Annual precipitation vs. the relative percentage of pollen of prairie forbs in modern pollen spectra from the northern midwest of North America (Bartlein et al., 1984). (From Bradley, 1999.) (d) Pedogenic (soil-formed) magnetic susceptibility vs. precipitation. The values come from scattered sites in the northern hemisphere and show a strong positive relationship up to c. 1500 mm per year. (Modified from Maher and Thompson, 1999.)

whole range of variability represented in the proxy record. In the first approach, the range of variability encountered during the relatively short calibration period will often be significantly less than that in the whole proxy record. Some degree of extrapolation of the calibration function will therefore be required. In the second approach, past periods may have no analogues in the present day, in which case the increased uncertainty in calibration can compromise attempts at quantitative reconstruction. Figure 4.5 illustrates the problems that can arise when values for a proxy lie outside the calibration space.

Ideally, reconstructions should be based on absolute values for each process or property, with realistic error estimates ascribed to each value. This is not always achieved. There is a spectrum of quantification ranging from the ideal to semi-quantitative, even entirely qualitative descriptions in which only non-numeric, relative statements (warmer/cooler; wetter/drier) can be applied. Moreover most, if not all, proxies are subject to uncertainties and biases that may be only partially captured in error analysis. For these reasons, it is always important to use as many independent sources of information as possible (see 4.6). A wider bandwidth spanning well-constrained possible values is always preferable to narrowly defined, but spurious precision.

The text that follows is organised, in so far as possible, by reconstructed environmental property (air temperature, precipitation, etc.), rather than by type of archive or proxy. This approach links the proxy records more directly to the themes outlined in Section 1.3. An exhaustive account of the full range of

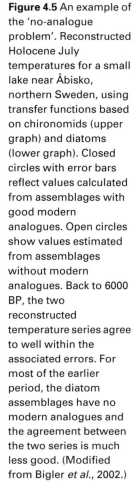

Figure 4.5 An example of the 'no-analogue problem'. Reconstructed Holocene July temperatures for a small lake near Åbisko, northern Sweden, using transfer functions based on chironomids (upper graph) and diatoms (lower graph). Closed circles with error bars reflect values calculated from assemblages with good modern analogues. Open circles show values estimated from assemblages without modern analogues. Back to 6000 BP, the two reconstructed temperature series agree to well within the associated errors. For most of the earlier period, the diatom assemblages have no modern analogues and the agreement between the two series is much less good. (Modified from Bigler *et al.*, 2002.)

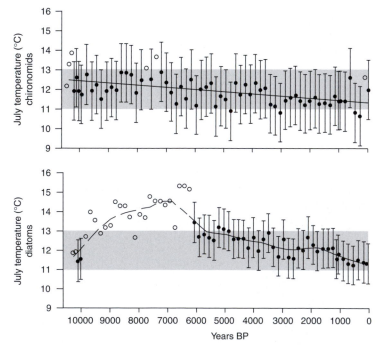

proxies used in Palaeoclimate reconstruction alone is well beyond the scope of this chapter. For this, readers are referred to Bradley (1999) and Lowe and Walker (1997). With regard to Palaeoclimate, the present account is necessarily rather brief and selective, but it includes, in addition, sections on the reconstruction of other environmental processes and properties reflecting hydrological and ecosystem changes in response both to climate and human activities.

4.2.2   Stable isotopes – a special case

The stable isotopes of oxygen, hydrogen and carbon in particular are used in a wide range of archives to provide insights into a remarkable variety of environmental processes. This reflects the fact that many processes, both biotic and abiotic, lead to fractionation whereby one stable isotope is sequestered, precipitated, metabolised or released preferentially. For example, in the atmosphere, on a global scale, the ratios between the stable isotopes of oxygen ($^{18}O/^{16}O$) and also those of hydrogen ($^{2}H/^{1}H$) reflect, to a first approximation, air temperature. There is therefore often a quantifiable link between air temperature and the stable-isotope ratios present in the water droplets that form and fall as precipitation, with lower/lighter ratios characteristic of cooler temperatures, and vice versa. On a smaller spatial scale, other factors influence the ratios, including the amount of prior precipitation, air mass source and seasonality. Incorporation of stable-isotope signatures into sedimentary archives involves additional fractionation processes and these have to be taken into account before climatic variables can be inferred from the stratigraphic record.

Variations in the carbon stable-isotope ratio ($^{13}C/^{12}C$) may also be indirectly linked to changing climate, though they are more often a function of a more complex set of processes. One of the ways in which $^{13}C/^{12}C$ ratios have been used is as an indicator of the source of carbonates and organic matter in sediments. They can also be used to identify the relative contribution of different types of source (marine versus terrestrial for example) to the carbon in atmospheric CO_2, both past and present. Variations in stable-isotope ratios are usually expressed as deviations (delta values) from an international standard, thus: $\delta^{18}O$, δD (deuterium, or ^{2}H) and $\delta^{13}C$. Some sense of the versatility of stable-isotope measurements as palaeo-environmental indicators emerges from the sections below and from reviews such as those of Bradley (1999), Jouzel *et al.* (2000), Cole (2003), Maslin *et al.* (2003), Leng (2003), Leng and Marshall (2004) and McDermott, F. (2004).

4.2.3   Proxy-based Palaeoclimate reconstructions: air temperature

Air temperature is the single most frequently reconstructed climatic variable, and public interest in 'global warming' gives it special significance beyond the scientific community. Attempts have been made to reconstruct air-temperature variations from many different archives. Stable-isotope measurements in high-latitude ice-cores are often used as a basis for inferring the temperature of the air masses through which the precipitation incorporated in the ice-core record has fallen (see Figure 4.4b). Since stable-isotope ratios in both oxygen (δ^{18}O) and hydrogen (δD) can be measured, a degree of mutual constraint is introduced, as well as additional insight into moisture source. The temperature reconstructions for ice cores from both Greenland and Antarctic cores are based on extensive calibration exercises (Jouzel *et al.*, 1997), despite which they have been challenged by recent studies. In the case of central Greenland, reconstructions from borehole temperatures recording the temperature of the ice itself as it changes down the hole, can, once calibrated (see summary in Alley, 2000), provide a smoothed record of past air temperatures. They show that the isotopically inferred difference in air temperature between the present day and the last glacial maximum in central Greenland is probably an underestimate by some 10 to 15 °C (Johnsen *et al.*, 2001). Borehole temperatures in rock and permafrost have also been used to reconstruct rather smoothed records of air temperature change spanning the last few centuries from a number of continental sites (see, e.g., Huang *et al.*, 2000). Relative temperature series have been derived from high-latitude ice caps away from Greenland and Antarctica by calculating the percentage of each ice-core segment recording summer melting and percolation of meltwater into the firn (Fisher and Koerner, 2003 and Figure 7.1).

Under favourable circumstances, the remains of fossil wood representing the former position of alpine tree-lines have been used to reconstruct past changes in temperature. Although the concept is rather simple, in practice it is complicated by a series of problems; for example, the degree to which present-day tree lines have been modified by human activities, the possibility that the highest fossil wood does not represent the highest trees for a given period, and the likely delay between climate change and tree migration.

Tree rings have become, for many parts of the world, the main source of information on climate change, partly because of the skill with which they can be used to reconstruct aspects of past climate and partly because they mostly provide absolute dates with annual resolution. By overlapping distinctive tree-ring sequences from living trees with those from fossil wood or archaeological sites, a few continuous palaeo-temperature reconstructions using tree rings have been developed spanning over 7000 years (e.g., Grudd *et al.*, 2002). Reconstructing long sequences poses additional calibration problems and requires special statistical treatment in order to retrieve a reliable record of both the high-frequency and lower-frequency variability (Briffa, 2000). The need to adjust tree-ring series for growth functions seriously complicates reconstruction of the lowest-frequency (e.g., millennial scale) variability. A further complication arises for some high-latitude series when the most recent trends are compared with temperature measurements (Briffa *et al.*, 1998; 1999). Since, where necessary, periods of overlap during the earlier part of the twentieth century can be used for calibration, this need not impair dendroclimatological reconstructions.

Some of the earliest work on dendroclimatology was carried out in the south-west of the United States where the main influence on tree growth is water stress. Under these conditions, tree rings can be used to reconstruct hydrological variability (see, e.g., Grissino-Mayer, 1996; Hughes and Funkhauser, 1998; Grissino-Mayer *et al.*, 2000; Stahle *et al.* 2000 and Section 7.8; Figure 7.10a)

Three types of evidence have been used for reconstructing Palaeoclimate from tree rings: ring width, late-wood density and stable-isotope composition. Indices derived from both ring width and late-wood density (Briffa *et al.*, 2002a; 2002b) in several high-latitude evergreen tree species have been used successfully to recon-struct palaeotemperature variations as part of an integrated northern hemisphere, circum-polar study. Tree-ring-temperature links have also been established for other types of tree, including several species growing in South America and Tasmania. Calibration is achieved by comparing tree ring indices to climate data from nearby meteorological stations for the same time interval (Figure 4.3a). Since the growth rings reflect metabolic activity during the growing season, most reconstructions relate to this period of the year. Jones *et al.* (2003) point out that during the period of instrumental records, winter temperatures have warmed more than summer temperatures in some regions. One implication of this is that using summer-related proxies such as tree-ring records as indicators of mean temperature over the whole year may rest on somewhat flawed calibration. It also follows from their analysis that warm-season temperatures rather than mean-annual temperatures may be a more realistic basis for testing model performance against some proxy-based palaeo-data. Variations in the stable isotopes of carbon, $\delta^{13}C$ (McCarroll and Pawelleck, 2001; McCarroll and Loader, 2004) and hydro-gen, δD (Lipp *et al.*, 1995) have both been used for palaeo-temperature recon-struction from tree rings, although their use has, so far, been more restricted and, in some cases, problematic.

An increasing number of studies derive air-temperature changes from the stable-isotope signatures contained in the calcite precipitated by cave water in the form of speleothems. In all cases, though, there is a need to understand and make quantitative adjustment for all the processes that intervene to modify a climate signature as water passes from the lower atmosphere, through soil and bedrock, to the point of calcite precipitation. These processes call for careful study site by site before the sequence of changes that is derived from speleothem analysis can be transformed into climate parameters (Lauritzen and Lundberg, 1999; McDermott, 2004).

Stable-isotope measurements of lake carbonates, both biogenic (e.g., mollusc shells and the tests of ostracods) and chemically precipitated, have also been used to infer air temperatures (Leng, 2003; Von Grafenstein *et al.*, 1999; 2000; Leng and Marshall, 2004). Using stable-isotope signatures in lake sediments for air-temperature reconstruction calls for calibration that takes into account not only atmospheric processes, but fractionation effects within the water column and, where appropriate, within the organisms whose remains are used – so-called vital effects. Recent studies have sought to base temperature reconstructions on stable-isotope ratios in organic carbon (Anderson *et al.*, 2001) and biogenic silica (Shemesh *et al.*, 2001). Other biological signatures in lake sediments – for example, the fossil remains of chironomids and diatoms – have been calibrated to air temperature with considerable success. Non-biological sediment properties, such as varve thickness, have also been used as climate proxies after time-series calibration (Overpeck, 1996; Francus *et al.*, 2002).

Lake sediments and peats also yield remains of the fossil Coleoptera (beetles) that have provided excellent records of past temperature variations. Atkinson *et al.* (1987) have reconstructed the mean temperatures of the warmest and coldest months for a variety of sites and time intervals in the British Isles using a 'Mutual Climatic Range' method.

The concentrations of noble gases relative to each other in groundwater provide information on the temperature at the water table where the gases were dissolved at the time of formation of groundwater. Studies of noble-gas ratios in groundwater dating from the last glacial maximum (LGM) have yielded some of the strongest indications of lowered tropical, continental temperatures at that time (Stute *et al.*, 1995; Stute and Talma, 1998).

Some of the methods for temperature reconstruction depend on proxies that can be encountered in more than one kind of archive. Pollen and spores are preserved in a wide variety of environments: peat, sediments both lacustrine and marine, buried soils and even loess and continental glaciers, albeit rather sparsely. Although changing pollen spectra relate primarily to changes in the species composition of the surrounding vegetation, which can reflect many more influences than climate, techniques have been developed for inferring climate from pollen records (see, e.g., Anderson *et al.*, 1991; Guiot *et al.*, 1989), even in regions that have been heavily impacted by human activity, as is the case in most of Europe. These methods work best when applied to periods during which the assemblage of pollen types recorded finds an analogue somewhere within the range of sites where contemporary pollen deposition has been sampled for calibration. Many pollen sequences, however, include periods with pollen assemblages that have no present-day analogues. In these cases, climate reconstructions, even where possible, are subject to greater uncertainties.

4.2.4 Proxy-based Palaeoclimate reconstructions: precipitation, surface moisture and precipitation minus evaporation (P − E)

Although palaeoclimatologists have often devoted most attention to temperature, in many of the most populated parts of the world, changes in water availability are much more important than temperature variability alone.

Changes in temperature and precipitation are often linked to changes in modes of variability that strongly influence both (see 7.5). The ideal is to create reconstructions that encompass temperature and either precipitation or the balance between precipitation and evaporation.

Several of the archives and proxies already referred to above, notably tree rings, speleothems and pollen records, provide reconstructions of past precipitation, surface moisture status, or P − E, but only in the case of ice cores that do not contain melt layers is there a direct measure of past precipitation. In this case, the changing accumulation rate expressed as water equivalent, reflects changes in precipitation at the site. Some of the most dramatic and widespread past changes in climate have been recorded as sudden, major changes in ice accumulation in central Greenland (Alley, 2000; Figures 6.3 and 6.4).

Less direct lines of evidence for variations in precipitation or P − E, include the following:

1. Water balance calculations. The level of water in closed-lake basins (those without an outflow) reflects total water volume, which, in turn, reflects the balance between input in the form of precipitation and inflowing run-off from the drainage basin, and loss through percolation and evaporation (Figure 7.2a). Changing water levels can be reconstructed from exposed strandlines and terraces and by careful stratigraphic study of sediments contained within the lake basin (e.g., Harrison and Digerfeldt, 1993). Although many factors influence all these terms, fluctuations in water level are clearly at least a qualitative expression of P − E. In some cases, reconstructions of P − E from high lake levels have been sufficiently quantitative to allow their use as tests of model simulations seeking to replicate past conditions (Kohfield and Harrison, 2000, and Section 7.4).

2. Even in the absence of physiographic and stratigraphic records of high and low stands, lake-level variation can be reconstructed from sedimentological (e.g.,Verschuren et al., 2000; Figure 7.10b), as well as various combinations of biological and geochemical evidence (e.g., Gasse, et al., 1987; Hodell et al., 1995). Whereas the stable-isotope ratios of oxygen and hydrogen (δ^{18}O and δD) are strongly linked to air temperature in many temperate and higher latitude locations, in the tropics and in most semi-arid and desert areas, they are more closely linked to changes in the sources and available volume of moisture. It follows that their varying proportions, as well as changes in δ^{13}C, can sometimes be used to reconstruct past changes in moisture sources and hydrological balance. This applies especially to speleothems (e.g., Bar-Matthews et al., 1997; Frumkin et al., 1999) and lake sediments (Talbot, 1990; Leng, 2003). In both types of archive, stable-isotope measurements are often combined with other geochemical studies. The signatures that are used to infer changes in P − E are usually ones indicative of the concentration or dilution of lake waters, or the source waters of the speleothem. Higher values of the heavier isotope generally reflect sources or archives that have undergone evaporative enrichment and/ or periods of lower rainfall. Fleitmann et al. (2002) used this approach to reconstruct variations in the strength of the monsoon system in Oman.

3. One ecosystem sensitive to changes in P − E is the type of peatland referred to as ombrotrophic. Many such peatlands have developed in damp, cool-temperate climates where there is an excess of precipitation. As a result of this, partially decomposed plant remains accumulate to the point where the growing vegetation is excluded from inflowing surface and groundwater and becomes entirely dependent on direct precipitation. In view of their sensitivity to changes in the degree of water-logging at the peat surface, the remains of the vegetation and associated micro-fauna preserved in the peat can provide a record of past surface wetness. The most successful studies have been based on the changing plant macrofossil content of the peat, or on the remains of testate amoebae, although several other approaches, including stable-isotope ratios and the degree of peat humification (decomposition), have also been used (Barber and Charman, 2003).

4. Just as reconstructions of past vegetation from pollen analysis can be used to reconstruct past temperature, in regions where the changing patterns of vegetation have been more responsive to changes in moisture regimes, they pollen analytical evidence can also be used to reconstruct past changes in precipitation (see Figure 4.4c). In some cases, parameters related to both temperature and precipitation, or P-E, have been reconstructed at the same time (e.g., Bartlein *et al.*, 1984; Watts *et al.*, 1996).

5. The buried soils (palaeosols) contained within loess sequences have also been used as a basis for reconstructing past changes in moisture regime. One of the most successful methods derives from an empirical relationship between soil magnetic susceptibility and rainfall. This has permitted reconstructions of the varying strength of the east Asian summer-monsoon system (Maher and Thompson, 1999 (see Figure 4.4d). Broadly similar results have also been obtained using changes in the relative proportions of extractable iron in palaeosols (Guo *et al.*, 2000). The accumulations of windblown loess that are separated by the palaeosol horizons provide complementary evidence on winter-monsoon variability in the form of changing particle size assemblages. These reflect varying wind strength as it affects both deflation in the source areas, and transport efficiency to the point of deposition (Xiao *et al.*, 1995).

4.2.5   The atmospheric concentration of greenhouse gases

Records of past atmospheric greenhouse-gas concentrations are held in air bubbles trapped within ice cores (Raynaud *et al.*, 2003). At any given time, in the uppermost layers of accumulating snow, the air is able to exchange with the atmosphere, but, as more layers accumulate, exchange with the atmosphere becomes progressively more limited until eventually the bubbles are occluded and all further exchange ceases. The mixing processes (first by convection, then by molecular diffusion) that occur during this time have the effect of smoothing the record and introducing a difference between the age of the ice and that of the gas in the bubbles it contains. The length of the delay between deposition and occlusion is controlled by the temperature and rate of accumulation of the ice and, for any given combination, it can be calculated with reasonable accuracy. Lack of precise knowledge about past temperatures and accumulation rates, however, means that calculations are subject to a degree of uncertainty. The time lag for ice formed in central Greenland is estimated to be just a few centuries, but at Vostok, where temperatures are lower and accumulation rates much slower, the delay is around 3000 years (Barnola *et al.*, 1991). The ability to make an accurate estimate of the delay becomes critical when any attempt is made to establish the sequence of events during the major changes recorded in ice-core records (see, e.g., Figure 6.5). The signatures of rapid changes in gas concentrations are also attenuated in the air-bubble record. For example, Spahni *et al.* (2003) calculate that in the Dome C core from Antarctica, the amplitude of the methane variations associated with rapid changes in climate around 8200 years ago (see 7.3) is attenuated by between 34 and 59%.

Of the main greenhouse gases present in the atmosphere CO_2, CH_4 (methane) and, more recently, N_2O (nitrous oxide) (Flückinger *et al.*, 1999; 2001; 2004; Sowers, 2001; Figure 7.13) have been measured in ice cores. As indicated in Chapter 3, reliable records of past CO_2 concentrations in environmental archives are almost entirely restricted to ice cores from Antarctica. Methane and N_2O have been reliably measured in both Antarctica and Greenland.

A final point to note in this section is that water vapour, the main greenhouse gas, does not leave a record of past variations in its atmospheric concentration. This has been a source of uncertainty in all attempts fully to model forcings and feedbacks in the past climate system. It is also one of the weapons used by global-change sceptics to cast doubt both on the interpretation of past climate changes and the development of future climate scenarios.

4.2.6   Other aspects of atmospheric chemistry

Many chemical species have been successfully measured in ice cores (Mayewski *et al.*, 1997) and in some cases it has been possible to apportion them to soluble or insoluble phases (e.g., Laj *et al.*, 1997). Chemical analysis can also be used in conjunction with continuous, high-resolution logging techniques (Wolff *et al.*, 1997) such as the electrical conductivity method (ECM) and dielectric profiling (DEP). The former responds only to varying acidity, the latter to acid, ammonium and chloride. Application of this array of techniques to the ice-core records from central Greenland has shed light on changing atmospheric-circulation patterns, aerosol composition and the history of volcanicity (see 4.3 and Figure 4.7b), biomass burning and dimethyl sulphide (DMS) emissions from marine biota.

4.2.7   Atmospheric dust

Three main types of archive have been used to reconstruct past atmospheric dust concentrations and rates of dust deposition. Ice cores from polar, temperate and tropical environments (Thompson, 1995; Thompson *et al.*, 2002) include dust particles, the concentrations of which may either be logged directly (Bray *et al.*, 2001) or estimated from calcium concentrations (Mayewski *et al.*, 1997; Rothlisberger *et al.*, 2002; Figure 6.6). The terrigenous component in deep-sea sediments formed in areas remote from riverine input is largely the result of atmospheric deposition. Provided this component of the bulk sediment can be separately quantified and the rate of sediment accumulation calculated, the dust flux to sites in the deep ocean can be determined. On land, *in situ* loess deposits, are, by definition, aeolian in origin. Once again, given adequate chronological control, rates of deposition can be calculated. By putting together measurements from these diverse archives, reconstructions of changing atmospheric-dust concentrations have now been produced (Harrison *et al.*, 2001; Claquin *et al.*, 2002).

4.3   Sources of evidence for patterns of external climate forcing

4.3.1   Astronomically controlled changes in solar irradiance

These changes are independent of any variations in solar output; they depend on changes in the Earth's orbit. Although the potential palaeoclimatic significance of these variations was first recognised in the nineteenth century, they are most often associated with the work of Milankovitch (1941) who elaborated the idea that they were of major importance in controlling the alternation between glacial and interglacial conditions. Unlike virtually all the forcing and feedback mechanisms considered in this chapter, reconstructing past changes in the orbital parameters does not depend on proxy records. The changes can be independently calculated, as can their effects on incoming radiation outside the atmosphere at any given latitude and for any season. The three orbital parameters (Figure 4.6a) have different periodicities. Precession cycles vary from 19 000 to 23 000 years, with a mean period of 21 700 years; obliquity, 41 000 years; and eccentricity, 100 000 years. Each has a characteristic signature at any given latitude, so their combined effect also varies with latitude. The fact that these changes can be reconstructed from theoretical considerations, together with the unambiguous way in which they can be recognised in many long-term records of climate change, gives them great significance both in considerations of climate forcing and in the development of chronologies (see 3.4.1 and 5.2.1).

4.3.2   Sub-Milankovitch-scale solar variability

Direct observations of total solar irradiance (TSI) at the top of the Earth's atmosphere have only been possible since the development of suitable satellites. They date from the late 1970s and thus span only two full 11-year (Schwabe) sunspot cycles (see Figure 4.6b). The record has been extended back to the time of the invention of the telescope in the early seventeenth century by analysis of sunspot observations (see Figure 4.6c). Extending the record further back in time involves using proxies for TSI in the form of cosmogenic radionuclides. Those most commonly used are ^{10}Be and ^{14}C. Variability in the latter is seen as a divergence from the normal radioactivity decay function through time (see Figures 4.6d and 4.6e, and Section 3.4.4). Although both proxies for past changes in TSI match direct observations over the last four centuries, the long-term relationship is complicated by the likelihood that changes in the shielding effect of the Earth's magnetic field and, in the case of ^{14}C, in ocean ventilation, may also have contributed to the measured variability. Both these qualify the notion that changes in cosmogenic radionuclides reflect solely changes in TSI. Until recently, it was thought that changes in the strength of the Earth's magnetic field had affected the flux of cosmogenic radionuclides only on timescales of several thousand years, and that once the deposition records of ^{10}Be or ^{14}C were de-trended for long-term effects, they could be used as proxies for TSI. Following the work of Wagner et al. (2000b), Snowball and Sandgren (2002), St-Onge et al. (2003) and Zillen (2003) it now appears quite possible that changes in the Earth's magnetic field have modulated cosmogenic nuclide production on much shorter timescales, of the order of centuries. This does not discount the contention that ^{10}Be and ^{14}C may serve as proxies for changes in the receipt of solar radiation in the lower atmosphere.

The possible effects of changing ocean ventilation on ^{14}C are different, for they would not affect receipts, but rather the balance between the radionuclide concentrations in the atmosphere and the oceans, independent of any changes in radiative flux. This becomes an important issue when evaluating arguments for a strong and persistent influence of changes in incoming solar radiation on surface climate (see 7.11.1).

4.3.3   Volcanism

Volcanic eruptions that result in the release of aerosols into the stratosphere, where they are widely mixed over the whole Earth, have major short-term effects on climate, mainly because of the resulting reduction in incoming solar radiation. The processes involved in the volcanism–climate linkage have been observed directly for the most recent major eruptions such as those of El Chichon in 1982 and Pinatubo in 1991. Many earlier explosive events are well documented in written records and oral history. For even earlier periods, direct evidence for past events can be reconstructed from acid layers in ice cores as well as from volcanic ash (tephra) layers in ice, sediments and peats. Indirect evidence comes from the effects of eruptions on tree growth. Putting together all the evidence for the history of volcanism and its impacts on climate involves tying many lines of evidence into a precise and detailed common chronology

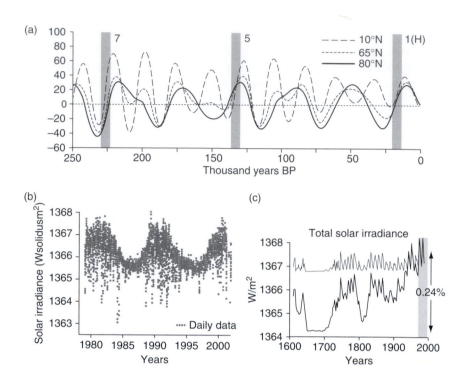

Figure 4.6 Solar variability: reconstructions of changes in solar radiation received by the Earth on a wide range of timescales. (a) Long-term variations in solar radiation resulting from the Earth's orbital changes. The three curves show changing solar radiation for July at three different latitudes. The changes result from the combined effects of three types of orbital changes: *eccentricity* (with a 100 000 year periodicity); obliquity (with a 41 000 year periodicity) and precession (with a periodicity mainly at 23 000 years). Note that the 41 000-year obliquity cycle dominates variability in solar receipts at high latitudes, but the 23 000 precession cycle has a stronger influence at low latitudes. The vertical lines mark the times at which the combined variability, paced by the 100 000 year eccentricity cycle lead to the onset of interglacial conditions during MIS 7, 5 and 1 (the Holocene) The importance of orbitally driven variability is introduced in Section 3.4.1 and considered further especially in Sections 5.2.1 and 6.4. (Modified from Bradley, 1999.) (b) Satellite-based records of changing solar irradiance at the top of the atmosphere over the last two 11-year (Schwabe) cycles, using daily data. (Adapted from Foukal, 2003.) (c) Variations in total irradiance from the early seventeenth to the late twentieth century. The thick line is derived by adding a long-term component determined by Lean *et al.* (1995) to the observations of sunspots since the development of the telescope (thin line). Note the periods of minimum values from AD 1645 to 1715 (the Maunder minimum) and during the early nineteenth century (the Dalton minimum). The timespan covered by (b) is shaded. (Adapted from Bradley, 1999.) (d) For the period from the early seventeenth century onwards, reconstructed changes in total solar irradiance are compared with variations in two likely proxies of solar variability, ^{14}C and ^{10}Be. The coherence between all three traces supports a longer-term reconstruction of past variability back to *c.* AD 900 on the basis of the ^{14}C and ^{10}Be changes shown. The timespan covered by (c) is shaded. (Adapted from Bradley *et al.*, 2003.) (e) Variations in ^{14}C activity (values inverted) over the last 12 000 years based on tree-ring measurements de-trended for radioactive decay and low-frequency changes in the Earth's magnetic field. Note that although Figure 4.6d encourages the view that residual $\Delta ^{14}$C may reflect changes in solar radiation, it is also affected by ocean ventilation, which may be at least partly responsible for some of the major changes, especially in the early part of the record (see 3.4.1).The time interval spanned by (d) is shaded. (Based on Stuiver *et al.*, 1998.)

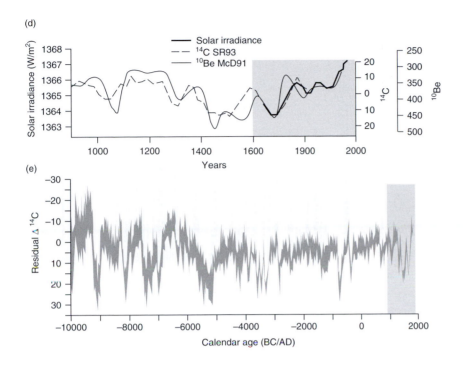

(Zielinski, 2000). Figure 4.7 shows reconstructions of explosive volcanicity over the last 500 and 9000 years.

4.4 Reconstructing past changes in the world's oceans

4.4.1 Past changes in global mean sea-level

On the timescales under consideration here, the main control of global mean sea-level has been changing ice volume. As ice forms it sequesters a higher proportion of the lighter isotope, which therefore becomes depleted in the seawater. The $\delta^{18}O$ values in marine organisms are thus increased during glacial intervals and depleted during interglacials. Changes in ice volume, hence global mean sea-level can therefore be inferred, to a first approximation, from changes in $\delta^{18}O$ in benthic foraminifera (Bard *et al.*, 1989). Similarly, changes in $\delta^{18}O$ in the bubbles contained in ice cores ($\delta^{18}O_{atm}$) are also related to changing global ice volume, hence sea-level. In both cases, the links between changing $\delta^{18}O$ and sea-level are complex, with many additional variables affecting the stable-isotope signatures (see 5.2.3 and Bradley, 1999, p 215; Ruddiman, 2003a). An alternative approach has been to infer global sea-level from dated coral terraces, once their heights have been adjusted for tectonic uplift (Chappell and Shackleton, 1986; Bard *et al.*, 1996). Labeyrie *et al.* (2003) present a synthesis of the various lines of evidence for changing sea-level over the last 450 000 years

Figure 4.7 Past volcanic activity reconstructed from observations and documentary evidence from AD 1500 onwards (a) and from SO$_4$ measurements for the last 9000 years (b). The shaded area shows the timespan common to both graphs. (a) The dust veil index (DVI) arising from volcanic eruptions is calculated using the location of the eruption, the estimated volume of dust produced and assumptions about distribution and residence times in the atmosphere. (Adapted from Bradley, 1999.) (b) The SO$_4$ peaks in the record from the GISP 2 ice core in central Greenland serve as a proxy for past volcanic activity during the Holocene (Zielinski *et al.*, 1994). (Adapted from Bradley *et al.*, 2003.)

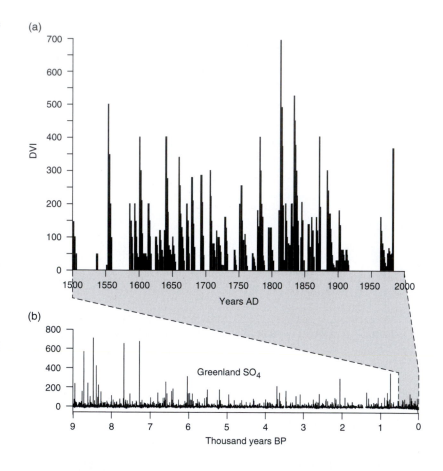

(Figure 5.1). More recently, Siddall *et al.*, (2003) have reconstructed global sea-level changes over the last glacial cycle using stratigraphic evidence from the Red Sea (Figure 12.6a). This record is detailed, continuous and, unlike most others, overlaps with reconstructions of mean sea-level made for the last 200–1000 years. In view of this, it is considered in 12.3.2 in relation to present-day sea-level trends.

4.4.2   Sea-surface temperature (SST) and salinity (SSS)

One of the main methods for reconstructing past climate change involves inferring past changes in SSTs from evidence contained in marine sediment cores and corals. The methods include the following:

1. Estimates derived from assemblages of the remains of marine organisms, most often foraminifera, the temperature ranges of which have been determined by calibration to modern species distributions (e.g., Imbrie and Kipp, 1971, and Figure 4.8). This approach assumes that the modern assemblages used in calibration represent the full

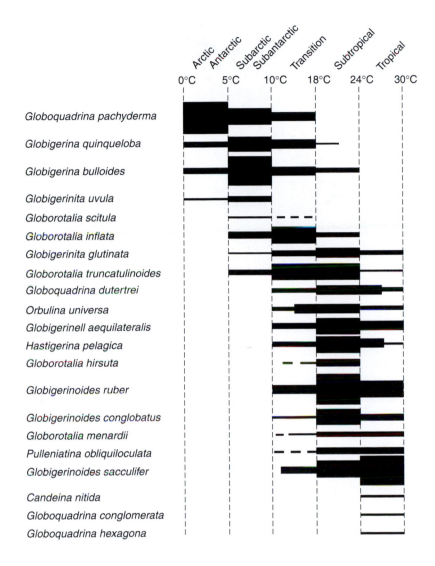

Figure 4.8 The present-day distribution of selected temperature-sensitive species of planktic foraminifera in relation to SST (Boersma, 1978); an example of the basis for reconstructing SSTs from foraminiferal assemblages in marine cores. (Modified from Bradley, 1999.)

range of past variability under study, that the species used have not undergone evolutionary changes that could have affected their temperature requirements and that the assemblages used in the calibration are in equilibrium with present-day conditions despite the likelihood that, because of the slow rates of accumulation in most marine sediments, they span a longer time interval than any available measurements of SST.

2. Alkenone palaeothermometry uses an index (U^k_{37}) derived from the response of certain coccolithophore species (mainly *Emilyania huxleyi*) to changes in water temperature. It is based on the relative proportions of different long-chain alkenones each of which contains a different number of carbon atoms. The index derived from the

Figure 4.9 The
relationship between
$\delta^{18}O$ and salinity in
modern ocean surface
waters (Broecker,
1989). (Modified from
Bradley, 1999.)

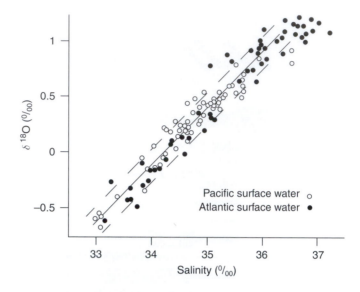

Figure 4.9 The relationship between $\delta^{18}O$ and salinity in modern ocean surface waters (Broecker, 1989). (Modified from Bradley, 1999.)

ratios can be directly calibrated to water temperature and the relationship is insensitive to changes in salinity or isotopic composition; nor does it degrade significantly over time (Müller *et al.*, 1998). It is therefore an extremely valuable and independent measure of past SSTs. Differences arise between alkenone-based reconstructions and those based on other methods as a result of differences in preferred water depths and in the seasonality of growth represented by the different organisms.

3. Stable-isotope ratios. The relationship between $\delta^{18}O$ and temperature in the sea is even more complex than in the atmosphere. The primary control, on the timescale of the Vostok record, is changes in ice volume (see 4.4.1). Because this signal is global, comparison between records from different sites can make it possible to extract other signals from changes in $\delta^{18}O$ values in marine sediments, but the variations are strongly influenced by salinity changes as well as by temperature. Whereas the global nature of the ice-volume signature makes $\delta^{18}O$ variations of great value in tuning the record in marine cores to the orbitally derived timescale (see 3.4.1), the capacity to record changes in salinity allows detection of major meltwater pulses associated with deglaciation.

Corals provide a second major source of evidence for past changes in SST and SSS. Cores from giant corals have been used to generate reconstructions of SSTs in tropical seas for periods of up to 350 years (Gagan *et al.*, 2000). Both $\delta^{18}O$ and element ratios (e.g., Sr/Ca and Mg/Ca) are commonly used and can be calibrated to both temperature (Figure 4.3b) and salinity variations (Figure 4.9) with considerable success. Similar reconstructions are possible for fossil corals and these have provided data on El Niño variability for several time-slices through the last glacial/interglacial cycle (Tudhope *et al.*, 2001).

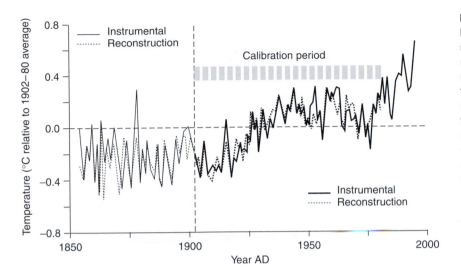

Figure 4.10 Comparison between measured and multi-proxy-based reconstructions of northern hemisphere temperatures for the nineteenth and twentieth centuries. The period 1902–80 is the calibration period, with dense instrumental coverage. The period 1854–1902 is an independent period for which less dense coverage by instrumental data is available. The close correlation between the instrumental and multi-proxy-based reconstruction for this earlier period strengthens the view that the reconstructions are accurate and can be patched onto the instrumental record for the last 20 to 30 years for which the full range of proxy data are not available. (Mann et al., 2000). (Adapted rom Bradley et al., 2003.)

4.5   Reconstructing past impacts resulting from climate change and human activities

Over the last decade or so there has been a strong tendency to extract palaeoclimatic inferences from every possible environmental archive and proxy. This is understandable given the degree of concern about future climate change. It has led to a huge increase in our knowledge about past climate variability, but it has had several unforeseen and less desirable consequences. Using biological proxies for Palaeoclimate precludes the possibility of reconstructing the impact of past climate change on plant and animal distributions and ecosystems without resorting to circular arguments. These issues have become of increasing concern in view of the likely impacts of projected changes in climate. There are signs that the need to consider impacts alongside independent proxy records for climate is being increasingly recognised. The chances of doing this are especially favourable when direct climate proxies and indicators of ecosystem responses coexist in the same archive (Amman and Oldfield, 2000; Jones et al., 2002).

Another consequence of the preoccupation with Palaeoclimate has been a tendency to ignore or find ways of discounting evidence for past human impacts on ecosystems (Oldfield and Dearing, 2003). As noted in Section 4.2.3, the use of contemporary European pollen analytical data for calibration to climate variables necessitated this. Now there is an increasing need to reconstruct the long-term history of human impact alongside that of climate change, especially in regions where human activities have had a significant effect on land cover for millennia. Many of the reasons for this are outlined in section 1.5. In addition, Ruddiman (2003b) and Ruddiman and Thompson (2001), by proposing close links between mid- to late-Holocene variations in methane and CO_2, and deforestation and cultivation by human populations (see 7.12), have reinforced the

Figure 4.11 Two examples of multi-proxy-temperature reconstructions for the Late-Glacial period from 13 500 to 11 000 BP. Kråkenes is a lake in western Norway, Gerzensee, in Switzerland (Al – Allerød; YD – Younger Dryas; PB - the Pre-Boreal period at the beginning of the Holocene). All proxies show a significant increase in summer temperature at the opening of the Holocene, though at each site the absolute temperatures for any given period vary with the proxy used. (Based on Lotter, 2003.)

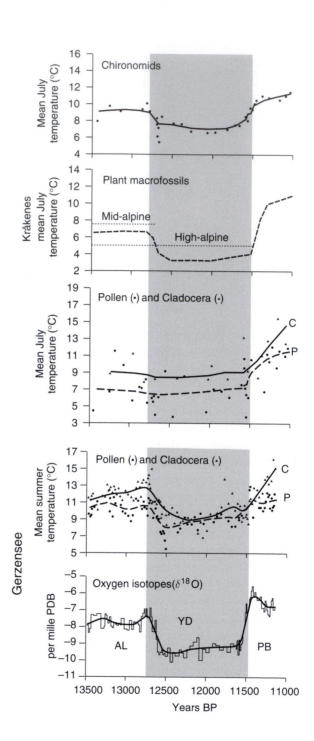

future importance of several complementary lines of research. These include efforts to calibrate pollen analytical data to changing landscape openness (e.g., Broström *et al.*, 1998b) as well as research attempting to trace the impact of human activities over large areas of Eurasia (e.g., Ren and Zhang, 1998). Human impacts on terrestrial ecosystems go well beyond changes in land cover and the growth of crops. They have involved dramatic changes in biogeochemical fluxes through a large number of interacting processes. Some of these are amenable to quantitative reconstruction using proxy evidence from sediment sequences, for example. Oldfield and Dearing (2003) outline several of the ways in which this kind of evidence can be used and Chapters 8 to 10 include examples that set recent observation-based evidence into the longer-term context provided by proxy-based reconstructions.

Many lakes and shallow sea environments have been heavily impacted by human activities over the last 200 years. Because instrumental and documentary evidence for the changing status of these types of ecosystem very rarely allows any kind of consistent quantitative reconstruction of impacts over the periods of critical change, techniques have been developed for reconstructing some of the key responses of aquatic ecosystems to human perturbations. Many of these involve transfer functions of the kind noted in Section 4.2.1 above. Chapter 10 considers these in more detail in Sections 10.2 and 10.4.

4.6   The multi-proxy approach

In this and the previous chapter, each archive, proxy and dating method has been discussed separately. In reality, the growing tendency is to use a range of proxies and at any given site (see, e.g., Mann, 2002a; Lotter, 2003). Figure 4.10 shows the skill with which multi-proxy-based temperature reconstructions for the northern hemisphere can capture changes in the instrumental record both for the period of calibration (1902–80) and for an earlier period (1854–1901) over which the two series are entirely independent of each other. Figure 4.11 shows the results of a multi-proxy approach to reconstructing changes in past temperature at individual sites in Norway (Kråkenes) and Switzerland (Gerzensee) for the late-Glacial period. Just as relying on a single proxy may tempt unrealistic precision, using only a single basis for constructing a chronology can also be misleading. Oldfield *et al.* (1995; 2003) include examples of chronologies derived from several independent lines of mutually constraining evidence.

In the next two chapters, we see how the many lines of evidence available are put to use to reconstruct the processes involved in past environmental changes at critical times over the last 420 000 years.

Chapter 5
Glacial and interglacial worlds

5.1 Key aspects of past variability

This and the next two chapters are concerned with Earth-system history over the last ~420 000 years, with special emphasis on the last ~130 000 years, the period since the beginning of the last interglacial. In the present chapter, we are concerned with the changes that are associated with the last four glacial cycles, up to the end of the last glacial maximum (LGM). For much of this period, the mean state of the climate and hence Earth-system components linked to climate, differed greatly from that in the recent past. Our main goal is to elucidate the combination of forcings and feedbacks responsible for the major changes observed, with a view to understanding better the sequences and synergies arising from their interactions, especially during periods of rapid change.

The high-latitude records show temperature variability to have been much greater during glacial periods than during interglacials, including the Holocene. By contrast, hydrological variability in lower latitudes has been extreme even during the Holocene. Variability of changing amplitudes, in both temperature and hydrology, is the norm throughout the period. This initial observation is of outstanding importance, for it tells us that even if we discount the likelihood of anthropogenic effects on climate, this does not dispose of future climate change. Scientists and policy makers across the whole spectrum, from the most sceptical about the impact of increasing atmospheric greenhouse-gas concentrations to the most firmly convinced, cannot afford to ignore all the evidence for natural climate variability. Even in the absence of any significant anthropogenic effect, it will continue into the future and play a vital role in human societies and human welfare. Equally, in so far as anthropogenic climate change is, or becomes, a reality, it will be superimposed upon, and hence interact with all the non-anthropogenic influences that give rise to natural climate variability. Looking to the future, the question is not – will climate change? It is rather – will future changes in climate differ from those that can be characterized from the record of the past; and if so, in which ways and by how much? This highlights the importance of past climate change, its causes, spatial expression, global and regional characteristics and its far-reaching consequences for ecosystems and people.

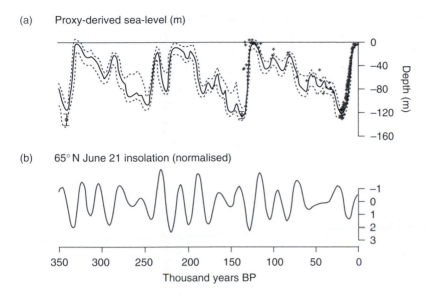

(a)   Proxy-derived sea-level (m)

(b)   65° N June 21 insolation (normalised)

350   300   250   200   150   100   50   0

Thousand years BP

Figure 5.1 Sea-level changes over the last four glacial cycles. Graph (a) is derived from proxy records and shows estimated uncertainties. Graph (b) shows the changes in June 21 insolation at 65° N. (Adapted from Labeyrie *et al.* (2003), which also gives the sources for each graph.)

Visual comparison of the trends from various archives spanning a range of latitudes in both hemispheres (see, e.g., Labeyrie *et al.*, 2003), shows that there is broad synchroneity in the *major* shifts of temperature between northern hemisphere sites and Vostok in Antarctica. In particular, the marine isotope stages considered as 'interglacials', (MIS 1, 5e, 7, 9 and 11) are common to both hemispheres, as are the overall trends between these temperature maxima. There is a shared tendency for a ragged decline from each interglacial peak to culminate in temperature minima shortly before the next steep rise to warm, interglacial conditions. Interglacial periods of peak temperature broadly comparable to those found during the Holocene span no more than 10% of the period, irrespective of the temperature series considered. There is no simple proportionality between the changing solar input signal at any latitude and the pattern of temperature change. Indeed, Wunsch (2004) shows that the fraction of Quaternary-climate variability attributable to orbitally controlled solar forcing never exceeds 20% and is usually much less. Nevertheless, a remarkably strong coherence in *phasing* can be recognised between orbitally driven changes in insolation and many long series of climate-linked changes, including the most truly global response, namely sea-level. Figure 5.1 confirms this coherence, to a first approximation, when changes in mean sea-level are compared with summer insolation at 65° N. It also gives the first clue regarding one of the likely major feedback processes in the climate system on orbital (multi-millennial) timescales. Periods of low summer insolation at this latitude correspond with periods of low sea-level, hence more extensive ice formation. The latter link is confirmed by the similar coherence between the insolation curve and the worldwide marine carbonate $\delta^{18}O$ variations that form the basis for the MIS chronology noted in Section 3.4.1. The pulse of the major oscillations from glacial to interglacial conditions appears to follow closely the 100 000 year cycle of changing orbital eccentricity, but since this alone was probably insufficient to provide the required forcing, it is likely that combinations of the other astronomical periods were involved (Santer *et al.*, 1993; Ruddiman and Raymo, 2003). One alternative hypothesis to orbital pacing, namely that the recurrent passage of the Earth's orbit through intergalactic dust clouds (Muller and MacDonald, 1997) may have reduced incoming solar radiation, has met with little favour. Whatever the initial trigger, the changes set in motion that led eventually to the onset of glaciations and glacial terminations must have included strong positive feedback within the Earth system.

In order to explore further the pattern of forcing and feedbacks responsible for the major climate shifts on orbital timescales, some key features of the Vostok record (Figure 5.2a) are now considered in more detail:

1. Even more dramatic than glacial inceptions are the glacial terminations. At each one, slow, smooth and rather modest changes in the amount and distribution of external energy input herald rapid and dramatic changes from full glacial to interglacial conditions. As with the periods of global cooling, the warming shifts are inconceivable without a range of feedbacks capable of dramatically amplifying the changes in incoming solar radiation.

2. Over the course of the four cycles represented in the Vostok core, atmospheric CO_2 concentrations peak during each interglacial around 280–300 ppmv, and methane values peak between 650 and 800 ppbv. The minimum values for CO_2 during glacial intervals fall consistently between 180 and 200 ppmv, and between 300 and 400 ppbv for methane. The narrow band within which the maximum and minimum values fall is remarkable, especially in the case of CO_2 concentrations, which invariably rise rapidly by *c.* 100 ppmv at each glacial termination. These observations imply that over the last four glacial cycles, atmospheric greenhouse-gas concentrations have remained within rather stable limits. This in turn suggests that in addition to the positive feedbacks operating during periods of rapid transition, there are also negative feedback mechanisms at work in the unperturbed Earth system that limit the range of variation. Even making the improbably conservative assumption of no further increases in atmospheric CO_2 and applying to present-day values a smoothing function to replicate the natural attenuation that occurs during the preservation of the trace-gas signal in the ice at Vostok (see 4.2.4), the atmospheric concentration of CO_2 already lies significantly outside the natural envelope of values (Raynaud *et al.*, 2003). Concentrations are now heading for levels that exceed the upper limits of natural variability in the Vostok record by values that are themselves in excess of the total range associated with glacial–interglacial cycles.

3. If we compare the main changes in trace-gas concentrations in the Vostok record with each other and with the sequence of temperature changes inferred from stable-isotope measurements in the ice, we see that despite some divergences in detail, the methane and CO_2 traces are broadly parallel throughout the sequence, and that both are tracked

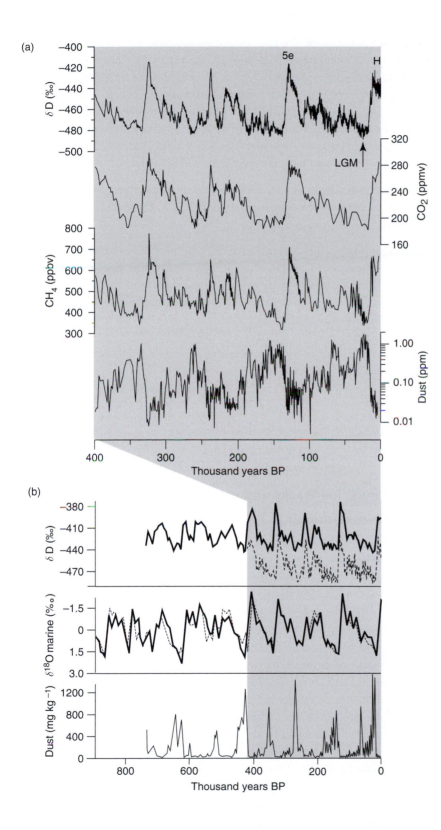

Figure 5.2 Long-term temperature, trace-gas and dust records from Antarctica.
(a) Greenhouse-gas and isotopically inferred temperature changes over the last 420 000 years recorded at Vostok. (Based on Petit *et al.*, 1999.)
(b) The record of δD (a temperature proxy) and dust mass at Dome C, Antarctica, plotted alongside marine-isotope records (δ^{18}O marine) that serve as a rough proxy for ice volume. The δD values, at Vostok are also shown for comparison. (Adapted from Delmotte *et al.*, 2004.)

by the temperature record. The trace-gas records are global in their implications, but the temperature record is specific to the source areas providing the moisture from which the Vostok ice is ultimately formed. Nevertheless, the main changes in isotopically inferred temperature parallel, to a first approximation, those that are based on long sediment records both from marine and terrestrial sequences from many parts of the world. The parallels between temperature and greenhouse gases at each glacial termination are especially significant in light of the inexorably rising greenhouse-gas concentration now and for the immediate future.

From these observations, we can identify a series of interlinked questions that arise once we consider the contemporary and future relevance of these results from Vostok:

1. What are the likely feedback mechanisms that have amplified external forcing at the onset of each glacial and interglacial and in what sequence did they operate? Specifically, can we establish from the precise phasing of events during glacial terminations, the stage at which increases in greenhouse-gas concentrations began and the likely role that their increase has played in triggering the shift from glacial to interglacial conditions?
2. What internal feedback processes serve to limit the amplitude of natural variability in trace-gas concentrations shown in the Vostok record?
3. What have been the major sinks, sources and fluxes for carbon during successive glacial cycles?

These are considered further in Sections 6.5 and 6.6.

5.2   Forcings, feedbacks and phasing

5.2.1   External forcing

Orbital changes can be computed accurately for each time interval, season and latitude (Berger, 1979; Berger and Loutre, 1991). Computing their net effect on energy balance at the top of the atmosphere for each latitude is also reasonably straightforward (Tricot and Berger, 1988). Translating these into net radiative forcing at the Earth's surface for each latitude is more difficult, since complex feedbacks operate to modify radiation receipts within and at the base of the atmosphere (see, e.g., Kutzbach and Guetter, 1986). Much more controversial is the issue of what other direct and indirect effects they (and indeed solar variability in general and on all timescales) may have on the Earth's atmosphere. This topic is considered further in Section 7.11.1, but for present purposes, it is sufficient to re-emphasise that all serious analyses of past-climate change stress the impossibility of accounting for the strength and speed of climatic response to solar forcing on orbital timescales without considering additional feedback processes operating as part of the interactive Earth system.

5.2.2   Feedback mechanisms

Feedback from the growth of ice sheets has already been hinted at as one of the key amplifiers of change during each phase of glacial inception (see 5.1). Several processes are involved. As snow and ice replace soil and vegetation over the land surface for progressively longer periods, the albedo changes; less heat is absorbed and more is radiated back into the atmosphere, reinforcing any cooling trend. The growing ice sheets enlarge the area of the arctic high-pressure system, displace the frontal systems equator-wards and increase meridional temperature and pressure gradients. These effects may be crucial in delivering heavy precipitation to the growing ice-sheet margins. As sea-ice forms, it increases surface albedo, seals off heat loss and gas efflux from the ocean over large areas, alters surface salinity, hence stratification, vertical mixing and circulation, and changes the pattern of primary productivity. Most of these processes reinforce any cooling trend that may have acted as the initial trigger. By contrast, as ice melts, a whole suite of changes in surface albedo, ocean and atmospheric circulation may follow, which tend to have the reverse effect.

Within the atmosphere, there are the, by now familiar, greenhouse gases, of which only the natural ones (e.g., water vapour, CO_2, CH_4, N_2O and tropospheric ozone) are of concern at this stage. Other natural trace gases are of indirect importance: for example, CO helps to determine the oxidising capacity of the atmosphere, hence the residence time of CH_4. Water vapour is by far the most abundant greenhouse gas. Although the total water-vapour content of the atmosphere is mainly controlled by temperature, few data are available to constrain past variations in atmospheric concentrations.

Aerosols (liquid or solid particles held in suspension in the atmosphere) also influence climate, both directly through their effect on radiative balance and indirectly through their influence on the formation, nature and persistence of clouds. Dust plays a vital role in delivering limiting mineral nutrients to those areas of ocean remote from other types of continental input. Much attention has been devoted to the possible influence of changing dust concentrations in the atmosphere and we shall consider evidence for its role in climate change in Sections 6.4.4. and 6.5.3. As noted in Section 4.2.6, the aeolian component of marine sediments can often be estimated geochemically, wind-blown deposits on the continents (e.g., loess) are widespread and increasingly well dated, and the dust content of glacier ice can also be quantified. Each type of palaeo-archive has been used to develop reconstructions of dust fluxes for key time intervals in the past. It has proved much more difficult to provide quantitative information on past changes in the secondary aerosols, mainly sulphur compounds. Dimethylsulphate (DMS) arising from marine biogenic activity may have an important role to play in providing condensation nuclei for cloud formation. Although one of its oxidation products, methanosulphonic acid (MSA) can be recovered from ice cores, it appears to be strongly affected by post-depositional changes.

Other feedbacks arise from both oceanic and marine processes. In many cases, these interact with each other and with the atmospheric variables already noted. Changes in atmospheric CO_2 and CH_4 can only be understood in the context of the carbon cycle and this involves a vast range of biogeochemical processes within the lithosphere, biosphere and hydrosphere. Figure 5.3 shows the main reservoirs that may serve to modulate fluxes of carbon on the timescales considered in this chapter. Even a cursory examination brings out the daunting complexity of the system. A couple of examples will suffice. Exchanges between the terrestrial biosphere and the atmosphere on a global scale reflect the aggregate balance between carbon sequestration through photosynthesis, release via respiration, storage through retarded decomposition and loss through export for each and all of the diverse range of biomes on Earth at any given time. Variations in any one of these processes will affect carbon exchange with the atmosphere. Moreover many of the processes are largely inter-related, hence not realistically considered as independent variables. They are all not only responsive to climate, but also capable of influencing climate through their effects on albedo and moisture flux. To add a further complication, changes in atmospheric CO_2 probably had an impact on vegetation independent of climate, since they may have a direct fertilising effect in some ecosystems and they also influence the competitive relations between species, especially between those with C3 and C4 metabolic systems. It is not surprising that estimates of the change in biospheric carbon storage between the last glacial maximum and the present day are highly varied (Figure 5.4) At least equally complicated are the exchanges between ocean and atmosphere. These involve physical processes (e.g., temperature changes and wind-generated eddy fluxes) and chemical gradients (differences in the partial pressure of CO_2) both of which govern gas exchanges across the interface. Changes in solubility, alkalinity and biological productivity within the ocean all affect the carbon reservoir in different components of the ocean–sediment system. Each one of these processes is modulated by complex changes in water chemistry. In addition, variations in carbon reservoirs and fluxes within the ocean may also be driven by changes in ocean circulation, which in turn responds to a variety of variables including atmospheric circulation, sea-ice formation and inputs of fresh water from the continents.

Accounting for changes in atmospheric concentrations involves reconstructing changes in the fluxes between reservoirs, but palaeodata often fail to provide this type of information. In the case of the marine component of the carbon cycle, for example, all it allows is some estimate of past inventories in each reservoir, usually with wide error bars (LeGrand and Alverson, 2001; Pedersen *et al.*, 2003) Bridging the gap between imperfect data and the requirements for reconstructing the fluxes that modulate changes in atmospheric concentrations necessarily falls within the remit of the modeller. In this most challenging area, the inferential framework currently falls far short of providing a context for rigorous hypothesis testing.

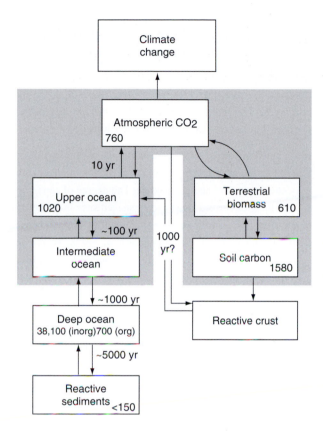

Figure 5.3 The major carbon reservoirs (with sizes included in the boxes and expressed in Gt C) involved in exchanges on millennial and sub-millennial timescales. Exchanges between them are thought to be the main ones involved in climate change. Directions of exchange and mean time constants are indicated by the thin arrows and the numbers alongside them. The reservoirs and fluxes within the shaded area are the ones of greatest relevance to the present account. (Modified from Pedersen *et al.*, 2003.)

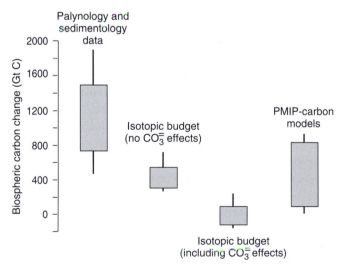

Figure 5.4 Estimates of the increase in biospheric carbon storage between the LGM and the present day. The range exceeds three orders of magnitude and illustrates well the uncertainties involved in characterising changes in the carbon cycle between glacial and interglacial conditions. One of the challenges posed by Ruddiman's hypothesis (see 7.12) is reconciling the incompatibility of the palynological and sedimentologically derived estimates with those based on models using stable-isotope data. (Adapted from Pedersen *et al.*, 2003.)

5.2.3   Phasing and the role of greenhouse gases in glacial inception and termination

Addressing this issue calls for the establishment of secure chronologies and between-archive correlations, so that the sequence of changes can be better portrayed. The main dating methods have already been outlined in the previous chapter, but several points are further highlighted here.

- As noted in Section 4.2.5, reconciling the chronology of trace gas-and ice-records involves calculating the effects of convection, molecular diffusion and gas-bubble occlusion on the trace-gas records. The degree of consistency between the trace-gas chronologies derived from widely spaced sites in Antarctica, with different rates of accumulation, as well as close agreement with quite independent estimates based on thermal fractionation (Severinghaus et al., 1998; Severinghaus and Brook, 1999) encourage the view that the current chronologies are sufficiently secure to allow valid comparison with their associated ice-derived records, but always within error limits that vary with core and depth.
- Synchronising ice-core records between Greenland and Antarctica using variations in atmospheric methane concentration (see 3.4.3) provides one of the vital keys to establishing the phasing of events between the two sets of polar archives.

There are still significant difficulties in establishing precise correlations between marine, terrestrial and ice-core records for some key time intervals, and points of unambiguous synchronisation are infrequent. One of the bases for proposed correlations between ice and marine cores arises from the inferred link between reductions in the strength of the Earth's magnetic field during brief magnetic excursions and the presence of peaks in the concentration of cosmic radioisotopes such as ^{10}Be and ^{36}Cl. The former can be detected in marine sediments, both in ice cores. Two peaks, corresponding to the Laschamp and Mono Lake palaeomagnetic 'events' (Laj et al., 2000; Wagner, 2000a), have been used to reinforce correlations between Greenland ice-core and North Atlantic marine-core sequences.

Currently available chronologies are adequate if the main concern is to follow the broad sequence of changes over the globe. As soon as questions of phasing during rapid transitions become important, the requirements in terms of accuracy and precision increase dramatically and the chronologies developed so far have not always kept pace with these requirements. Changing ice volume (hence global mean sea-level), air temperatures, atmospheric greenhouse-gas concentrations and dust content are the main variables considered in studies aimed at reconstructing the phasing of changes during major transitions between glacial and interglacial intervals. Close correlation between polar-ice-core and marine-sediment chronologies is thus especially vital for establishing the phasing of the major feedback mechanisms operating during periods of rapid change.

Changes in ice volume are usually inferred from dated stratigraphic evidence of sea-level changes, or from changes in $\delta^{18}O$, either in marine foraminifera ($\delta^{18}O_{sw}$)

or in the oxygen preserved in the gas bubbles trapped in ice ($\delta^{18}O_{atm}$). Provided non-global temperature effects can be excluded, $\delta^{18}O_{sw}$ may be used as a direct proxy for ice volume. These and other considerations led Shackleton *et al.* (2000) to propose a revised Vostok chronology based on tuning changes in $\delta^{18}O_{atm}$ to orbital variations, using the closely dated GISP 2 ice core to establish the phase relationship between the precession and obliquity components of the $\delta^{18}O_{atm}$ signal, and orbitally driven changes in insolation. Support for a chronology based on $\delta^{18}O_{atm}$ is provided by Bender (2002). He tested such a chronology by comparing it with that derived from a comparison between changes in O_2/N_2 ratios within the Vostok ice bubbles, and orbitally controlled summer insolation at $78°\,S$. The rationale for this lies in a proposed link between summer insolation and the physical properties of the ice grains that control fractionation between O_2 and N_2 during bubble closure.

Ruddiman and Raymo (2003) proposed a further refinement of the Vostok chronology. They pointed out that variations in $\delta^{18}O_{atm}$ depend partly on $\delta^{18}O_{sw}$, partly on the offset caused by marine and terrestrial photosynthesis, the so-called 'Dole effect'. This leads to a systematically varying offset between the two sets of values and produces probable errors in the Shackleton (2000) chronology. According to Ruddiman and Raymo the offset results from the dominance of $\delta^{18}O_{atm}$ variations by changes in the strength of monsoon regimes during periods when ice sheets are small, and by changes in ice volume when ice sheets are large. Ruddiman and Raymo's chronology is obtained by tuning the variations in CH_4 to the 23 000-year precession cycle (Figure 5.5). Strong support for this approach comes from the correspondence between the timing of the late-glacial CH_4 maximum and that of maximum July insolation at $30°\,N$, mainly in response to obliquity changes. The link between methane and obliquity arises from the behaviour of tropical monsoon systems, which Ruddiman and Raymo regard as a major response to obliquity-linked orbital forcing, as well as the dominant control on variations in atmospheric CH_4. Their chronology is close to that of Shackleton (2000) and both imply significant revisions to the 'GT4' Vostok timescale (Petit *et al.*, 1999) for the period prior to 250 000 years.

Ruddiman and Raymo point out that their methane-based chronology gives much more consistent phasing between Vostok CO_2 and $\delta^{18}O_{sw}$. They have

Figure 5.5 The coherence between insolation changes and atmospheric methane over the last 350 000 years forms part of the basis for Ruddiman's proposed revision of the Vostok chronology (see 5.2.3). The stages at which the conjunction of orbital changes leads to the development of interglacial MIS 9, 7e, 5e and 1 are marked. Dark shading within the Holocene denotes the divergence between methane concentrations and insolation (see Figures 7.14 and 7.15). (Modified from Ruddiman, 2003a.)

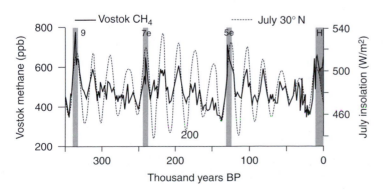

examined its implications for the possible role of CO_2 in ice-volume changes. They first propose corrections to the SPECMAP timescale and also allow for the probability that the SPECMAP $\delta^{18}O$ signal lags true ice volume by around 1500 years. Using these adjustments, they suggest that there was little, if any, lead between CO_2 and ice volume at the 41 000-year period. From this, they infer that CO_2 is unlikely to have acted as an independent forcing mechanism on this timescale, but rather as positive feedback. A small lead remains a possibility on the 23 000 year period, leaving the question of forcing or feedback more open. On the 100 000-year period, their analysis retains a lead of c. 5000 years for CO_2. They qualify possible support for a forcing role for CO_2 on this timescale by pointing out that $\delta^{18}O$ signals in marine sediments are complex and should not be taken as exact proxies for ice volume. In conclusion, they state: 'We can rule out the possibility that CO_2 lags behind ice volume at any of the orbital periods. But we cannot distinguish between the possibility that CO_2 leads ice volume, in which case it is an active part of the forcing of the ice sheets, or is in phase with ice volume, in which case it plays the role of an amplifying positive feedback to ice-volume changes.' Their conclusions rest on a careful evaluation of both the chronologies and the proxies upon which the reconstructions of processes and phasings depend. With respect to the 100 000-year period, their conclusions do not differ significantly from the earlier ones of Petit *et al*. (1999): 'These results suggest that the same sequence of climate forcing operated during each (glacial) termination: orbital forcing ... followed by two strong amplifiers, greenhouse gases acting first, then deglaciation and ice-albedo feedback.'

Both Bender (2003) and Ruddiman (2003a) explore further the likely implications of their new Vostok chronologies in the context of a wide range of evidence from polar, marine and terrestrial palaeo-records. Bender evaluates a huge volume of empirical evidence for the possible role of the biosphere in climate change on glacial–interglacial timescales. Paying particular attention to albedo effects and the possible role of dust and marine productivity, he summarises many lines of evidence in support of significant biosphere feedbacks of both kinds, but considers that conclusive proof of a large role for the biosphere remains elusive. Several of his arguments are consistent with a feedback via dust deposition and southern ocean productivity (see 6.5.3).

Ruddiman (2003a; 2004) compares his reconstructions with earlier views of the nature of orbital forcing, beginning with CLIMAP (1981). Using the methane-based chronology, he expands on the conclusions in Ruddiman and Raymo (2003) and proposes that:

- The main external forcing of major changes in climate and ice volume, at least over the last two to three glacial cycles is linked to the 41 000-year and 23 000-year obliquity and precession cycles acting in combination. The 100 000-year eccentricity cycle serves simply to 'pace' these as drivers of glacial–interglacial transitions by reinforcing their combined effects at eccentricity maxima and minima.

- Both CH_4 and CO_2 provide essential feedbacks that amplify solar forcing on orbital timescales. The former amplifies the 23 000-year interglacial signal during precession-driven insolation maxima, the latter provides negative feedback during 41 000-year obliquity-driven minima.
- The combined effects of insolation forcing and greenhouse-gas modulation, both positive and negative, generate a net response in terms of changing ice volume that matches empirically reconstructed changes in ice volume within the rather wide uncertainties that still exist.

Ruddiman's suggestions do more than revise the chronology. Feedbacks from changing greenhouse-gas concentrations are an inherent property of the scheme he proposes.

In summary, the arguments used to establish chronologies, infer sea-levels and reconstruct the phasing of major changes through glacial–interglacial cycles are complex and still being refined. Despite this, none of the analyses quoted suggest that ice-volume changes predate those in atmospheric greenhouse-gas concentration. All portray greenhouse gases as playing either a significant forcing or, more probably, a crucial feedback role in driving or amplifying the changes in ice volume and climate.

Caillon *et al.* (2003b) outline a possible way of overcoming uncertainties in the relative phasing of ice-temperature and gas-derived changes in the Vostok core by using the isotopic composition of argon in the gas bubbles as a proxy for temperature changes across the glacial–interglacial transition that took place around 240 000 years ago. Although the precise mechanism whereby the isotopic composition of argon serves as a palaeo-temperature proxy is not fully understood, this approach promises to obviate one of the key sources of uncertainty in phasing, since it should allow both temperature and atmospheric greenhouse-gas concentrations to be calculated from the same material. Using methane and $\delta^{18}O_{atm}$ variations as indicators of northern hemisphere changes, they suggest that the increase in CO_2 post-dated the first signs of Antarctic warming by around 800 ± 200 years, but clearly precedes northern hemisphere deglaciation. Their results strengthen other indications that:

- CO_2 acted as a feedback, after initial insolation forcing.
- The changes in atmospheric-CO_2 concentrations reflect processes in the Southern Ocean.
- Southern hemisphere warming and outgassing of CO_2 from the ocean preceded northern hemisphere deglaciation.

Finally, it should be noted that likely feedbacks from changes in terrestrial vegetation and soils during glacial inceptions and terminations have received very little attention by any of the authors cited so far. Meissner *et al.* (2003) have carried out simulations using an Earth-system model coupled to land-surface and dynamic global-vegetation models. They claim that terrestrial feedbacks during glacial inception double atmospheric cooling and reduce meridional overturning in the North Atlantic.

Some integrative hypotheses that build on the main conclusions outlined above are considered in Section 6.5, once the transition from glacial to Holocene conditions has been considered in more detail.

5.3   Marine isotope stage 5e and the last (Eemian) interglacial

As the last, best-dated and most intensively studied full interglacial, the 'Eemian' has received special attention. The demonstration that the Holocene temperature record in central Greenland showed much less variability than that which characterised the preceding glacial period has led to speculation as to whether this was a general characteristic of all interglacials. This, in turn, has provoked questions concerning the degree of variability during the Eemian. The term Eemian was coined to represent an interglacial defined on the basis of a period of consistently high tree-pollen percentages in European pollen diagrams. Some of the controversy surrounding the issue of Eemian variability has probably arisen as a result of problems of timing and definition. Until quite recently, the terms Eemian and MIS 5e were often used interchangeably. As both Shackleton *et al*. (2003) and Tzedakis (2003) have confirmed, there is a significant difference in timing between the two. The marine isotope sub-stage is best characterized by a plateau in marine $\delta^{18}O$ values, representing the highstand of sea-level, and dated radiometrically to between *c*. 128 000 and 116 000 years BP. Shackleton *et al*. (2003) report on a core off the coast of Portugal containing both marine and terrestrial records. They thereby show that the opening of the Eemian, as defined by pollen analysis, postdates the mid-point of the MIS 6/5e transition by some 6000 years and the beginning of $\delta^{18}O$ 'plateau' values (and the associated peak in sea-level) by some 2000 years. There was a similar lag at the end of MIS 5e, with forest vegetation surviving in Portugal until *c*. 110 000 years BP, well after the onset of continental ice accumulation in North America. Tzedakis (2003), whose chronology for the Eemian in southern Europe is in good agreement with that of Shackleton *et al*. (2003), is able to show that the end of forested conditions in *northern* Europe came much earlier, around 115 000 years BP and thus almost coincided with the end of the $\delta^{18}O$ 'plateau'.

These results may be compared with those from a series of marine cores from different latitudes in the North Atlantic summarised in Labeyrie *et al*. (2003). They show that above *c*. 60°N sea-surface temperatures (SSTs) fell rapidly between *c*. 125 000 years and 115 000 years BP, roughly in line with the decline in summer insolation at that latitude. By contrast, SSTs around and south of *c*. 40° N remained high beyond 110 000 years BP. Over this period, the winter insolation curve for 20° N, which is the reverse of the summer receipts at 65° N, matches the low latitude SSTs rather well and suggests that external insolation may have played a role in the steep north–south gradients at the end of the Eemian as detected in both continental and marine records. These steep gradients appear to have resulted in changes in both marine and terrestrial biomes that were asynchronous from north to south.

Well resolved and accurately dated records from MIS 5e/Eemian times are less common in other archives and from other parts of the world, but a comparison may be made with the record of climate change in loess sections from the western part of the Chinese Loess Plateau. Here, aeolian deposition persists through the periods of palaeosol formation that mark interglacial stages. This has the effect of expanding the stratigraphic record from these periods and greatly reducing pedogenic processes that overprint the climate signature in buried soils further east. Using particle size as a proxy for the strength of the winter monsoon, and frequency dependent susceptibility as a proxy for the summer monsoon, Chen *et al.* (2003) show that there is no evidence in their sections for significant variability during what they define as MIS 5e.

From both the Chinese and European evidence, as well as more widespread research on marine $\delta^{18}O$ values and sea-level changes summarised by Shackleton *et al.* (2003) we may draw several conclusions:

- There is no conclusive evidence either for strong temperature variability or for major changes in the east Asian monsoon system during MIS 5e as currently defined.
- Perspectives on the 'last interglacial' depend both on location and on the types of archive and proxy under study. Spatial variations in insolation, latitudinal gradients and possible lags in response all influence the nature of the record at any given site.
- The changes recorded during the Eemian/ MIS 5e show links to the pattern of solar forcing and associated latitudinal gradients in insolation receipts that were distinctive and specific, making it quite unrealistic to consider that period as an analogue for the Holocene.

Not only did the pattern of radiative forcing differ greatly from that during the Holocene, the apparently uninterrupted transition to interglacial conditions in the northern hemisphere at the beginning of MIS 5e was very different from the case at the opening of the Holocene. Simple analogues from the past, if they exist, need to be sought with more caution and discrimination. Marine isotope stage 11 most closely corresponds with the Holocene in terms of the pattern of external radiative forcing. The close attention currently being paid to this period in the Dome C record (EPICA community members, 2004) and elsewhere promises to bring new insights.

The steep latitudinal gradient in insolation and SSTs already noted during the later part of MIS 5e and beyond, into the late Eemian, spanning the beginning of MIS 5d, is believed to have been critical for glacial inception. Not only did the low summer insolation at high latitudes favour the development of ice sheets, the steep gradient probably enhanced the transport of heat and moisture to the growing ice margins, bringing increased snow deposition. Recent modelling (Khodri *et al.*, 2001) tends to support this inference and highlights in addition the likely importance of feedbacks from sea-ice formation and albedo changes on land (De Noblet *et al.*, 1996). In simulating glacial inception around 120 000 BP, Wang and Mysak (2002) conclude that two additional feedback mechanisms may also be critical: the orographic effect of increased elevation as the ice sheet grows, and the role of freezing rain and re-freezing meltwater.

5.4   Rapid climate oscillations during the last glacial period

One of the most dramatic findings to emerge from the analysis of ice cores from central Greenland was evidence for rapid oscillations of isotopically inferred temperature, with an amplitude around 50 % or more of the total range between glacial minima and the Holocene maximum. Only after the results were replicated in several cores and their imprint discovered in marine-sediment sequences in the North Atlantic and elsewhere were they accepted for what we now know them to be: major reorganisations of the Earth system impacting climate over much of the globe. They are usually referred to as Dansgaard/ Oeschger (D/O) events or oscillations. The challenge of unravelling their origins and dynamics has been given new urgency through the realisation that at least one of the processes involved – freshening of North Atlantic waters leading to the weakening or total collapse of North Atlantic deep water (NADW) formation and of the meridional overturning circulation that carries warm surface water into the North Atlantic – may occur in future as a result of high-latitude climate changes already detectable at the present day. These changes could eventually have dramatic consequences (see Vellinga and Wood, 2002).

Figure 5.6 shows the trace of the D/O oscillations as recorded in the GISP 2 ice core compared with selected sequences from lower latitudes and the isotopically inferred temperature record from Byrd in Antarctica. This and associated records from other ice core, terrestrial and marine sequences permit a series of broad generalisations:

● The D/O oscillations, some 24 in all, are most typical of the period between *c.* 60 000 and 15 000 B,P although there are somewhat similar events both before and after.
● The amplitude of warming and cooling during each event is at least of the order of 10 °C in central Greenland (Severinghaus *et al.*, 2004).
● Shifts in atmospheric methane concentrations are in phase with the oscillations to within *c.* 30 years (see, e.g., Flückiger *et al.*, 2004). The warm limb of the oscillations coincides with higher methane values, pointing to widespread changes in terrestrial ecosystems during each oscillation.
● At least some, if not all of the highest and broadest peaks in isotopically inferred temperature correspond with well studied interstadials (relatively mild intervals within the overall glacial period) identified from pollen records in Europe and elsewhere.

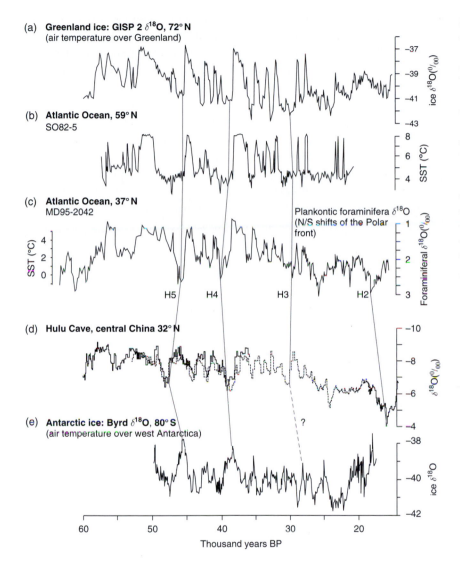

Figure 5.6 Proxy records of climate change 60 000 to 10 000 BP from different archives and latitudes. The vertical lines show inferred correlations corresponding to Heinrich layers in the North Atlantic. Note the episodes of clear antiphase relationship between the northern-hemisphere temperature records and that in the Byrd ice core from Antarctica. (Adapted from Labeyrie *et al.*, 2003, which also gives the sources for each graph.)

- Six of the deepest 'troughs' correspond with characteristic layers of glacial detritus rafted by icebergs to locations far south in the North Atlantic: the so-called Heinrich layers (Heinrich, 1988).
- The D/O oscillations during most of this period have a rather regular periodicity centred around *c*. 1500 years from peak to peak, whereas the Heinrich events are not so regularly paced.
- Steep increases in temperature to peak values follow most of the Heinrich events, after which the amplitude of the oscillations declines, giving rise to cycles of oscillations (so-called Bond cycles) beginning with the strongest, then tailing off towards the onset of the next cycle (Bond and Lotti, 1995; Bond *et al.*, 1997; 1999).

- The characteristic signature of these D/O oscillations is clearly discernable in marine cores throughout the northern hemisphere, even at sites as remote as the Santa Barbara Basin off the west coast of California (Behl and Kennet, 1996; Hendy *et al.*, 2002; Seki *et al.*, 2002; Figure 5.7). The comprehensive survey of sites recording D/O type oscillations around the globe by Voelker *et al.* (2002) also includes marine cores from low latitude sites in the southern hemispheres.

- Parallel oscillations are detectable in stalagmites from the Hulu Cave in central China (Wang *et al.*, 2001) and the Moomi Cave on the island of Socotra in the Indian Ocean (Burns *et al.*, 2003), both of which link warm phases in Greenland to increased precipitation at the cave sites. There are also parallels with sequences from the western part of the Chinese Loess Plateau (An and Porter, 1997; Chen *et al.*, 1997; Fang *et al.*, 1999; An, 2000), where cool phases in Greenland appear to correspond with increased wind strength and probable dessication. One problem brought to light by the independent chronology developed for the Moomi Cave stalagmite is a significant disagreement with the central Greenland ice-core chronology. The level of detailed coherence between the two records virtually precludes any significant time lag between them. For the period dated in central Greenland to between 37 000 and 53 000 the U/Th chronology for the Socotra stalagmite suggests that the ice-core ages are too young by some 3000 years (cf. Shackleton *et al.*, 2004).

- At least the strongest of the D/O oscillations are detectable in the ice-core records from Antarctica, but synchronisation of the chronologies using the methane record (see 3.4.3) confirms that they are in antiphase when compared with the temperature record from Greenland.

Figure 5.7 Rapid and spatially coherent climate variability during the last glaciation (MIS 4, 3 and 2), including Dansgaard/ Oeschger oscillations and Heinrich events. The two upper graphs show high resolution changes in SST reconstructed from alkenone records, alongside parallel changes in δ^{18}O in planktonic foraminifera, from the Santa Barbara Basin off the coast of Southern California. The lower graph plots reconstructed changes in δ^{18}O over the same time interval in the GISP 2 ice core from central Greenland. Dansgaard/ Oeschger oscillations are numbered from 1 to 21, Heinrich events from H1 to H5. YD – Younger Dryas; B/A – Bølling/ Allerød. (Adapted from Seki *et al.*, 2002.)

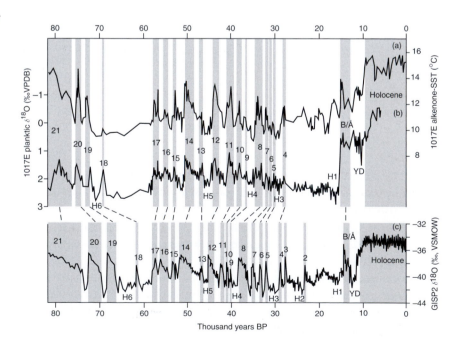

Even at the level of simple description, these oscillations are immensely impressive. The challenge is to develop testable hypotheses that can account for their dynamics. Over the last few years, something approaching consensus has begun to emerge with respect to the major cold excursions associated with Heinrich layers. Although there is some indication of possible non-synchroneity in response from north to south (Chapman *et al.*, 2000), the southern spread of ice-rafted detritus (IRD), confirmed by lithological and geochemical analysis of North Atlantic sediments, points to a major collapse of the Laurentide ice sheet, leading to a massive release of icebergs. Both the empirically demonstrated antiphase relationship with the record in Antarctica, where gradual warming coincides with the temperature minima in Greenland, and a variety of model simulations, are consistent with the inference that the Heinrich layers correspond to periods during which meridional heat flux associated with the formation of NADW at high latitudes was, at the very least, significantly reduced if not completed cut off. This led to a cooling in the North Atlantic as the surface flow of warm water from the south ceased, and to a corresponding build up of heat in the South Atlantic – the so-called bipolar seesaw (Broecker 1998; Stocker, 1998). There is also strong agreement that the likely cause of the shut down in NADW formation was a reduction in salinity leading to the failure of dense, sinking water to develop. Much less agreement surrounds the reasons for the freshening of the North Atlantic during these events. Ice-sheet dynamics – the so-called binge–purge hypothesis (MacAyeal, 1993) – is one possible process. A recent study, using a fully coupled three-dimensional model, has reinforced the credibility of this explanation by successfully simulating quasi-periodic large-scale surges from the Laurentide ice sheet, with dynamics and periodicities broadly consistent with the empirical evidence (Calov *et al.*, 2002). Broecker *et al.* (1990) proposed an internal oscillator in the thermohaline circulation (THC) linked to salinity variations, but the rationale for this hypothesis has tended to weaken in recent years in the light of the growing body of empirical evidence. Kaspi *et al.* (2004) suggest that feedback from rapid, synchronous and extensive reductions in sea-ice extent may provide an effective amplifying mechanism during the abrupt warming that took place at the end of each cold limb of the cycle. The authors claim that this mechanism is sufficient to amplify significantly quite small changes in THC as the atmosphere is warmed through the changes in albedo and insulation that result from sea-ice melting. Several authors have invoked major changes in the routing of melt-water from the Laurentide ice sheet via the St Lawrence rather than the Mississippi drainage system (see, e.g., Clark *et al.*, 2001).

Clark *et al.* (2002) postulate three modes of ocean circulation in the Atlantic: a modern (interglacial) mode, an intermediate, glacial, mode and a Heinrich mode in which warm-water flow into the North Atlantic is strongly curtailed. They propose that switches between the latter two modes may result from quite small changes in freshwater input, of the order of 0.1 Sverdrup ($1 \, Sv = 10^6 \, m^3 \, s^{-1}$).

It follows from their analysis that the sensitivity of the North Atlantic to perturbation by hydrological changes may be strongly dependent on the mean state of the climate, indeed the Earth system as a whole at the time.

Both Claussen *et al.* (2003) and Stocker and Johnsen (2003) have developed models designed to simulate the full range of behaviours associated with D/O oscillations and Heinrich events. Both favour the notion of a bipolar seesaw mechanism as the dominant process involved. In the conceptual model proposed by Stocker and Johnsen, the seesaw oscillation is dampened and modulated by a heat reservoir in the Southern Ocean. Their model allows them to predict from the GRIP temperature series in Greenland, the pattern of temperature oscillations recorded in the Byrd ice core, which provides a higher resolution record for the last glacial period in Antarctica than does the Vostok core. From this, it is possible to calculate the correlation coefficient between the simulated and the measured variations at Byrd. For the period from 35 000 to 65 000 years BP, the correlation is almost 0.8 when the timescale on which the heat reservoir operates is set between 1000 and 1500 years – a credible value for glacial times in the Southern Ocean. The model thus simulates the temperature variations at the Byrd Station with remarkable skill. It does not require a radical difference in mechanism between the Heinrich events and the D/O oscillations of lower amplitude. The model further suggests that the smaller D/O oscillations, as well as the Heinrich events, should have a corresponding antiphase signature in Antarctica. Recent synchronisation of the GRIP and Dome C records by means of [10]Be during D/O event 10 confirms that at this time a minor response can indeed be detected in Antarctica (Caillon *et al.*, 2003a).

Claussen *et al.* (2003) force their intermediate complexity CLIMBER model with freshwater pulses into the North Atlantic to trigger Heinrich events. For the D/O oscillations of smaller amplitude, they follow Rahmstorf and Alley (2002) and Rahmstorf (2003) in suggesting that the rather regular periodicity of the oscillations, along with evidence for the, albeit strongly dampened, persistence of this periodicity over longer periods (Bond *et al.*, 1997) points to regular external forcing. Rahmstorf and Alley (2002) liken the processes involved to the phenomenon known as stochastic resonance, whereby a weak external signal combines with internal variability to force the system over a critical threshold, with a periodicity in phase with that of the external forcing. Their model also generates credible patterns of variability that replicate many of the spatial and temporal characteristics of climate variability between 60 000 and 30 000 years BP.

Taking these modelling studies together, along with those of Rahmstorf (2002) and Schmittner *et al.* (2002; 2003), the issue of whether the difference between Heinrich and D/O events is one of degree, or of significantly different processes, remains, for the moment unresolved; however, Elliot *et al.* (2002) clearly show that the two types of event have different effects on the North Atlantic, with only the former demonstrably reducing deep-water formation, despite the fact that the D/O events are marked by oscillations in sea-surface temperatures (SSTs) and iceberg flux.

One of the features of the D/O oscillations not obviously related to the above explanations is the extent to which the signal is propagated atmospherically over what is probably the whole of the northern hemisphere. Labeyrie *et al.* (2003) integrate atmospheric processes in their conceptual model of D/O oscillations. They also note that methane varies synchronously with temperature throughout the sequence, consistent with pollen analytical evidence for responses in the terrestrial biosphere. The same authors infer a direct coupling between atmospheric circulation, coastal ice sheets and ice shelves. During warm, interstadial phases, they envisage rapid snow build-up along ice-sheet margins as a result of their juxtaposition with the northern extension of warm water. Expansion of the ice mass would lead to enlargement of the polar vortex and a gradual cooling. As a result, penetration of warm waters northwards would be curtailed and ice-sheet growth inhibited. This trend is seen as eventually culminating in the waning of coastal ice sheets and ice shelves, the diminution of the polar vortex and the onset of a new cycle. Ocean, atmosphere and cryosphere are linked in this scheme, but no numerical model is offered that purports to simulate it. Many questions remain regarding, for example, the possible role of the tropics (Stott *et al.*, 2002) and the degree of ice melting required to generate the recorded sea-level variations during the strongest oscillations. The summary by Sarnthein *et al.* (2002) outlines some of the key ways in which further progress will be made. Throughout their analysis they stress the need for more secure and finely resolved chronologies that allow precise synchronisation of ice and marine cores and terrestrial records. To this end, the revision by Shackleton *et al.* (2004) of the GRIP/GISP 2 timescales goes a long way towards harmonising records from both poles and from key marine and terrestrial sequences for the period back to 80 000 BP.

The present state of knowledge may be summarised as follows:

- Some modification of the bipolar seesaw mechanism whereby changes in the strength of meridional overturning circulation in the Atlantic generate opposing modes of heat exchange between the North Atlantic and the Southern Ocean comes closest to explaining the growing array of evidence for rapid climate oscillations between 60 000 and 15 000 years BP.
- The intervals of most extreme cooling in the northern hemisphere correspond with warm intervals in Antarctic ice cores.
- The model that most convincingly simulates the succession of anti-phase linkages between Greenland and Antarctica also predicts a weaker antiphase response in Antarctica to the periods of less extreme northern-hemisphere cooling. Evidence is beginning to emerge in support of this.
- There is, at the very least, a credible case for inferring an external forcing trigger as the mechanism responsible for the timing of the full sequence of D/O events, though its nature is not known and the existence of such a mechanism remains unproven.

- The one generally accepted mechanism for initiating the cooling limb of the most extreme oscillations involves a freshening of the surface waters of the North Atlantic in the wake of a surge of the Laurentide ice sheet and a strong southerly extension of the area over which icebergs carry and deposit mineral detritus.
- The existence of the oscillations appears to have been linked to the mean climate state during the period between the initiation of post-Eemian glaciation and the period of the last glacial maximum.
- The oscillations appear to originate in the Atlantic, but their impacts, propagated through changes in both the atmospheric and ocean circulation, are essentially global; however, the boundaries between areas over which temperature changes are in or out of phase with the Greenland record are not yet fully defined.
- There is a growing consensus that freshening of the North Atlantic is a key process generating each of the periods of strong northern-hemisphere cooling, though the mechanisms responsible for this and the size and rate of freshening remain uncertain and probably variable.

Finally, there is strong evidence to suggest that D/O variability is not unique to the last glacial period. In this regard, the high-resolution methane record from Vostok, published by Delmotte *et al.* (2004), is especially significant. The methane record reflects northern-hemisphere and low-latitude changes in terrestrial biota and soils, and is shown to undergo high-frequency variability within the range 350–600 ppbv, with a typical duration of 1500 to 3000 thousand years. This constitutes clear evidence for Dansgaard/Oeschger variability in Antarctica during all four glacial periods represented. Moreover, the phasing of the methane variability relative to isotopically inferred temperature suggests that the bipolar seesaw mechanism already outlined above prevailed throughout.

The evidence so far falls some way short of allowing rigorous testing of models equally applicable to present-day, projected-future and appropriate-past boundary conditions and forcing. Nevertheless, the information available already provides important constraints on models designed to assess the probability of a shut-down of the meridional overturning circulation in the future (see 13.3).

For the moment, we leave the enigma of rapid climate oscillations and consider conditions at the LGM, but in 6.4, analysis of the sequence of events associated with the transition from the LGM to the Holocene will once more include some detailed consideration of abrupt changes.

5.5 The last glacial maximum (LGM)

Taken literally, the LGM must represent the time when sea-level was at its lowest, hence ice volume at a maximum. This can be either identified and dated directly from stratigraphic evidence, or inferred from benthic $\delta^{18}O$ values (see 4.4.1). Since in many cases, palaeo-temperature reconstructions at a given site do not precisely parallel these global changes, the first step must be to define the global signature of the literal LGM chronologically at sites where it can be established unambiguously. Mix *et al.* (2001) suggest a calibrated age of between 21 000 and 22 000 years BP. In many cases, though, the term is used rather less scrupulously, since chronological control for a given sequence of changes may not be adequate. A date range of between 19 000 and 23 000 years probably brackets the period of glacial minimum prior to the first rise in sea-level at 19 000 years. Broecker (2000) gives a comprehensive summary of some of the main features of the LGM environment as reconstructed from empirical evidence.

Establishing the conditions at the LGM are important for several reasons (Mix *et al.*, 2001):

- The period identified was one of relatively stable climate compared with the preceding MIS 3 and the succeeding transition to the Holocene.
- The main boundary conditions – orbital configuration, continental geography, atmospheric trace-gas concentrations, sea-level and ice-sheet areas and volumes – are reasonably well known.
- It is the only glacial maximum within reach of both ^{14}C and U-Th dating.
- The period has become a major target for model simulations, partly as a result of the growing body of empirical evidence relating to the period. It provides a crucial test of the ability of models to cope with boundary conditions very different from those at the present day.

The focus on the LGM has a long history, beginning with the CLIMAP (Climate Long-range Investigation, Mapping and Prediction) study that took place in the 1970s and was published in 1981.

By now a plethora of models of varying complexity exist for the LGM, including a set of simulations carried out as part of the Paleomodel Intercomparison Project (PMIP) (Pinot *et al.*, 1999). Although the majority of recent models and a wide range of empirical data suggest that the CLIMAP reconstruction may have underestimated the reduction in temperature in the tropics at the LGM, there is a good deal of disagreement between them. This has provided further stimulus to expand and refine the evidence for conditions during the LGM.

Some of the more robust findings so far are as follows:

- Sea-level at its minimum stood at around −136 m, exposing large areas of continental shelf to sub-aerial processes.
- Although some areas remained moist and both forests and wetlands survived across a range of latitudes, in general the area covered by steppe, tundra and xeric vegetation was much greater in extent than during the late Holocene (Prentice *et al.*, 1993; Petit-Maire, 1999). This was partly the result of a cooler and drier climate, partly a response to lowered CO_2. Cowling and Sykes (1999) outline the case for viewing the fall in CO_2 as strongly influencing the balance between C3 and C4 plants, though Huang *et al.* (2001) conclude from their studies that climate, rather than reduced CO_2, is the

major factor responsible for the differences in vegetation between the LGM and the present day.

- The difference in temperature between the LGM and the present day in tropical areas has been a topic of long-standing disagreement. Continental archives record a decrease averaging within the range 4 to 7 °C while recent estimates of the decrease in SSTs mainly fall between 1 and 4 °C (Kohfield and Harrison, 2000). The general consensus is that the temperature difference was much greater at high latitudes. One effect of this was to increase meridional-temperature and -pressure gradients, especially in the northern hemisphere (e.g., Rind, 2000). It seems quite likely that continental temperatures were more depressed than those over the oceans; also that altitudinal temperature gradients were steeper.
- Dust deposition globally was around an order of magnitude greater at the LGM than today (Harrison *et al.*, 2001; Mahowald *et al.*, 1999).

Kohfield and Harrison (2000) evaluate the skill with which current models are in agreement with data for the LGM. They show that simulations with computed rather than prescribed CLIMAP SSTs capture the pattern of terrestrial biomes more successfully. Feedback from terrestrial vegetation was not included in the models they evaluated. The account by Mix *et al.* (2001) of a range of models applied to the LGM shows how it is relatively easy to generate similar outputs with radically different model structures. Overall, we are left with the conclusion that a fully effective synergy between data and equilibrium models for the LGM is still some way off. It will depend on the development of models with a more comprehensive and realistic range of feedbacks, as well as on significant improvements in palaeo-data. One recent study (Kim, 2004) using an atmosphere –ocean– sea-ice climate-system model compares the modelled cooling effects of atmospheric CO_2 concentrations with those of ice-sheet topography and other changes. Kim concludes that reduction in atmospheric CO_2 concentrations was responsible for around 60% of the total reduction in temperature at the LGM.

Chapter 6
The transition from the last glacial maximum to the holocene

6.1   The temperature record at each pole

The 10 000 years that intervene between the LGM and the opening of the Holocene period have come under intense scrutiny from almost every point of view. They represent the most recent and most accessible example of dramatic global warming. Consequently, both the dynamics of Earth-system changes over this period and the response of ecosystems to these changes (in so far as these can be separately distinguished as 'responses' within such an interactive system) are of outstanding importance.

Changing temperatures can be inferred from δ^{18}O and δD measurements in the ice itself (see 4.2.3) The initial warming corresponds with the early stages of a rise in orbitally driven insolation at $60°$ N that continued into the early Holocene, peaking at 10 000 years BP (Figure 6.1). In Antarctica the isotopically inferred temperatures begin to rise beyond the preceding range of variability around 19 000 years BP. The only significant (and unexplained) exception is in the record from the Siple Dome on the coast alongside the southern Pacific (Taylor *et al.*, 2003), where a sudden rise in surface temperature of some 6 °C is recorded as early as 22 000 years BP, followed by a decline, before temperature once more increases, rather more steadily, from *ca.* 18 000 BP onwards (Ahn *et al.*, 2004). In all the ice-core records from Antarctica, temperatures increase smoothly right through into the early Holocene, with only one 'set back', the so-called Antarctic cold reversal (ACR). On the Vostok chronology of Petit *et al.* (1999) the ACR spans some 2000 years, beginning around 14 500 years BP. Ahn *et al.* (2004), who link data from several Antarctic cores to the GISP 2 timescale, show the ACR beginning around 14 800 BP and continuing to 13 000 BP. Discrepancies of the order of a few centuries still persist between the chronologies applied in current literature to ice cores from Antarctica spanning the last deglaciation (see, e.g., Figures 6.5 and 6.6). The summary by Ahn *et al.*, building on the higher resolution sequence from Siple Dome, points the way to harmonizing these more closely both with each other and with the records from central Greenland.

The latest evidence from the Dome C core (EPICA community members, 2004) points to remarkable parallels between this, the last glacial termination,

Figure 6.1 Through the LGM to the opening of the Holocene. Changes in summer insolation, Arctic and Antarctic temperatures, atmospheric CO_2 and relative sea-level from 26 000 to 10 000 BP. HL –Heinrich layer; MPW – melt-water pulse; YD – Younger Dryas. (Adapted from Labeyrie et al., (2003), which also gives the sources for each graph.)

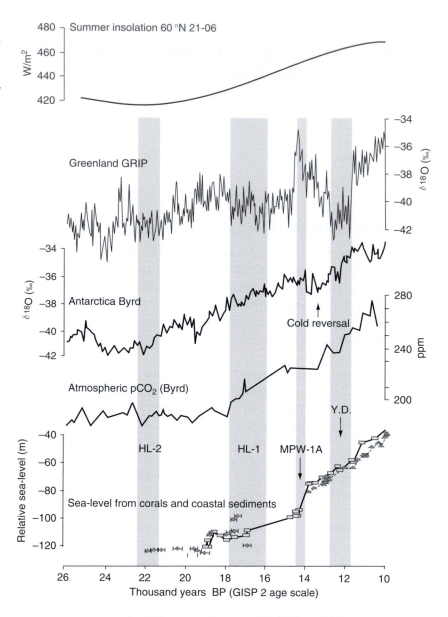

and that occurring just before the start of the Vostok record (termination 5, at the transition from MIS 12 to 11). Not only are the absolute values for many of the properties measured on either side of the two transitions very similar, there appears to be a similar 'cold reversal' during an early stage in MIS 11.

In Greenland, it is difficult to choose a point in the stable-isotope sequences where the first significant increase begins. There is a element of personal judgement involved in attaching any real significance to the initial, somewhat inconclusive increase around 19 000 years BP, prior to the cooling associated with

Heinrich layer 1. Most authors consider the opening of the Bølling/Allerød (B/A) interstadial at some time between 14 500 and 14 700 years BP as the first clear sign of northern-hemisphere warming outside the range of preceding variability. After the steep, rapid increase in temperature at the opening of the (B/A) interstadial, temperatures declined in a ragged fashion. There was then a cool interval between 12 800 and 11 600 years BP, known as the Younger Dryas (Y/D).

Comparing the inflexions on the curves of rising temperature from each set of polar records between the LGM and the opening of the Holocene, they are seen to be broadly in antiphase. Thus, the warm B/A interstadial between c. 14 600 and 12 800 BP corresponds closely, in terms of timing, with the ACR in Antarctica. This has generally been regarded as a reflection of the bipolar seesaw effect described in Section 5.4. Morgan *et al.* (2002), however, raise the possibility that at the Law Dome site, the ACR may have begun around 15 000 years BP, a few centuries before the start of the B/A. If subsequent research confirms this version of the relative timing of these two events, the results would cast doubt on the bipolar seesaw, in its simplest form, as a satisfactory explanation for all the opposed changes in temperature between Greenland and Antarctica.

6.2   Spatial patterns of temperature change

The sequence of temperature changes inferred from the stable-isotope records in central Greenland ice cores has come to serve as a template for a wide area on either side of the North Atlantic. Figure 6.2 shows how closely coherent the detailed records of changing climate are from Greenland eastwards into western and central Europe and southwards at least as far as the coast of Venezuela. Whatever the causes of the variability recorded, the spatial propagation of the signatures must have involved atmospheric processes (cf. Mikolajewicz *et al*, 1997). Indeed, the main sequence of changes, well known from Europe since the late 1920s and named after type-sites in Denmark, can be detected in a great diversity of marine and continental archives and proxy records across most, if not all, of the northern hemisphere (Mikolajewicz *et al*., 1997; Broecker, 2000; 2003; Rahmstorf, 2002). Much the same can be said for the earlier D/O oscillations already considered (see 5.4), but continental archives with a full and detailed record of these earlier events are much less common than are the records from the LGM–Holocene transition.This is mainly because of the more widespread survival of archives postdating the LGM, especially in previously glaciated regions.

The temperature change during the final transition from Younger Dryas to early-Holocene conditions in central Greenland was accomplished in a matter of decades (Figure 6.3). Its amplitude was originally estimated from stable-isotope measurements as around 10 °C, but more recently, borehole-temperature records (Cuffey *et al*., 1995; Dahl-Jensen *et al*., 1998) suggest that this is almost certainly an underestimate by some 10 to 15 °C. Estimates of the parallel temperature changes in western Europe vary with the site and proxy used, but generally

Figure 6.2 The detailed coherence of selected records of climate change around the North Atlantic, from Greenland, Europe and the coast of Venezuela. (a) Part of the grey-scale trace from the Cariaco Basin, reflecting changes in the strength of the northeast trade winds (Hughen *et al.*, 1996). (b) The $\delta^{18}O$ record (bulk carbonates) from Hawes Water, northwest England (Marshall *et al.*, 2002), reflecting changing temperature. (c) The $\delta^{18}O$ record from GRIP, central Greenland (Johnsen *et al.*, 2001). (d) The $\delta^{18}O$ record from Leysin in central Switzerland tuned to the GRIP ice core timescale (Schwander *et al.*, 2000).

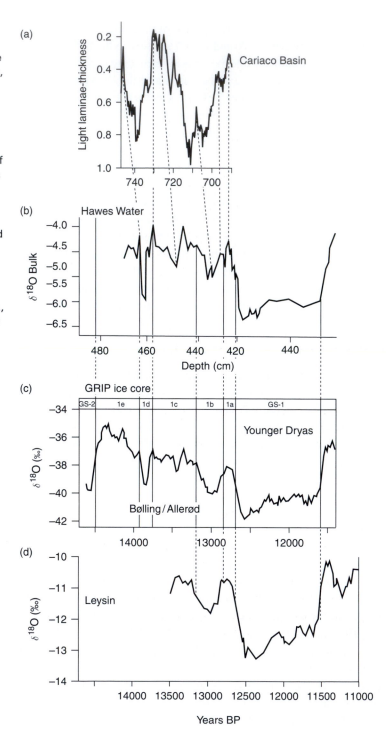

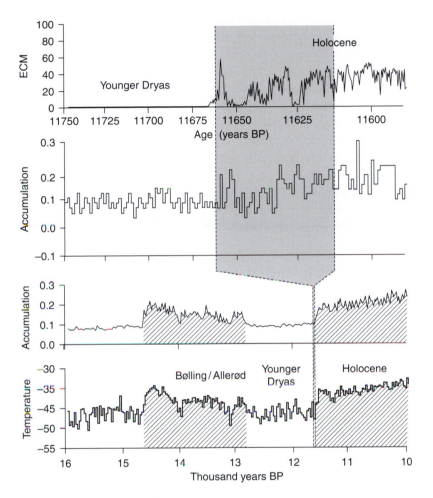

Figure 6.3 A detailed record of ice accumulation, electrical conductivity (ECM) and isotopically inferred temperature in central Greenland for the late-glacial and early Holocene. The upper two graphs are expansions of the lower timescale around the Younger Dryas–Holocene transition. Note the flickering nature of variability across this transition, as well as the amplitude and sudden impact of each step in the switch. The whole transition occupies around 50 years. (Modified from the PAGES website; See www.pages-igbp.org.)

range from 4 to 10 °C, at sites from northern. Norway to central Switzerland (Ammann, 2000; Birks *et al.*, 2000; Jones *et al.*, 2002; Marshall *et al.*, 2002). Estimates of the change in sea-surface temperatures (SSTs) across this transition in the North Atlantic and neighbouring seas range up to 10 °C (Karpuz and Jansen, 1992). Both alkenone and δ^{18}O records from the Adriatic point to a shift of around 8 °C in SSTs there at the opening of the Holocene (Ariztegui *et al.*, 1996). In the case of the closely linked changes recorded in the Cariaco Basin (Hughen *et al.*, 1996), the shifts in sediment type that track the climate express changes in productivity linked to differential upwelling as a result of modulation of the northeast trade winds.

There is no simple boundary between the latitudes dominated by the northern-hemisphere sequence of changes and the Antarctic one. Terrestrial records in the southern-hemisphere include some from both South America (Hajdas *et al.*, 2003) and New Zealand that claim to detect a phase of cooling during the Younger Dryas time interval (see Rahmstorf, 2002). Using the deuterium-excess (δD) record from

Dome C, Stenni *et al.* (2001) inferred the likely existence of a cold reversal in the Southern Ocean closer to the age of the Younger Dryas than to the well-documented ACR. By contrast, Lamy *et al.* (2004) show that the record of surface-water temperature changes in marine sediments off the coast of southern Chile follow the timing recorded in Antarctica. The issue has become more complicated as a result of data from Japan. Although the pattern of change recorded there parallels, in broad outline, that found on either side of the Atlantic, the varved sediments from Lake Suigetsu in Japan suggest dates for the main transitions that differ by several centuries (Nakagawa. *et al.*, 2003). The authors conclude that the sequence of changes in Japan reflects both solar insolation changes and Atlantic influences. These results caution against accepting correlations that are not based either on an extremely high level of signal coherence (cf. Figure 6.2), or on accurate, precise and independent chronologies.

6.3   Changes in continental hydrology

Figure 6.4 shows a more extended record of isotopically inferred temperature changes alongside changes in the rate of ice accumulation through this period in the GRIP icecore. The periods preceding the B/A and the YD are both character-ized by low rates of accumulation. Sharp increases mark the onset of the B/A and the Holocene. Markgraf *et al.*, (2001) use lake-level variations to infer changes in moisture regime during this period from a suite of sites along the western side of the

Figure 6.4 Ice accumulation and isotopically inferred temperature in central Greenland over the last 17 000 years. Note the sudden warming transitions around 14 6000 and 11 600 years BP, as the opening of the Bølling/ Allerød and the Holocene, respectively, and the reduced variability during the Holocene (modified from the PAGES website. (See www.pages-igbp.org).

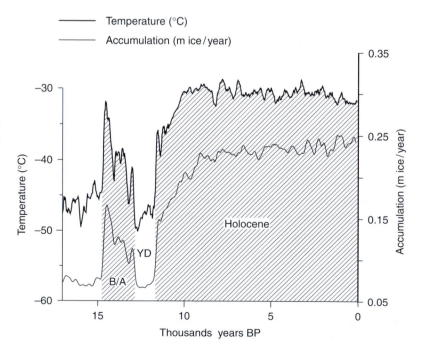

Americas. Their results point to a generally dry and cold pre-Holocene at high latitudes, as well as in equatorial regions, with moister conditions generally more prevalent in temperate latitudes. In equatorial Africa, sediment records from Lake Victoria suggest that it experienced quite severe desiccation at least once and perhaps twice between the end of LGM and the B/A; there was also a less severe phase of lower lake levels during the YD (Stager, *et al.*, 2002).

6.4 Dynamics, forcing and feedbacks

6.4.1 Introduction

During the course of the complex and interrupted rise in temperature between the LGM and the opening of the Holocene, the potential feedback processes involved include the following:

- The waning of continental ice sheets and sea-ice extent.
- Increases in atmospheric greenhouse-gas concentrations (CO_2 recorded reliably only in Antarctica, CH_4 in both Greenland and Antarctica).
- Sharply declining rates of dust deposition recorded in ice cores from both polar regions, from the tropics (Thompson, 1995) and in loess sections (An, 2000).
- Changes in ocean circulation.
- Changes in vegetation cover and soil moisture.

Each is considered briefly below, then questions of phasing are addressed

6.4.2 Ice volume

For the LGM–Holocene transition, the shrinking of the northern hemisphere ice sheets can be tracked from geomorphological and stratigraphic evidence allowing identification of ice extent at any given period (e.g., Clark *et al.*, 2001). Global ice extent can be derived from the evidence for global sea-level rise (Figure 6.1) During the period from the LGM to the opening of the Holocene, sea-level rose by just over 60 m, with the steepest rise, termed melt-water pulse (MWP) 1A, beginning just before 14 000 years BP and coinciding with the early part of the B/A interstadial. The evidence from the Barbados coral sequence (Fairbanks, 1989), and sediments off the coast of northwestern Australia (Yokohama *et al.*, 2001) and from the Irish Sea basin (Clark *et al.*, 2004) shows a smaller, but significant, increase of some 10 m, around 19 000 years BP.

6.4.3 Greenhouse gases

Monnin *et al.* (2001) subdivide the increases in CO_2 and CH_4 concentrations into four stages (Figure 6.5). During the first stage, from *c.* 17 000 to 15 400 years BP, there are steep, synchronous increases in both trace gases. The estimate by Monnin *et al.*, for the lag between the temperature and the CO_2 rise is 800

+/−600 years, in good agreement with estimates from other Antarctic ice cores. During the second stage, between 15 400 and 13 800 years BP, CO_2 values rise more gently and CH_4 concentrations hardly change. Around 13 800 years BP, at or just after the beginning of the ACR (correlated with the B/A by CH_4 concentrations) there is a sudden jump in both sets of values, after which they remain stable until 12 300 BP. Between 12 300 and 11 200, CO_2 values rise steeply once more, but there is a trough in CH_4 concentrations, which only recover to early-Holocene values at the end of this phase. The CH_4 concentrations correlate this trough with the YD (note that, as shown on Figure 6.5, the Dome C chronology appears to differ from the GRIP timescale by *c.* 500 years). Comparison with the δD temperature record gives a maximum correlation ($r = 0.94$) with CO_2 if the CO_2 values are shifted back by 410 years. This is so close to the range of uncertainty in estimates of the age difference between the ice and the enclosed gases that no significant delay between temperature and greenhouse-gas increases can be established with certainty. The higher-resolution record from Siple Dome (Ahn *et al.*, 2004) suggests a time lag of around 210 to 330 years. Their results make it highly unlikely that the increase in CO_2 values predated that in temperature and make a short lag probable, though not certain. A close link between the two is once more strongly indicated.

Figure 6.5 The record of atmospheric greenhouse-gas concentrations over the LGM–Holocene transition at Dome C, Antarctica. The four stages in the increase of greenhouse-gas concentrations are discussed in Section 6.4.3. ACR – Antarctic cold reversal; YD – Younger Dryas; B/A – Bølling/Allerød (modified from Monnin *et al.*, 2001). Note the offset between the Dome C and GRIP chronologies indicated by the sloping lines near the foot of the graph.

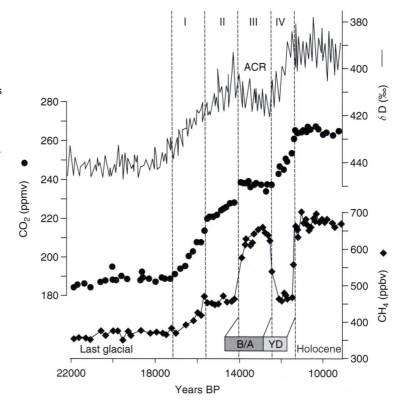

The evidence from Dome C suggests a similar phasing between temperature and CO_2 during the transition between marine isotope stage (MIS) 12 and 11 (EPICA community members 2004), though the increase in CH_4 during the earlier deglaciation begins after the start of the CO_2 rise. Some decoupling between CO_2 and CH_4 changes is thus apparent in both transitions, though the pattern is not the same in each. The decoupling is consistent with the view that whereas changes in CH_4 reflect mainly responses by terrestrial vegetation and soils in the northern hemisphere and low latitudes, changes in CO_2 are mainly linked to ocean processes. It is now well established that the increase in CH_4 at each glacial termination recorded in the Vostok ice core lags isotopically inferred temperatures by around 1000 years (Delmotte *et al.*, 2004).

6.4.4   Dust deposition

Remarkable data sets are available from both central Greenland and Antarctica using Ca concentration as a proxy for dust deposition, or detailed microparticle measurements (Figure 6.6). These latter are available for the last 27000 years from Dome C (Delmonte *et al.*, 2002) and, at lower resolution for the whole of the last 740 000 years (EPICA community members, 2004).

In Greenland, values decline between 20 000 years and 18 000 years BP, plateau for a while, then peak around 17 000 to 15 500 years, after which they fall almost to zero at the beginning of the B/A. They only increase again at the opening of the YD around 12 800 years BP, after which they decline to negligible levels at the opening of the Holocene. The records from Vostok (Petit *et al.*, 1999) and Dome C (Delmonte *et al.*, 2002; Rothlisberger *et al.*, 2002) show over an order of magnitude decline between *c*. 18 000 and 11 000 years BP. Loess deposition and transport out over the Sea of Japan and the Pacific declined steeply over the same period (Tada *et al.*, 1999; Irino *et al.*, 2001). Between 18 000 and 17 000 years BP, dust deposition onto the Huascaran ice cap in Peru also fell by an order of magnitude (Thompson, 1995). Although the precise phasing of declines in dust deposition varies between sites, there is a common overall trend irrespective of location. In central Greenland and at Dome C in Antarctica, where the records for the period of deglaciation are temporally highly resolved, dust deposition and temperature are strongly anti-correlated at every stage. Similar rapid falls in dust concentration accompany all of the eight glacial terminations recorded in the long sequence from Dome C.

6.4.5   Ocean circulation

Several lines of evidence summarized by Clark *et al.* (2002) suggest that during the whole period of deglaciation, each major step in the sequence of changes

Figure 6.6 Ice-core records of changing dust concentrations and temperature during the LGM–Holocene transition. The GRIP, Huascaran and Vostok graphs are from Raynaud *et al.* (2003), giving the sources for each graph; the Dome C, Antarctica, graphs are from Delmonte *et al.* (2002). At each pole, dust (or Ca) concentrations decline as temperatures increase. The steep decline at Huascaran falls within the same time interval. ACR – Antarctic cold reversal; YD – Younger Dryas. Note the offset between the Vostok and Dome C chronologies.

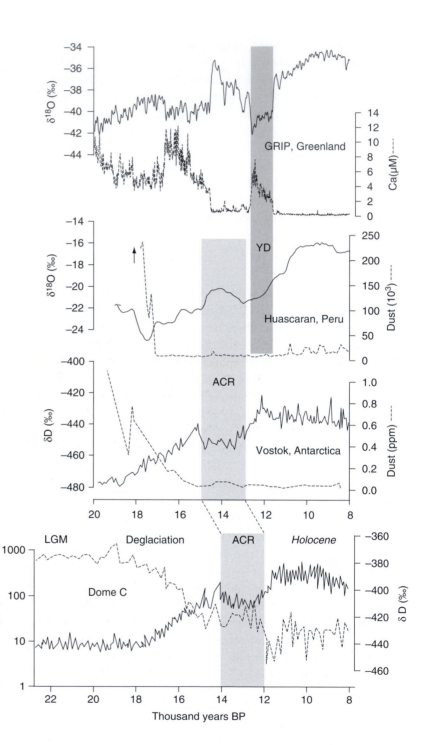

from the dip in temperature in central Greenland contemporary with Heinrich 1 (18 000–16 000 years BP) through the B/A and YD, into the opening of the Holocene, involved a reorganisation of ocean circulation. These reorganizations included major fluctuations in the strength of North Atlantic deep water (NADW) formation, as was the case during the earlier Heinrich events and intervening interstadials. Much research has focused on two themes:

- The sequence of meltwater discharge events around the margins of the Laurentide ice sheet and associated changes in ocean stratification and circulation in the North Atlantic.
- Temperature, stratification, circulation and productivity changes in the Southern Ocean.

Combining the results presented by Teller *et al.* (2002) and Aharon (2003), it now seems highly likely that the main trigger for the onset of the YDs was sudden freshwater dischargee from glacial Lake Agassiz. The discharge was routed into the North Atlantic via the Great Lakes St Lawrence valley, providing a freshwater cap to the North Atlantic sufficient to disrupt NADW formation. Prior to the onset of the YD, discharge from the ice margin was mainly routed southwards via the Mississippi valley. Up until the beginning of the B/A, peaks in freshwater discharges into the Gulf of Mexico had coincided with periods of cooling in Greenland, suggesting that they may have played a role in changes in NADW formation. Subsequently, peak discharge via this southerly route into the Gulf of Mexico roughly coincided with the start of the B/A interstadial – a period of dramatic warming in Greenland and around the North Atlantic. The discharge peak appears to have accounted for around 50 % of the rise in global mean sea-level during melt-water pulse (MWP) 1A. This conflicts with the proposal by Weaver *et al.* (2003) that MWP 1A was mainly the result of melting of the Antarctic ice sheet.

Interest in the role of the Southern Ocean as a driver of climate change during deglaciation has been sharpened by the model simulations of Knorr and Lohmann (2003). They showed how a gradual warming of the Southern Ocean and a reduction in sea-ice leads to changes in ocean circulation that can suddenly give rise to a switch from the glacial to interglacial mode of circulation in the North Atlantic. Their model generates a rise in SSTs in the North Atlantic of up to 6 °C over a few decades – a result compatible with the sudden rise in temperature associated with the opening of the B/A, around 14 600 years BP. In their model, this process can even overcome the effects of meltwater discharge into the North Atlantic. These results therefore help to reconcile the apparent paradox of a large freshwater flux into the Gulf of Mexico linked to MWP 1A falling within the early stages of the B/A interstadial. Linking these several lines of evidence together, it is now possible to propose a coherent, albeit still speculative, sequence of events in the Southern Ocean and North Atlantic during the period of deglaciation (Stocker, 2003 and Figure 6.7). The existence of a possible parallel with the ACR during the deglaciation immediately preceding MIS 11 (EPICA community members, 2004) opens up the possibility of a broadly similar sequence of changes during that earlier transition.

The only insolation changes required in the scheme outlined in Figure 6.7 are those linked to smooth orbital forcing. On the other hand, van Geel *et al.* (2002) make the case that since the sudden and wide-scale climate changes around the time of the B/A–YD transition coincide with a strong anomaly in atmospheric ^{14}C concentrations, the cooling may have been at least in part linked to a strong reduction in insolation. Solar forcing of this kind is implicit in Rahmstorf's (2003) review linking the YD to earlier Dansgaard/Oeschger oscillations. Although neither Marchal *et al.* (1999), nor Clark *et al.* (2002), discount the possibility, the balance of evidence may be swinging away from sudden solar

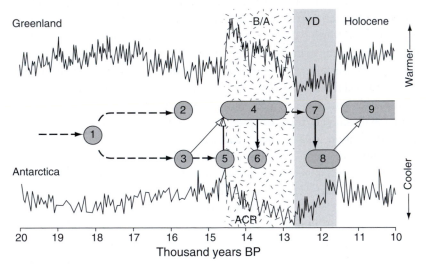

Figure 6.7 A schematic model of the likely sequence of changes in climate and circulation in the Atlantic during the transition from full glacial to Holocene conditions. (1) Increased solar radiation triggers the onset of the transition. (2) Freshwater from the melting of northern ice sheets shuts down the meridional overturning circulation (MOC) in the Atlantic (Heinrich 1). (3) As a result, the bipolar-seesaw mechanism enhances warming in the southern hemisphere, which quickly turns on the MOC once more, by surface advection of saline water (4) as well as a discharge of melt water in the south (5). As a result, the north now warms rapidly (the Bølling/Allerød interstadial) and the seesaw mechanism leads to cool conditions (the Antarctic cold reversal) in the south (6). Melting of the northern ice sheets accelerates, leading to a second abrupt cooling in the north (7), which once more stimulates warming in the south (8). Finally, this turns on the MOC once more, leading to the sudden warming (9) at the opening of the Holocene. Solid lines with black arrowheads represent the bipolar-seesaw mechanism. Lines with clear arrowheads indicate the mechanisms thought, on the basis of modelling, to be responsible for changes initiated in the southern hemisphere and eventually espressed as the resumpyion of the MOC at the opening of the Bølling/Allerød interstadial and the Holocene. (Modified from Stocker, 2003.)

forcing at this time. Delaygue *et al.* (2003) use a dynamic model of intermediate complexity to show that, provided the additional effects of an increased oceanic reservoir age for ^{14}C are taken into account, simulated changes associated with a meltwater discharge are probably sufficient to account for the anomaly in atmospheric ^{14}C concentrations.

6.4.6   Changes in vegetation cover and soil moisture

A much more comprehensive global survey of vegetation changes during this time interval is needed before the possible role of feedbacks can be confidently analysed. Adequately dated terrestrial sequences outside Europe and north America are still too sparse. In general terms, the northern hemisphere cold intervals before and after the B/A appear to have been characterised, with some exceptions in monsoon-dominated regions, by more widespread grassland, steppe and tundra communities, under drier as well as colder conditions. In so far as changing atmospheric CH_4 concentrations during the period indicate anaerobic decomposition in terrestrial biomes, moister soil conditions appear to have prevailed during the warm intervals, both in tropical and high-latitude environments (Monnin *et al.*, 2001). If the model-based conclusions reached by Meissner *et al.* (2003) concerning land-surface feedbacks during glacial inception have comparable implications during deglaciation, there is an urgent need to build vegetation and soil feedbacks into models simulating deglaciation.

6.4.7   Questions of phasing

From the above account, we can draw several tentative conclusions:

- Increases in atmospheric-greenhouse-gas concentrations probably began a few centuries at most after the first indications of rising temperatures in most cores from Antarctica. A small rise in global mean sea-level ($\sim$10 m), implying some ice melting, probably preceded this, but the main rise in sea-level (hence rapid ice melting) post-dated it by 2000 to 3000 years. There is still much room for improvement in establishing more precisely the phasing of different forcings, feedbacks and responses during deglaciation.
- As in the case of the strongest oscillations during MIS 3, the antiphase relationships in inferred temperature between Greenland and Antarctica during the period point to a major role for ocean circulation in driving the changing patterns of response at high latitudes in each hemisphere.
- The interruptions to northern-hemisphere warming during Heinrich 1 and the YD intervals are most probably linked to surges in fresh-water input to the North Atlantic, the first via the Gulf of Mexico, the second via the Great Lakes–St Lawrence route. Both involved complete or partial collapse of the Atlantic meridional overturning

circulation, with consequent warming in the South Atlantic. The steep warming trend in the northern hemisphere that came at the end of each coincided with the swift resumption of the circulation (McManus *et al.*, 2004).

- The sudden onset of northern hemisphere warm conditions at the opening of the B/A and the Holocene are thus linked to a rapid development of the 'modern' mode of NADW formation.

- There appear to be significant differences between Heinrich 1 and the YD. For example, the latter coincides with much higher and quite steeply rising levels of atmospheric CO_2 concentrations. The reasons for these differences remain unresolved, though changed conditions in the Southern Ocean may have been involved.

- The close antiphase relationship between changes in dust deposition and isotopically inferred temperature, in ice cores from both poles, suggests the possibility of a significant role for atmospheric dust as an agent in climate change during deglaciation.

- The parallels between higher atmospheric concentrations of CH_4 and evidence for warmer, moister conditions in the northern hemisphere during deglaciation (and vice versa) point to changes in terrestrial vegetation and soil-moisture status as the main driver of the recorded changes in methane concentrations (but see 6.5.4).

- Modulation of atmospheric CO_2 changes during deglaciation appears to be linked to changes in the world's oceans.

- Such changes, by modifying ocean ventilation, are likely to have been the main, though not necessarily the only, processes generating the changing concentrations of atmospheric ^{14}C during the period.

- This consideration, coupled with the increasing skill with which models are able to replicate some of the key features of the Earth-system changes recorded during deglaciation, without invoking external forcing, tends to cast doubt on, but does not yet preclude, the possibility that sub-millennial changes in external forcing play a major role in the rapid oscillations characteristic of deglaciation.

6.5   Transitions and feedbacks – deglaciation in the context of longer-term changes

6.5.1   Challenges to modelling

It is clear from all the references to modelling above, that despite significant progress, providing coherent, well-constrained and testable models of all the dramatic changes in Earth-system function that took place between the beginning of MIS 5d through to the opening of the Holocene remains a formidable challenge. There are several underlying reasons for this (Clark *et al.*, 2002; Labeyrie *et al.*, 2003):

- Combining the degree of complexity and the spatial and temporal resolution required for comprehensive modelling on the one hand, with the length of model run needed to capture even the rapid transitions on the other, is still out of reach. Compromises have to be made and these involve simplifications that limit the value of the model.

- Proxy data on many of the key variables are too sparse and too uncertain to provide an adequate basis either for initialising models or for testing their performance.
- There are significant mismatches between the timescales on which abrupt changes occurred (years to decades) and those on which transient states were sustained (centuries). Moreover, the periodicities of variability do not match any of the currently known timescales of ocean and atmosphere variability.

Only by integrating cryospheric and terrestrial biome changes with atmospheric and oceanic processes and linkages, and by generating interactions that include strong hysteresis, can models be produced that replicate aspects of the recorded variability in a credible way. The models generated simulate alternations in thermohaline circulation between preferred modes but they are still some way from allowing confident detection of the threshold values that can tip the system between modes under any given mean state. This limits both their explanatory power for past variability and their possible predictive power for likely future changes.

6.5.2   Positive and negative feedbacks

We now return to the three questions posed earlier at the end of Section 5.1. These concerned:

- The likely sequence of feedback mechanisms and the role of greenhouse gases, especially in the shift from glacial to interglacial conditions.
- Processes limiting the amplitude of natural variability in trace gas concentrations through the last four glacial cycles.
- The major sinks, sources and fluxes for carbon during successive glacial cycles.

There is evidence that changes in all the feedback mechanisms already discussed took place shortly after the initial change in solar radiation, believed to be the trigger for the onset of deglaciation. Attempting to quantify their relative importance depends on models that are still subject to great uncertainty. Increases in greenhouse gases probably followed initial temperature increases by a few hundred years. The most recent results cast some doubt on the earlier assertion that they invariably preceded significant ice melting and sea-level rise. Resolution of this question awaits better correlations and chronologies. There is general agreement that major changes in CO_2 concentrations during glacial–interglacialcycles reflect, above all, exchanges with the marine reservoir. The arguments for this are summarized by Pedersen *et al.* (2003). Unfortunately, these same processes are poorly constrained by existing data. This means that it is still impossible to make an informed choice between the various alternative hypotheses developed to account for the variations in the exchange of CO_2 between the ocean and atmosphere (LeGrand and Alverson, 2001). Pedersen *et al.* (2003) point out that several of the suggested processes may operate together, either in sequence or in parallel, to generate the vast reorganisations of the carbon cycle that we see during the major transition. All these considerations impede realistic modelling of the role of CO_2 in the sequence of changes involved in the last glacial terminations.

None of the studies summarised here questions the feedback role that increasing concentrations of atmospheric greenhouse gases played during glacial terminations. Few, if any, models of glacial–interglacial transitions fail to highlight greenhouse-gas feedbacks and none of the evidence available so far seriously challenges the inference by Lorius *et al.* (1990), on the basis of the early records from Vostok, that the sensitivity of temperature to a doubling of CO_2 is in the region of 3–4 °C, a value consistent with physical theory and within the range of sensitivity used by the IPCC (see 13.1.3).

The processes responsible for limiting the extreme values of greenhouse-gas concentrations during interglacial maxima and glacial minima have not yet been elucidated. One scenario has been outlined by Falkowski *et al.* (2000) and Scholes (2002). Their conceptual model envisages that during each of the recurrent glacial minima, the productivity of terrestrial ecosystems in the cold, dry glacial climate dropped to the point where loss of carbon through respiration balanced uptake through photosynthesis. Consequently, plant growth was maintained at a minimal level. In the ocean, there was less carbon release to the atmosphere than today, partly through reduced thermohaline circulation, partly as a result of enhanced biological productivity. Large quantities of organic carbon were thus stored in sediments and inorganic-carbon remains in deep waters. The authors regard this state as quasi-stable and linked to an atmospheric CO_2 concentration of around 180 ppm. Emerging from it towards interglacial conditions requires the combination of forcings and feedbacks already outlined above, beginning with the initial insolation trigger. Transient equilibrium at peak values during interglacials is seen to reflect a balance between increased ocean productivity in areas of upwelling, and outgassing of CO_2 driven by increased solubility. The increased productivity is a result of long-term input of nutrients remobilised from land surfaces. The peak values of about 280 ppm reflect the point at which biological uptake of CO_2 both on land and in the oceans is balanced by the ocean-to-atmosphere flux. There are problems with several aspects of this model (see, e.g., Bender, 2003), but, with all its limitations, it serves to indicate some of the possible linkages between systems that may be involved in regulating the carbon budget of the Earth.

An alternative view of the linkages between the various components of the Earth system during periods of transition envisages a crucial role for atmospheric dust both as a fertiliser in parts of the ocean where it is the key limiting nutrient and as a feedback mechanism in the atmosphere through its effects on radiative balance.

Bopp *et al.* (2003) use an ocean biogeochemistry model and palaeo-data to suggest that an explicitly simulated dust field for the LGM can provide enough iron to the Southern Ocean to stimulate a diatom bloom sufficient to lead to a global loss of carbon to the deep ocean of around 6%, through the operation of the 'biological pump'. This, in turn, would generate a draw-down in atmospheric CO_2 of 15 ppm. Higher atmospheric-dust content at the LGM results from the interaction of several processes. The continental areas prone to dust entrainment are enlarged as a consequence of reduced C_3 plant growth, which in turn results from lower atmospheric CO_2 and drier climatic conditions (Mahowald et al., 1999). Stronger winds and a reduced hydrological cycle, as a result of lower precipitation and cold oceans, lead to increased atmospheric transport. The evidence for dust deposition in an extensive data set of marine and ice-core records is consistent with the level of enhancement in dust deposition generated by the simulated dust field of Bopp et al. Equally, the spatial pattern of changes reconstructed from marine-sediment proxies for productivity is consistent with the simulated changes in export production to the deep ocean as a result of the fertilising effect of the enhanced dust deposition generated by the model. Thus, the nature of the terrestrial biosphere in combination with changed atmospheric processes generates enhanced dust entrainment, which in turn drives enhanced marine primary productivity, which in turn leads to the sequestering of carbon in the deep ocean via the biological pump. Calvo et al. (2004) suggest that dust may also have changed the pattern of biological productivity by providing an input of silica to the open ocean. Their results from the Tasman Sea show that heavy dust deposition was associated with a switch from dominantly coccolithophore to diatom productivity. Whereas under dominance by coccolithophores the effects of the biological pump are offset by calcite secretion, in the case of diatom dominance, there is a clear net draw-down of CO_2. They infer that this type of change may have contributed to the lowering in atmospheric CO_2 during glacial periods.

Changes in terrestrial and marine biota are both essential elements in the schemes of interaction outlined above on the basis of Bopp et al. (2003) and by Calvo et al. (2004). Bopp et al., whose simulation is broadly consistent with existing palaeo-data from both marine (cf. Sigman and Boyle, 2000) and ice-core records, estimate that the maximum impact of high dust deposition on atmospheric CO_2 concentrations must be rather less than 30 ppm. Their conclusions are broadly compatible with those of Watson et al. (2000), Ridgwell and Watson (2002) and Ridgwell (2003). Ridgwell's analysis suggests that the 'dust–iron–marine productivity–carbon export to the deep ocean' hypothesis (see Figure 6.8) could account for about a third of the temperature variability recorded in Antarctica on glacial–interglacial timescales.

6.5.3 Iron fertilisation experiments and the 'biological pump'

The question of iron fertilization and ocean productivity – one of the key elements in the above hypothesis – has been addressed through direct experiments in areas of the ocean where productivity is low despite high nitrate concentrations in the water.

Figure 6.8 A schematic diagram of the feedbacks in the climate system involved in the hypothesis of iron fertilisation activating the 'biological pump' in the oceans. At the onset of glaciation, lower sea-levels, drier climatic conditions and reduced plant cover all contribute to increased dust supply, which serves to fertilise areas of the ocean where productivity has been limited by iron. Increased productivity leads to a draw-down of carbon to the deep ocean and a reduction in atmospheric CO_2. During deglaciation, the reverse suite of interactions lead to higher atmospheric CO_2 concentrations (Modified from Ridgwell, 2003.)

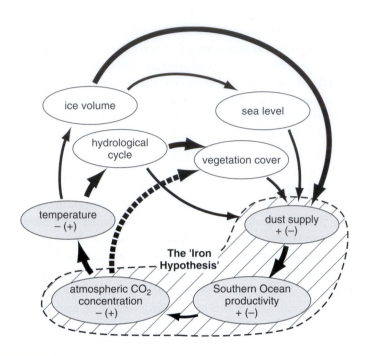

Coale *et al*. (1996) carried out the first, IRONEX, experiments in the equatorial Pacific, since when the SOIREE (Boyd *et al*., 2000) and SEEDS (Tsuda *et al*., 2003) experiments have been conducted in the Southern Ocean and subarctic Pacific, respectively. These studies confirm that iron fertilization can generate strong, rapid increases in primary productivity. Tsuda *et al*. recorded a shift in diatom population towards dominance by larger species, associated with 2.6 doublings of the population per day during their experiment. These results suggest that the conditions generated by the high productivity they recorded would favour rapid sedimentation of organic matter on a massive scale. This, along with the shift in dominant species, is a key process in the 'iron-enrichment' hypothesis as applied to glacial times. Nevertheless, the effectiveness of the mechanism as a major factor in the glacial (or future) carbon cycle remains in doubt, since none of the experiments directly confirm export of the resulting dead biomass to deep waters in large quantities. Boyd *et al*. (2004) included a sediment-trapping strategy in their SERIES iron-enrichment experiment in the Gulf of Alaska. They conclude that only a small proportion of the particulate carbon produced during the iron-induced phytoplankton bloom actually reached the deeper layers of the ocean. Much of the rest was re-mineralised by bacterial action or grazed by animal plankton. They also found that productivity was limited by the availability of silicic acid as well as iron. As the results of more iron-fertilisation experiments are published, the danger of extrapolating temporally beyond the timeframe of short experiments and spatially beyond the bounds of the experimental area become increasingly apparent.

Bishop *et al.* (2004), on the basis of the SOFeX experiment in the Southern Ocean, demonstrated not only increased productivity but also enhanced export of carbon at least to a depth of 100 m. They claim that their results reinforce the view that iron fertilization may have enhanced the biological pump in the Southern Ocean during the LGM. The complementary study, by Buesseler *et al.* (2004) within the framework of the same experiment, suggests that the level of carbon export to deeper waters is much the same as for natural blooms and casts doubt on the effectiveness of iron fertilisation as a response to future increases in atmospheric CO_2.

6.5.4   Other processes

Harrison *et al.* (2001) and Claquin *et al.* (2002) also consider the effects that enhanced atmospheric-dust loading at the LGM may have had on net radiative forcing. Following Overpeck *et al.* (1996) Claquin *et al.* agree that the effect would have been positive over ice sheets, but their simulation suggests that this was more than compensated by a large and negative effect in the tropics. Their model suggests that the overall effect may have been broadly comparable to that of the reduction in atmospheric CO_2. Although sensitivity tests suggest that this is a robust conclusion, there is still much doubt as to the effect on the actual energy balance at the Earth's surface.

Although the simulations outlined above claim consistency with available palaeo-data, it is important to note that in a situation where data are sparse, subject to significant, often unquantified uncertainties, and sometimes contradictory, claims of consistency often fall short of confirmation. For example, Maher and Dennis (2001), claim that the phasing of changes in dust flux and CO_2 around the time of the penultimate deglaciation, together with the low dust flux to the Southern Ocean during the period, both run counter to the hypothesis that dust was a major controlling influence on atmospheric CO_2 concentrations. Through analyses of ice-core N_2O concentrations and isotopic ratios over the last 106 000 years, Sowers *et al.* (2003) infer that there has been little change in the relative proportions of terrestrially derived and marine N_2O, a finding that tends to cast doubt on changing marine productivity as the dominant cause of changes in atmospheric-CO_2 concentrations between glacial and interglacial times. Furthermore, Caillon *et al.* (2003b), on the basis of their timing of changes during termination III, the previous deglaciation, favour a link between CO_2 and marine processes that involves a longer time lag than the 'iron fertilisation hypothesis' would imply.

That other ocean processes than dust deposition were involved in glacial–interglacial changes in atmospheric-CO_2 concentrations is implied by the indication that the combined dust effects were unlikely to account for more than around 30% of the recorded changes in CO_2. Sarmiento *et al.* (2004) and Ribbe (2004), in his comments on the Sarmiento *et al.* paper, point to a dominant role for the Southern Ocean as a driving force in ocean-productivity changes on a global scale. Modulation of the role of the Southern Ocean as a wide-scale supplier of nutrients, through changes in ocean circulation, would therefore seem to be another candidate for contributing to changes in CO_2 on glacial–interglacial timescales.

That the increase in atmospheric CO_2 concentrations during glacial–interglacial transitions coincides with strong evidence for increasing productivity by terrestrial ecosystems is one of several reasons for regarding ocean as the main player in modulating CO_2 changes. One of the contributing processes was probably the development of coral reefs in shallow sea-shelf regions as sea-level rose to cover these and reef growth became more widespread. The carbonate production involved in reef development leads to CO_2 release. Vecsei and Berger (2004) evaluate the evidence for the timing and effects of the process. They show that the influence was significant and was probably strongest during the early Holocene, between 9000 and 6000 BP, when the eustatic sea-level rise had slowed down. It is therefore not likely to have been a major influence on either the earliest increases in atmospheric CO_2 concentrations, or the later ones that have been linked by Ruddiman (2003b) to human activities (see 7.12).

The changes in methane concentration during the last deglaciation (Figure 6.5) are usually interpreted as reflecting mainly changes in the extent of tropical wetlands, with northern hemisphere boreal peatlands playing a subsidiary role. The claim that these sources have controlled atmospheric methane on longer timescales has been challenged by Kennett et al. (2000; 2003). They propose a major role for methane hydrates. These occur in situations where organic carbon is effectively trapped and retained under high pressure and/or at low temperatures. Vast quantities are trapped in solid and probably gaseous form under Arctic permafrost. They also occur more widely in organic-rich marine sediments. Destablisiation and release of methane from these deposits may have been triggered by changes in sea-level, water temperatures or ocean circulation. Evidence from the Santa Barbara Basin (Kennet et al., 2000; Hinrichs et al., 2003) suggests that hydrates were released into the Gulf of California during Dansgaard/ Oeschger oscillations and deglaciation. This led Kennett et al. (2003) to propose that similar processes were more widespread and may have contributed significantly to the rise in atmospheric-methane concentrations during deglaciations. This remains a controversial hypothesis. Maslin and Thomas (2003) attempt to establish the maximum contribution this process could make to changes during deglaciation by using the stable-isotope ratios of carbon to constrain the sources of carbon. They accept the likelihood of hydrate release, but conclude that it was unlikely to have contributed more than 30% of the additional methane released at the time and reaffirm the view that terrestrial sources dominated. It follows from a wide range of model simulations for other time intervals that these and many other terrestrial feedbacks were important components of Earth-system change during deglaciation, but as yet it is difficult to characterise and quantify them fully, partly because of the problems involved in generating transient simulations that capture all the major features of the Earth system during the period.

6.6   Glacial–Interglacial changes and the modern world

The differences between the glacial world and the late Holocene, and between the rapid warming that succeeded the last glacial maximum and that recorded in the recent past and projected for the future, make it difficult to draw conclusions of direct, quantitative application to the present day from the research outlined in this and the preceding chapter. Some points are clear and non-controversial, for example, the value of reconstructing the LGM as reliably as possible in order to provide empirical data against which to measure the success of model simulations under different boundary conditions. The situation is much more complex when we consider the need to understand processes and process interactions. Some of the key changes that accompanied the major reorganizations of the Earth system described above – massive continental deglaciation in the northern hemisphere and an order-of-magnitude shift in atmospheric-dust concentrations, for example – are not currently relevant. Yet some of the processes with which these changes are linked – freshening of ocean waters in the North Atlantic, or changes in biological productivity in the oceans – may be of considerable importance in the future. Some of the tempting, apparent analogues for a warmer world, MIS 5e, for example, are clearly inappropriate, because it can be shown that they were the product of a different combination of forcings and feedbacks. Yet some of the processes involved – changed latitudinal temperature gradients and ocean circulation – may be key factors in the future. The argument that we can ignore evidence from glacial and interglacial times because things were so different then, misses vital points arising from the degree of inter-relatedness and non-linearity inherent in the Earth system:

- The value of future projections in many aspects of environmental change will depend on their ability to define thresholds and responses in complex, non-linear and highly interactive systems. The only realistic tests for the models used will be in their ability to capture amplifying or self-regulating feedbacks and resulting non-linear behaviour in the past. This and the previous chapter include many examples of this type of behaviour on a wide range of temporal and spatial scales.
- The present account, despite uncertainties regarding precise phasing, reinforces the view that CO_2 and methane played an important role in the sequence of amplifying feedbacks that led to global warming at the end of successive glaciations.
- Although data–model comparisons addressing both time-slice states and transient changes are still far from satisfactory in many cases, the progress achieved over the last decade is outstanding and ongoing. The fact that the models involved are dealing with a range of processes only some of which apply directly to the present and future, in no way limits the importance of developing, testing and improving them as tools designed to simulate key features of the Earth system as it changes through time.

Chapter 7
The Holocene

7.1 The transition to the Holocene

As the previous chapter has shown, the ice-core records from each of the poles provide contrasted templates for the transition from glacial to interglacial, Holocene conditions. In Greenland, the final transition appears to have been completed within a few decades, with most of the change in isotopically inferred temperature taking place in two jumps, each lasting less than a decade (Figure 6.3). The remarkable coherence between the Greenland records and those from Europe and around the North Atlantic (Figure 6.2) suggests that similarly rapid changes took place over a wide area. The question of just how widespread this pattern of change was is not yet fully resolved, but there is well dated evidence to suggest that it extended south as far as Patagonia (Hajdas *et al.*, 2003). Denton and Hendy (1994) claimed that the sequence of changes at the end of the last glacial in New Zealand are synchronous and parallel to the northern-hemisphere pattern, though not all lines of evidence accord with this view (McGlone, 1995).

Although the beginning of the Holocene is marked by a sharp boundary in many archives, it was also a point in a long period of transition involving a whole sequence of Earth-system changes that continued for several thousand years. The seasonality and spatial distribution of external insolation continued to change throughout the whole of the Holocene. At the end of the Younger Dryas, global ice volume was still sufficiently large for sea-level to lie some 60 m below its present-day level. It took another 5000 years for it to approach its present level. De-glaciation brought in its wake a sequence of changes in the terrestrial biosphere, replacing ice and tundra with forests over vast areas. This process also took up to thousands of years. In the account that follows, it is important to distinguish, wherever possible, changes that were mainly forced by processes linked to the ongoing transition from glacial to interglacial conditions, from those that reflect variability within the range of boundary conditions similar to the pattern prevailing in the late Holocene, immediately prior to major human impacts. The latter have distinctive importance, as they provide the best evidence for the dynamics, spatial expression and amplitude of the natural variability that will interact with any anthropogenically driven future changes. Fortunately, the range of archives and proxies available for study, the quality of chronological control, and the density of sites investigated increase greatly as we approach the present day.

In Chapters 5 and 6, much of the text was built round a sequence of major reorganisations of the Earth system that were, in broad terms, either globally synchronous, or at least temporally coherent, despite differences in the ways in which they were expressed in different parts of the world. The Holocene requires a rather different approach. The central-Greenland isotopic record shows a dramatic decline in variability at the opening of the Holocene. This has led some to see the climate of the last 11 000 years as rather invariant. By contrast, the hydrological record from low latitudes shows massive changes during the Holocene. The Holocene is not so much a period of low variability as one during which variability was less to do with temperature and was more spatially diversified, hence less easily characterised by global generalisations. Even in higher latitudes, where climate variability was undoubtedly less than during glacial periods, it was still significant in human terms. In lower latitudes, where the most readily recognisable changes were hydrological, these were of quite outstanding human significance.

The text that follows first outlines the types of changes taking place during the Holocene as a whole by presenting examples of Holocene climate change inferred from a variety of proxy records from different parts of the world. This provides a short descriptive framework within which to explore some of the key themes that link the patterns of observed change to the processes responsible. These form the main body of the chapter. The final section seeks to review these from the standpoint of dominant forcing and feedback mechanisms.

7.2   Patterns of overall climate change during the Holocene

There is a huge diversity of proxy records of climate change during the Holocene. In the text that follows, we begin by presenting typical proxy records from latitudinal extremes. The high-latitude records are reflecting, above all, changes in temperature and seasonality. The low-latitude records are largely reflecting major changes in precipitation minus evaporation (P−E). Although temperature changes may have contributed to the dramatic fluctuations, the main variable appears to have been hydrological. Adopting these two contrasted templates of Holocene variability at the outset is a simplification of much more complex spatial patterns, linked in part to orbital forcing. The long-term patterns of hydrological variability in the tropics form a counterpoint to the high-latitude glaciations. Instead of being paced by the 100 000-year eccentricity cycle, they are much more directly responsive to the 23 000-year precession cycle.

Figure 7.1 shows a range of proxy temperature records from high-latitude sites in the northern hemisphere. Whether we consider isotope values in

Figure 7.1 Circum-Arctic-Holocene temperature records from ice cores, marine and lake sediments. (a) Changing solar insolation at 65° N. (b) Pollen-based reconstruction of changing temperature in northern Finland (from Heikkila and Seppä, 2003: see Figure 4.4a) (c) Sea-surface temperatures for North Atlantic sites around Iceland (from Anderson *et al.*, 2004). (d) The eight curves shown are $\delta^{18}O$ records from ice cores from Arctic Canada and Greenland (from Fisher and Koerner, 2003). (e) The percentage melt recorded in the Agassiz ice for Ellesmere Island (from Fisher and Koerner, 2003). (f) The temperature history reconstructed from borehole temperature measurements at Greenland Summit (from Fisher and Koerner, 2003.)

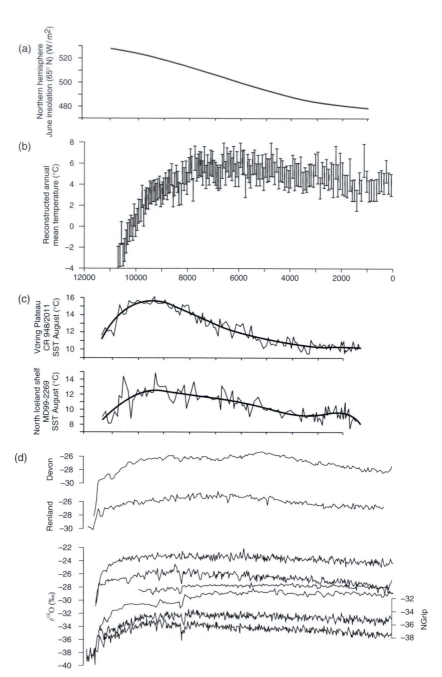

Greenland and Canadian Arctic ice, summer ice-melt layers, pollen-based, temperature reconstructions speleothem signatures, SSTs or marine diatom-based reconstructions, most show a fair measure of agreement, with a steep rise during the first 1000–3000 years and evidence for peak-Holocene temperatures

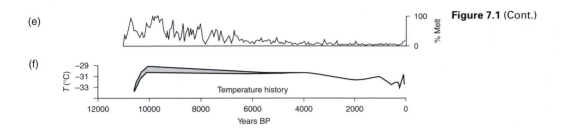

Figure 7.1 (Cont.)

between 10 000 and 6500 years BP. Several show a sharp drop in values around 8200 years BP and this is one of the themes considered in the next part of the chapter, as is the record of and response to rapid warming at the onset of the Holocene.

Figure 7.2 presents a range of low-latitude climate records from south America, Africa and southeast Asia. All show a strong degree of hydrological variability with a tendency, except in the south-American example, for wetter conditions during the first half of the Holocene, followed by evidence for varying degrees of desiccation thereafter. Evidence for extensive lakes and a rich flora and fauna confirms that the Sahara experienced a relatively humid climate for most of the Holocene up until at least 6000 years BP. Indeed, a broad area across the whole width of Africa shows similar trends, with increasingly variable, but mainly high lake-levels until around 6000–5500 years BP, followed by an irregular, pulsed decline. An episode of high dust deposition around 4000 years ago, recorded in ice cores from Mount Kilimanjaro in Kenya (Thompson *et al.*, 2002), points to severe desiccation around that time. The dynamics of these dramatic hydrological and ecosystem changes in the Sahara/Sahel region is one of the themes considered later, in Section 7.4.

Moving across into the monsoon-dominated regimes of east and southeast Asia, Morrill *et al.* (2003) summarise a wealth of evidence showing that there, too, the early to mid Holocene was a period with generally more humid conditions than those prevailing after 4500–5000 years BP. In South America, some of the most dramatic evidence for Holocene-climate change comes from the Altiplano and lower, desert regions of northern Chile. Here, conditions were drier during most of the early to mid Holocene, between 8000–9000 and 4000 years BP (Nunez *et al.*, 2002; Rowe *et al.*, 2003).

An extensive review of Holocene-climate variability over the whole globe is well beyond the scope of this book. Some additional sense of the range of variability recorded can be gained from the summaries in Mackay *et al.* (2003) and in the results from the PAGES pole–equator–pole (PEP) transects, (Markgraf, 2001; Dodson *et al.*, 2004; Battarbee *et al.*, 2005). The mid Holocene, immediately before the main changes in hydrology noted in this section has, like the last glacial maximum (LGM), become a key target for testing models (see Figures 2.2 and 2.3).

Figure 7.2 Low-latitude Holocene hydrological changes. (a) Changing water levels in lake Abhé, Ethiopia. (Modified from Gasse, 2000.) (b) Indicators of southwest monsoon variability in India. (Modified from Overpeck *et al.*, 1996.) (c) Evidence for hydrological changes from lake sediments and the Sajama ice core, central Andes and northern Chile. (Modified from Nuñez *et al.*, 2002.)

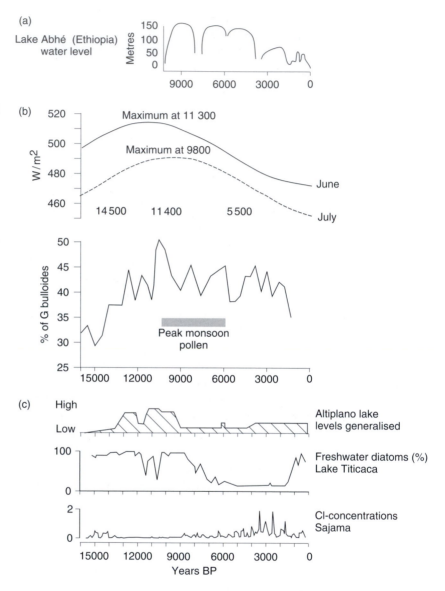

7.3   The early Holocene: delays, interruptions and biosphere feedbacks

This section highlights evidence for delayed response to orbital forcing, the importance of terrestrial-biosphere feedbacks at high altitude and the record of interruptions in the glacial–Holocene transition.

Figure 7.1 demonstrates the degree to which high-latitude northern-hemisphere temperature records peak in the early part of the Holocene and tend to decline steadily thereafter. To a first approximation, they mirror the pattern of solar radiation received during summer months around 65° N. There is a significant delay between summer solar receipts, which peak between 12 000 and 10 000 years BP, and the temperature maxima in the proxy records shown. In a survey of some 140 environmental archives of Arctic Holocene temperature change between 0° W and 180° W, Kaufman *et al.* (2004) show that peak temperatures, which reached 1.6 ±0.8 °C above mean twentieth-century levels, were time-transgressive. In northwest Canada and Alaska, peak temperatures were reached between 11 000 and 9 000 years BP, whereas in eastern Canada, they were delayed by up to a further 4000 years. This the authors attribute to the lingering effects of the Laurentide ice sheet. Although summer solar irradiance peaked between 12 000 and 10 000 years BP, winter irradiance at that time was at a minimum over the same latitudes. The likely impacts included much stronger seasonality. It is therefore important to note the extent to which the proxies used in Figure 7.1 and in the estimates of Kaufman *et al.* (2004) are likely to have been more responsive to summer than to winter conditions. The sea-surface temperatures (SSTs) in the North Atlantic show a greater difference between the early-Holocene maximum and the present day. Anderson *et al.* (2004) estimated this to have been as much as 4–5 °C along the eastern flank close to Norway, and around 2 °C on the western margin, close to Greenland. Their high-resolution marine records also point to shorter-term fluctuations of anything up to 3 °C, depending on location.

It was during the early-Holocene period of warmer summers and stronger seasonality that taiga and boreal-forest communities replaced ice and tundra as plants and animals migrated to recolonise the land previously either ice covered, or in the grip of periglacial conditions. The pollen records from high latitudes during the early to mid Holocene provide a chronicle of this recolonisation, as well as information on climatic conditions prevailing at the time. One of the surprising conclusions from pollen-based climate reconstructions for northern Eurasia is that not only summers but also winters appear to have been warmer during the early to mid Holocene 'optimum' than they are at the present day (Cheddadi *et al.*, 1997; Prentice *et al.*, 2000; Kaplan *et al.*, 2003). This rather unexpected finding has become known as the 'biome paradox'. The explanation for the warmer winters is believed to lie in the climate feedback from the terrestrial biosphere. As forest replaced landscapes previously blanketed in winter snow, the albedo of the land surface changed dramatically, especially during winter (Figure 7.3). Much less heat was reflected back into the atmosphere, much more retained in the soil. Model simulations, using an atmosphere–ocean model coupled with a BIOME vegetation model suggest that the effects of vegetation feedback, reinforced by synergy between vegetation and ocean influences, continued to enhance cool-season temperatures through into the mid Holocene (Wohlfart *et al.*, 2004).

All the high-latitude records show quite strong, sub-millennial scale variability in the early Holocene. One of the most distinctive features shared by many of the records is a sharp dip in temperature lasting for about a century

124   The Holocene

Figure 7.3 The role of biosphere (albedo) feedbacks during the early-Holocene transition from tundra to taiga and cold deciduous forest, inferred from modelling studies by Foley *et al.* (1994) and TEMPO members (1996). The inferred effect of the feedbacks included a >3 °C increase of spring temperatures as a result of the snow-masking effect of the forest. (Modified from Overpeck *et al.*, 2003.)

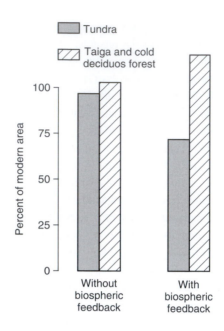

around 8200 years BP (Figure 7.4.). This 'event' appears to have been quite widespread and is well recorded in eastern and central North America and much of Europe (Alley *et al.*, 1997; von Grafenstein *et al.*, 1998), as well as the Cariaco Basin (Hughen *et al.*, 1996). It appears to correspond with a desiccation interval in parts of tropical Africa. Barber *et al.* (1999) ascribed the 8200 BP event to the sudden drainage of glacial lakes Agassiz and Ojibway via the Hudson Strait into the North Atlantic. They date the drainage, resulting from the melting of the Laurentide ice sheet that had previously dammed the lakes, to *c.* 8470 calendar years BP. This date is only a few decades earlier than the date of the onset of cooling recorded in the GRIP and GISP 2 ice cores. It is thought that the freshwater pulse reduced surface salinity in the northwest Atlantic, the effect of which was to reduce the formation rates of intermediate water in the Labrador Sea, as well as North Atlantic deep water (NADW). This in turn would have led to a reduction in northward transport of heat associated with the surface limb of the meridional overturning circulation (MOC) in the North Atlantic. The future significance of the 8200 BP event lies in the demonstration that a sufficiently large freshwater pulse is quite capable of disrupting ocean circulation and climate even under interglacial conditions.

The 8200 BP event is not the only such period of early-Holocene cooling recorded in Europe. For example, an earlier 'preboreal oscillation' dated to between 11 360 and 11 000 BP has been well documented in lake-sediment isotope records (e.g., Schwander *et al.*, 2000; von Grafenstein *et al.*, 2000) and a further cool oscillation to around 9200 BP by several authors.

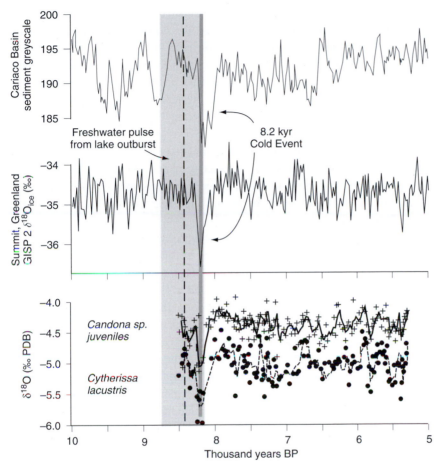

Figure 7.4 The record of the 8200-year event in central Greenland (GISP 2), southern Germany (Ammersee) and off the coast of Venezuela (Cariaco Basin) (Modified from Hughen *et al.*, 1996, data at www.ngdc.gov.paleo and von Grafenstein *et al.*, 1998.)

The above examples illustrate the extent to which, during the smooth, simple pattern of orbitally driven solar forcing, the resulting climate during the early Holocene in the northern hemisphere was strongly influenced by the dwindling Laurentide ice sheet, the discontinuous effects of its breakdown on ocean circulation and the feedbacks resulting from the replacement of ice and tundra by forested landscapes. Modelling or explaining the pattern of climate change during the period is not practical unless feedbacks from changes in the cryosphere, ocean circulation and terrestrial biosphere are taken into account.

7.4   Major hydrological changes – a green then a brown Sahara

One of the most striking climate shifts has been a tendency for many areas in low latitudes to experience a drier climate from the mid Holocene onwards. Rain-bearing summer-monsoon systems were generally stronger during the first half

(Morrill *et al.*, 2003). This can be partly explained by the interaction between the generally higher summer temperatures in the northern hemisphere and the effects of more rapid seasonal heating on land than over the oceans, where warming is delayed and dampered by the greater thermal inertia. These factors, acting in combination, created steeper thermal and pressure gradients between oceans and land, thus strengthening the onshore circulation associated with summer-monsoon climates. The orbital forcing that reinforced the monsoon systems in the northern hemisphere had the reverse effect in the south, leading to drier conditions in the monsoon-dominated regions of southern Africa and more especially South America. Enhancement of the monsoon in northern Australia during the early to mid Holocene appears to have been linked to strong ocean feedback, which overcame the effects of orbital changes (Liu *et al.*, 2004).

The contrast between early- and late-Holocene climates extended into some of the world's great deserts that currently lie beyond the influence of monsoon systems. Of these, the Sahara has received most attention. There is compelling evidence from lake-levels, from fossil- and pollen-based reconstructions of faunal and vegetation distributions and from archaeological sites, that during the first half of the Holocene much of the Sahara was well vegetated, with widespread freshwater lakes and abundant water-loving animals. It was also extensively settled by human populations. Clearly, the climate was dramatically different. Early attempts to model this different climate, using changed solar forcing alone, failed to generate the required moisture over the region. More recent simulations come much closer to including the required changes. They show that feedbacks from both the terrestrial biosphere (Broström *et al.*, 1998a) and ocean (Kutzbach and Liu, 1997) are required and that the two feedbacks actually interact synergistically (Braconnot *et al.*, 1999). Model simulations also suggest that the presence of partially vegetated wetlands may have been an important factor in maintaining the early-Holocene climatic and hydrological regimes (Carrington *et al.*, 2001). By now, coupled atmosphere–ocean simulations of the 'Green Sahara' (e.g., de Noblet-Ducoudré *et al.*, 2000) provide a much improved match with the empirical evidence.

Simulating the rapid shift from moist to arid conditions across the Sahara around 5500 BP also requires a model that incorporates ocean–atmosphere–vegetation interactions and feedbacks. The summer-radiative forcing undergoes a smooth decline, but the modelled monsoon system and related vegetation response in the coupled model show a sharper decline between 6000 and 5000 BP (Claussen *et al.*, 1999). This closely matches empirical evidence for aridification in the form of greatly increased aeolian dust flux to the site of a marine core off the west African coast (deMenocal *et al.*, 2000; Figure 7.5). The sequence of lake-level changes summarised by Gasse (2000) for lower-latitude sites (6° N to 18° N) also show steep mid-Holocene declines, but these tend to be somewhat later, between 4500 and 4000 BP, as does the evidence for dessication

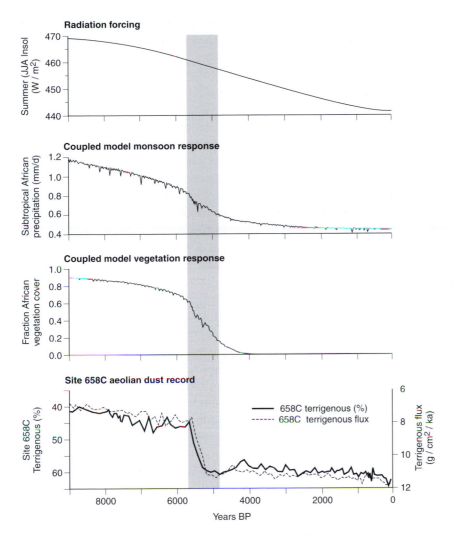

Figure 7.5 The Mid Holocene browning of the Sahara. The plots show how the CLIMBER2 intermediate complexity model, with fully coupled ocean, atmosphere and terrestrial vegetation, using the radiation forcing, simulates rather sudden precipitation and vegetation changes around 5500 BP (Claussen *et al.*, 1999) that coincide remarkably well with the aeolian dust record (scale inverted) in a marine core off the coast of west Africa. (Modified from deMenocal *et al.*, 2000.)

in the form of a major peak in dust deposition in the Kilimanjaro ice-core record (Thompson *et al.*, 2002). The above account follows the cited literature in ascribing all the recorded changes to biophysical processes. There is a case for considering whether or not, in some regions, human activities may have contributed to the environmental changes that marked the transition from moist to arid conditions.

The example of the Sahara further illustrates the extent to which any explanation of major Holocene-climate changes requires internal feedbacks from oceans and the land surface to be linked to external forcing. What both empirical reconstructions and models have so far failed to address effectively is the nature of the dynamic links between the pattern of centennial- to millennial-scale variability in high- and in low-latitude proxy records during the first half of the Holocene. Equally, the full range of empirical evidence for climate change at regional level during the first half of the Holocene is often difficult to synthesise objectively, even for a region as closely studied as western Europe. This is partly because of limitations in chronological control, partly through the way in which different archives and proxies record different aspects of the climate system, and partly because many of the proxies used respond to different thresholds with a good deal of inertia and non-linearity.

Over virtually the whole of the northern hemisphere, proxy records identify an early- to mid-Holocene period with temperatures, on average, higher than today. The timing varies with location and type of proxy, but the concept of a climate 'optimum' or 'hypsithermal' period is very general. Just as this was preceded by irregular warming, it follows that climates have, since the so-called 'optimum', tended overall to be somewhat cooler. In considering the last c. 3000 years, three themes – modes of variability, the changing incidence of climatic extremes and the evidence for wide-scale climate changes over the last one to two millennia – merit special attention before considering the broader question of forcing and feedbacks.

7.5   Modes of variability

During the later part of the Holocene, over the last three to four thousand years, ice extent and sea-level have remained approximately as they are now. Orbital forcing has changed smoothly towards present-day conditions and atmospheric greenhouse-gas concentrations have varied relatively little, compared with the amplitude of pre-Holocene and post-AD 1700 changes. The pattern of external forcing and major internal feedbacks has therefore been quite close to the immediately pre-industrial pattern. In consequence, the special interest in the period is rather different from that arising from earlier times, when boundary conditions were distinctively different. It lies not in testing models under different conditions from present, but rather in understanding the nature of natural variability, and its spatial and temporal expression, under pre-industrial conditions. The main reasons are twofold. This is likely to be representative of the variability that will interact with anthropogenic forcing over the next few centuries. It is also the 'background noise' against which any signal of anthropogenic forcing will have to be detected.

One of the most common ways of studying natural variability during the late Holocene is in terms of climate modes. These 'modes of variability' appear to be the direct result of interactions, mainly between atmospheric and ocean processes, within specific regions of the Earth's surface. They appear to capture a large percentage of recent climate variability and therefore provide a realistic framework for study. Each mode has a relatively well-defined spatial domain within which some of the physics underlying the variability is reasonably well characterised. The dominant modes also have effects well beyond their immediate sphere of action. To gain some sense of their role in present-day and recent climate, it is necessary to outline the physics of the processes generating the mode of variability, consider its variability in time and space and illustrate the wide ranging effects, as they vary on different timescales. In global terms, the most important of these modes of variability, the El Niño southern oscillation (ENSO), arises largely from the interaction between ocean and atmospheric processes in the tropical Pacific.

7.6   El Niño southern oscillation (ENSO)

7.6.1   The nature of ENSO

Under normal conditions, there is a strong temperature and sea-level pressure gradient across the equatorial Pacific from east to west, with the Indo-Pacific warm pool (IPWP), as it is known, carrying a layer of shallow surface water at or just above 28 °C. By contrast, along the western margins of the Pacific, cold upwelling waters against the coast of tropical South America are often as cool as 23 °C. Strong convection over the IPWP draws in moisture and forms part of a circulation system within the tropical Hadley cell, known as the Walker circulation. The surface limb of this is a westward-flowing stream of air feeding the strong convection over the IPWP. This pattern leads to heavy precipitation over the IPWP, which, in turn, tends to stabilise the shallow surface layer of warm water. By contrast, dry conditions prevail along the coast of South America along with high ocean productivity in that region, in the cold upwelling water.

Under El Niño conditions, the westward flow of air over the sea surface weakens and the centre of convection linked to the warm pool shifts eastwards into the central Pacific. The eastward displacement of the warm pool is linked to changes in ocean stratification, with the thermocline shallowing to the west and deepening to the east. Thus, the cold upwelling water along the coast of Ecuador and Peru is replaced by warmer water, which in turn tends to weaken the zonal surface temperature gradient and the westward surface air flow. Marine productivity declines catastrophically along the coast of South America and rainfall increases. Over the area, which, under normal conditions, lies within and around the IPWP, there is a sharp decline in rainfall. Figure 7.6a shows a thumbnail sketch of the main changes involved.

Excursions from the 'normal' state can also occur with the opposite physical characteristics, and these are known as La Niña events. Cold water spreads as a tongue from the eastern edge of the Pacific westwards and the Walker circulation is strengthened. This reinforces cold upwelling on the coast of Ecuador and Peru where drought tends to be exacerbated. Conversely, there are increases in precipitation around the region of the IPWP.

The physics of ENSO is well known to the point where early diagnosis of El Niño events can often be made with some success. The temporal behaviour of the oscillation can be reconstructed from instrumental records of the sea-level pressure

Figure 7.6 (a) Thumbnail sketch of El Niño, showing SSTs (darker= warmer), atmospheric circulation and the changing depth of the thermocline (from http://ocean world.tamu). (b) ENSO variability reconstructed from the tropical Pacific Maiana Atoll coral $\delta^{18}O$ measurements and from instrumental series. Note the rather good match between the proxy (coral stable-isotope) record and the instrumental reconstructions; also the changing frequency and amplitude of El Niño over the last 160 years. Shading distinguishes an early period during which the dominant interval between El Niño episodes is a decade or more, and the latest period when the interval averages three to four years. (Modified from Urban *et al.*, 2000.)

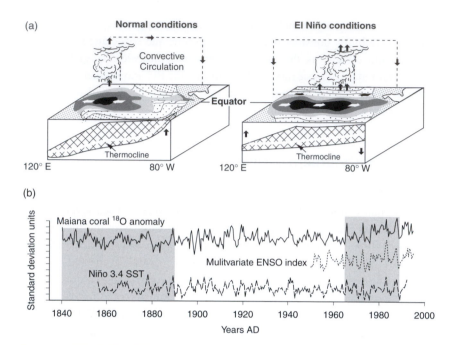

difference between Tahiti and Darwin that allow calculation of a southern-oscillation index (SOI). Extending this further back in time requires analysis of proxy records from within or close to the 'home territory' of the oscillation.

7.6.2   The variability of ENSO in the time domain

Figure 7.6b shows a reconstruction of the SOI, based on instrumental records, together with an extension of this back to the early nineteenth century derived from analysis of coral from Maiana Atoll in the central Pacific (Urban *et al.*, 2000). These records can be analysed for changes in the dominant periodicity of ENSO. At the beginning of the period, ENSO events tended to occur every ten years or so, whereas towards the end, the dominant frequency was closer to four years. The pattern of temporal variability is marked by periods of stability with relatively simple frequency spectra, and intervening transitions with more complex patterns. Just as the frequency of ENSO variability has changed over the last two centuries, so has the strength.

From the above, it is clear that ENSO is not a 'stationary' oscillation with similar amplitude and frequency through time. Both amplitude and frequency vary on decadal to centennial timescales. On the longer timescale of the Holocene as a whole, many proxy records from the areas influenced by ENSO (summarised in Gagan *et al.*, 2004) suggest that the mode of variability emerged in something comparable to its recent form some 5000 years ago. One of the most complete records comes from analysis of varved lake sediments from a site

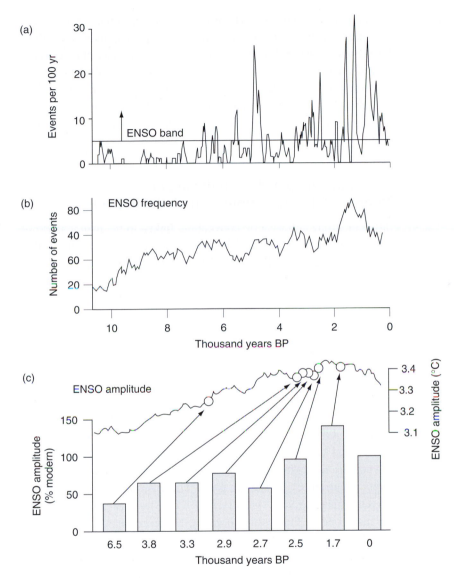

Figure 7.7 ENSO temporal variability through the Holocene. (a) ENSO events per hundred years reconstructed from a laminated lake-sediment record in Ecuador. (Adapted from Moy *et al.*, 2002.) (b) Modelled-ENSO frequency for the region of the eastern Pacific (Clement *et al.*, 2000). (Adapted from Gagan *et al.*, 2004.) (c) Modelled-ENSO amplitude (as for b), compared with records of the relative amplitude of $\delta^{18}O$ variability reconstructed from coral remains from the Huon peninsula, Papua New Guinea (Tudhope *et al.*, 2001) (Adapted from Gagan *et al.*, 2004.)

in southern Ecuador (Moy *et al.*, 2002). They find that there are virtually no signs of ENSO variability as they define it (two to eight-year frequency) prior to 6800 BP, rather few signs before 5000 BP, then sporadic peaks thereafter with the highest incidence of events between 1800 and 1200 BP. Putting these results together with those summarised from the Pacific itself by Woodroffe *et al.* (2003) and Gagan *et al.* (2004) (Figure 7.7a), and with model simulations (Figure 7.7b and c) it seems reasonable to conclude that ENSO variability has itself varied on all timescales from decadal to multi-millennial, a conclusion further reinforced by analyses of fossil corals of even greater age, each spanning short episodes in the late Pleistocene (Figure 7.7c; Tudhope *et al.*, 2001). From this we may infer that there are complex and as yet poorly understood links between ENSO variability and both external forcing and mean climate state; moreover that in order to gain any sense of the likely future role of ENSO, it will be necessary to develop and evaluate models designed to test hypotheses relating to ENSO variability on timescales longer than the last two or three events for which a wide range of instrumental data is available from both the atmosphere and the oceans.

7.6.3   The variability of ENSO in the spatial domain – teleconnections and global impacts

The generation of strong atmospheric convection, linked to the maintenance of high SSTs in the area of the IPWP, is but one of the processes ensuring that ENSO variability has a widespread impact on climates and environmental processes well outside the tropical Pacific region. These include the extent of Antarctic sea-ice, Atlantic-ocean circulation and the inter-annual variations in atmospheric-CO_2 concentrations. There is evidence to suggest that variability in the tropical Pacific leads global temperature with sufficient consistency to serve as a predictor of future global temperature changes (Bratcher and Giese, 2002).

Over the last few decades, ENSO events have generally been associated with drought over much of eastern and central Australia as well as in southeastern Africa and Madagascar, in addition to similar effects around the IPWP region itself. By contrast, wetter conditions are often experienced in the Gulf states of the USA. Although La Niña conditions are not simply the opposite of El Niño, they have generated damaging floods in central and eastern Australia. The contrast between the two extremes is well recorded in the luminescence bands of corals growing in the Great Barrier Reef. These closely track freshwater runoff from Queensland rivers (Isdale *et al.*, 1998). During the twentieth century, the main one of these, the Burdekin, has experienced, inter-annual fluctuations from 5% to 750% of the median values. Hendy *et al.* (2003) used luminescence banding to reconstruct Burdekin discharge since the early seventeenth century and showed that the strength of the link between river run-off and the SOI varies through time. It is significant prior to 1920 and after 1960, but between 1920 and 1950, it breaks down.

The record of varying monsoon strength preserved in the Dasuopu ice core from the southern edge of the Tibetan Plateau (Thompson *et al.*, 2000) shows a strong link with ENSO over most of the last five centuries, with ENSO events correlating with weaker monsoons. This coupling appears to have broken down since 1976 (Kumar *et al.*, 1999).

Moving further afield, ENSO events have been linked to drought in the United States. Figure 7.8a shows how the spatial domain within which this teleconnection is expressed varies depending on the time interval chosen. Between 1840

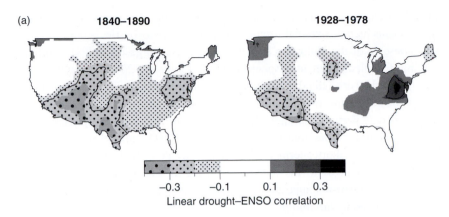

(a)

1840–1890   1928–1978

Linear drought–ENSO correlation

−0.3   −0.1   0.1   0.3

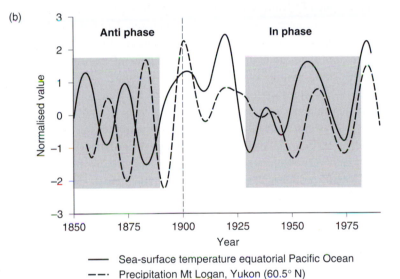

(b)

Anti phase   In phase

Normalised value

Year

—— Sea-surface temperature equatorial Pacific Ocean
—-— Precipitation Mt Logan, Yukon (60.5° N)

Figure 7.8 Changes through time in El Niño teleconnections in North America.(a) Maps of drought–ENSO correlations for the periods 1840–90 and1928–78, using tree-ring indices of drought (Cook *et al.*, 1999) and the coral record of ENSO by Urban *et al.*, (2000) . Correlations are significant at 90% within contoured areas. Note the changes in the phase of correlation in the eastern USA between the two periods. (Modified from Labeyrie *et al.*, 2003.)(b) Correlations between equatorial Pacific SSTs and precipitation in the Yukon, northwest Canada since 1850. During the nineteenth century the two series are anti phase; subsequently they are in phase. Shading shows the periods represented in a. (Adapted from Moore *et al.*, 2002.)

and 1890, the correlations between ENSO and drought are both stronger and more extensive than between 1928 and 1978. The sharpest contrast is in the northeast, where a strong positive correlation during the earlier time interval is replaced by an equally strong negative correlation. Similar examples come from the northwest Pacific (Moore *et al.*, 2001; Figure 7.8b).

From the above examples, it is apparent that when the time perspective is lengthened from decades to centuries, by using both instrumental and proxy records for ENSO variability, not only do the amplitude and frequency vary, but also the teleconnections. The mechanisms responsible for this variability are not known with any degree of certainty. One possibility is that under differing mean conditions, the feedback processes responsible for the short-term variations between El Niño, 'normal' and La Niña conditions can reinforce longer-term persistence in any given state. Additional processes may be involved, however, including changes in ocean stratification. Moreover, the extent to which the mechanisms include external forcing, or are entirely stochastic, is not clear. On the timescale of the Holocene as a whole, models suggest a link with orbital forcing. Coupled ocean–atmosphere models (Liu *et al.*, 2000; Clement *et al.*, 2000) forced by orbitally driven changes in seasonal insolation show an increase in El Niño events during the Holocene, with a peak between 3000 and 1000 years ago. This only partially parallels the palaeo-record as it fails to replicate either the suddenness of the transition to stronger ENSO variability in the mid Holocene, or the shorter-term variability already described. As in the case of the dramatic changes in the Sahara–Sahel region during the mid Holocene, additional mechanisms appear to be required in order the simulate the actual record of changing variability.

The extent to which El Niño is amenable to prediction as distinct from early diagnosis is a matter of debate. Kessler (2002) has taken the view that El Niño events are not simply generated by internal dynamics as part of a self-sustained oscillation. He sees them as being more in the nature of disturbance-linked events requiring some kind of externally derived stimulus. One candidate is volcanic activity. By comparing reconstructions of El Niño variability with independently developed chronologies of volcanic activity since AD 1649, Adams *et al.* (2003) show that a volcanic eruption in the tropics roughly doubles the likelihood of El Niño in the following winter. In contrast to the above authors, Chen *et al.* (2004) suggest that El Niño is largely driven by internal dynamics, and claim a high level of predictability for their model when it is applied to the temporal sequence of El Niño events over the last 148 years. Their study, which uses SSTs to initialize the model, relates to the prediction of individual and mainly major El Niño events up to two years in advance. The degree of success varies with the time interval covered and is highest for the strongest events. It does not address the issue of changes in the frequency and amplitude on decadal to millennial timescales.

The pervasive influence of ENSO variability on the functioning of the Earth system as a whole will demand further attention in Section 8.2, which outlines the factors modulating atmospheric greenhouse-gas concentrations on an inter-annual basis at the present day.

7.7   The North Atlantic (NAO) and other oscillations

The NAO is usually defined in terms of the pressure gradient between the high-latitude low pressure centred around Iceland to the north and the high-pressure system present in the region of the Azores or Gibraltar to the south (Hurrell *et al.*, 2003 and Figure 7.9) Instrumentally derived reconstructions currently span the period from AD 1821 to the present day (Vinther *et al.*, 2003). Several proxy-based reconstructions have been proposed, the longest well-validated one reaching back to AD 1675 (Luterbacher *et al.*, 1999; Schmutz *et al.*, 2000). More

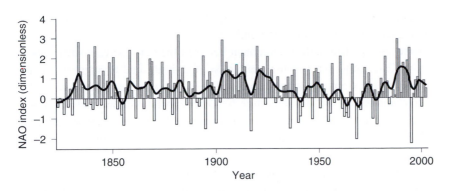

Figure 7.9 The winter NAO index from 1820 to the present day. (From www.cru.uea.ac.uk.)

recently, Mann (2002b) has proposed a cold-season temperature-based NAO reconstruction. Once calibrated, this allows the use of temperature-linked proxies to extend the reconstruction period back to AD 1750.

The north–south pressure gradient used to define the NAO exerts a strong control on the strength of the westerly circulation that in turn dominates the climate of western Europe. The effects of variations in this pressure gradient are most noticeable from November to April. A high NAO index brings strong westerly circulation, with wetter, milder and generally stormy winters. A weak NAO index is characterised by colder, more 'continental' winters in western Europe and a greater seasonal-temperature contrast. It is the consistency of this relationship that allowed Mann (2002) to define NAO in terms of a cold-season temperature pattern. The influence of the NAO stretches far beyond western Europe. The NAO strongly influences surface-air and sea-surface temperatures, especially in winter, in North America, Eurasia and as far south as the Mediterranean. The NAO has also been a dominant influence on changing winter stream-flow in the Middle East over the last *c*. 150 years (Cullen *et al*., 2002). Wang *et al*. (2003) demonstrated a discernable influence of the NAO on air temperatures as far afield as the Tibetan Plateau. Ocean currents and sea-ice extent in the North Atlantic and even thermohaline circulation are affected by the NAO, which may, in this way, influence wider areas on longer timescales.

The physics of the NAO and the mechanisms responsible for it are less well known than for ENSO, though the general view is that it results largely from stochastic interactions within the atmosphere. Ogi *et al*. (2003) claim that one external factor influencing these is the eleven-year solar cycle. Using the NCEP/NCAR reanalysis series (see 4.1.1) they suggest that the influence of NAO on summer climate is strongest at solar maximum. They infer that feedbacks from snow and ice cover, after strong NAO impact during the previous winter, play an important role in the link between NAO and summer temperatures, though the processes involved in the link to solar variability are not well understood. As with ENSO, there is some indication that early diagnosis may be possible. Saunders and Qian (2002) show that for the period 1950/1 to 2000/1, North

Atlantic SSTs in early November were quite good predictors of the sign of the winter NAO over the following months.

There is less than perfect agreement between the various reconstructions of the NAO spanning the last few centuries, though all tend to suggest that the variability is essentially unpredictable. The winters between 1988 and 1995 were characterised by unusually high NAO index values and the late twentieth century may prove to have experienced a degree of variability not replicated in earlier periods covered by currently available instrumental and proxy reconstructions. No characteristic periodicities have been convincingly identified for the NAO. It remains possible that the factors responsible for it are diverse and largely independent of each other. One possibility is that the NAO is coupled in some way to an oscillation in Arctic pressure and circulation. The Arctic oscillation (AO) involves changes in the position of the Arctic high-pressure system from the polar region to a zone around the pole at $c.$ $55°$ N. In its positive phase, the expanded high-pressure system strengthens westerly circumpolar circulation and storminess.

Decadal to millennial variability is a characteristic of late-Holocene climates throughout the world; the current tendency to describe the variability detectable in the instrumental and calibrated, recent proxy records in terms of modes, with more or less definable spatial domains and physical regimes, reflects the need to organise what otherwise would be a highly disorganised system. Within the extra-tropical Pacific region, strong decadal variability is detected in both instrumental and proxy records. The Pacific decadal oscillation (PDO) is characterized as an alternation between 'warm' and 'cold' regimes. During warm regimes, the western and central areas of the north Pacific remain cool while the eastern Pacific is warm. These conditions are reversed during each cool phase. During the twentieth century, the different regimes have tended to persist for 20 to 30 years with relatively abrupt shifts between. There is some indication that the PDO was weaker during the nineteenth century than in either the twentieth century or the second half of the eighteenth century (Gedalof et al., 2002). As in the case of the NAO and AO systems, research currently falls short of providing either a convincing physical forcing mechanism for the mode of variability, or explaining its changing spatial expression through time. It also seems unlikely that the PDO is an adequate paradigm for explaining all the main features of twentieth-century climate variability in the Pacific region, especially since 1999. Bond et al. (2003) claim that the variability is essentially stochastic and that predictability on a multi-year timescale is minimal.

For any given period, there is evidence for a whole variety of teleconnections between modes of variability but, as with the modes themselves, the patterns change through time (see, e.g., Wang et al., 2003). One important implication is that attempts to model or predict the behaviour of and teleconnections between modes of variability, on the basis of short-term records, will inevitably be flawed. The whole area provides a wealth of research challenges that form one of the key themes in the quest to develop better predictive ability at regional level under the changing patterns of forcing and feedbacks that lie ahead.

7.8   Examples of changing extremes

Changes in the magnitude, frequency and duration of extreme weather conditions – droughts, floods, cyclones and the like – are of crucial importance for human societies.

Several tree-ring (Grissino-Mayer, 1996; Stahle *et al.*, 2000; Knapp *et al.*, 2002) and marine-sediment (Biondi *et al.*, 1997, 2000) studies identify a period of persistent and extensive drought in North America during the sixteenth century (Figure 7.10a). By now there is evidence for several periods of multi-decadal drought over the last 1600 years in various parts of the west and southwest of the United States (summarized in Bradley, 2003). The reconstruction of drought by Benson *et al.* (2002) from stable isotope analyses of the sediments of Pyramid Lake, Nevada, provides a high-resolution record spanning the whole of the Holocene. They found that the duration of multi-year droughts ranged from 20 to 100 years, and the time intervals separating droughts varied from 80

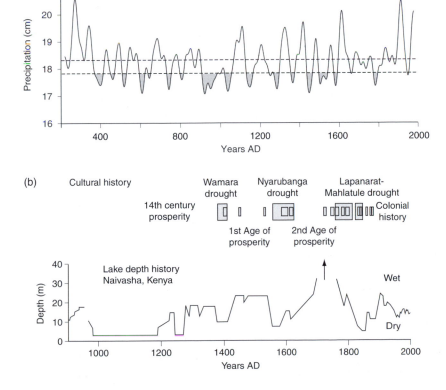

Figure 7.10 Records of changing drought incidence in the late Holocene. (a) Smoothed record of changing precipitation for southern Nevada over the last 1700 years reconstructed from a tree-ring network. Note the periods of severe multi-decadal drought prior to AD 1600. (Adapted from Hughes and Funkhauser, 1998.) (b) The changes in the level of Lake Naivasha, Kenya, over the last thousand years, inferred from sedimentological changes and biological evidence. Each episode of lowered lake-level corresponds to a period of drought recorded in the colonial and oral history of the region. (Based on Verschuren *et al.*, 2000.)

to 230 years. The period most prone to drought conditions was between 6500 and 3800 years BP.

Most of the intervals of severe desiccation referred to above, were they to recur in the future, would have severe consequences for ecosystems and for human populations. One of the clearest links is that between drought and forest-fire incidence. Wildfires require both a build up of fuel and a period of desiccation. There is therefore a strong link between climate variability and fire incidence (Swetnam, 1993). So far, there are few indications of the possible forcing and feedback mechanisms responsible for these recurrent droughts beyond clear links with the variability of ENSO and, in the case of the Pacific northwest of the USA, the PDO (Knapp *et al.*, 2002). Schubert *et al.* (2004), in a model-based study of the interactions between ocean, atmosphere and land that led to the American dust bowl, an extensive and persistent drought that affected the Great Plains during the 1930s, also pointed to anomalously cool SSTs in the tropical Pacific as a primary cause, with feedbacks from the land surface enhancing the precipitation deficit during the summer months. The record by Verschuren *et al.*, (2000, Figure 7.10b) from an entirely different climate regime shows similar indications of severe and persistent drought during the last millennium, though in that instance and several others noted below (Figure 7.12) there are strong indications of a link to solar variability.

A growing number of studies include attempts to reconstruct long-term records of other types of weather extreme, partly to address the issue of whether current incidence lies outside the range of natural variability (see especially 12.1). Knox (2000) presents a detailed analysis of palaeoflood magnitude and frequency throughout the Holocene for the Upper Mississippi drainage basin. He links the shifts in flood regime to changes in climate that alter the frequency of the dominant air-mass trajectories over the region. Using storm-generated inwash layers in the sediments from 13 lakes in the northeast of the United States, Noren *et al.* (2002) reconstructed the incidence of major rainfall events in the region over the last 13 000 years. Periods of storminess last for around 1500 years, with their peaks *c.* 3000 years apart. They link storminess in the region to changes over a much wider area and conclude that changes in the AO (see 7.7) are involved. Macklin and Lewin (2003) provide a comprehensive analysis of flood incidence in British rivers for the whole of the Holocene. For the last 4000 years, they link major flood episodes to independent evidence for shifts to wet conditions derived from peat stratigraphy (see 4.2.4). Clear links with climate change are apparent despite the fact that changes in land cover as a result of human activities significantly altered the sensitivity of river basins to climatic events. A possible link between floods and solar variability has been proposed by Schimmelmann *et al.* (2003), who infer a 200-year periodicity in California from a sequence of flood layers in the sediments of the offshore Santa Barbara Basin. As well as correlating these with many other sequences from the northern hemisphere and as far afield as South America, they suggest that the trigger for the events lies in the 208-year solar cycle. By contrast, Camuffo *et al.* (2000) discount the suggestion of a persistent and statistically significant link between solar activity and the record of sea storms during the last millennium in the Adriatic and western Mediterranean.

The above studies lead us to the conclusion that with few exceptions – the Sahel–Sudan region may be one (Hulme, 2001; Nicholson, 1982) – recent instrumental records fail to capture the full range of variability in the incidence of extremes during the late Holocene. They highlight the limitations of magnitude/frequency estimates based on the short span of instrumental records, a point made strongly by Nott and Hayne (2001) who demonstrate from palaeo-records that the recurrence interval of 'super-cyclones' along the Great Barrier Reef is an order of magnitude shorter than had previously been calculated using the period of instrumental measurements.

The above results also emphasise the need to take into account the full range of natural variability in developing future projections, irrespective of any anthropogenic effects on climate. What the records so far fail to do, for the most part, is provide clear evidence for the combination of forcings and feedbacks responsible for shifts in the magnitude and frequency of extreme-events. Even links between extreme-event incidence and mean conditions are not always clear. In the case of cyclone frequencies off the north-Queensland coast, the results actually suggest an insensitivity to changes in mean SSTs over the last 5000 years (Hayne and Chappell, 2001). In other studies there are tantalising, but disparate suggestions, with little or no sense of an emerging body of theory. This must be borne in mind in considering future projections of changing extremes (14.6 and 14.7). In this whole area of research the data so far provide all too little empirical constraint on models.

7.9   The Medieval Warm Period and the Little Ice Age

Two of the few generalisations to emerge from the complex records of variability briefly considered above are that all modes show changes in behaviour, often linked to shifts in the mean state of the system, and that these shifts are sometimes also expressed as changes in the teleconnections between systems over wide areas. Have these changes been sufficiently coherent to generate changes of broad regional, hemispheric or even global extent over the last one to two thousand years?

Many studies, beginning with the pioneering work of Lamb (1965), have claimed to detect a broadly parallel pattern of change over large areas of the northern hemisphere during the last millennium. Lamb first presented evidence for a climatic optimum during medieval times, now usually referred to as the Medieval Warm Period (MWP), or Epoch (MWE), followed by what has become generally known as the Little Ice Age (LIA). On balance, the evidence available has failed to confirm the existence of a temporally coherent MWP that spanned the same time interval over a wide area (Hughes and Diaz, 1994). On

Figure 7.11 Proxy-based
reconstructions of global-
temperature changes since
AD 200. Prior to AD 1000
only the reconstruction by
Mann and Jones (2003) is
shown. Thereafter, a range
of reconstructions is shown
against the background of
the uncertainty range in the
reconstruction by Mann
et al. (1999). The
observational record
begins in the mid
nineteenth century.
(Compiled from Mann and
Jones, 2003 and Bradley,
2003.)

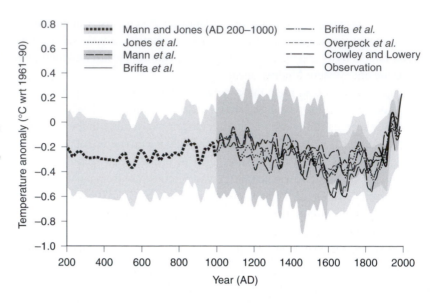

the other hand, reconstructions of past temperature variability over the last thousand years suggest that there were periods in the early part of the millennium, prior to *c.* AD 1300, that experienced generally warmer conditions over much of the northern hemisphere. Reconstructions based on documentary or proxy records point to warmer conditions during this time interval, at least as far south as the Mediterranean and north Africa (Schilman *et al.*, 2001), south and west as far as the Cariaco Basin off the coast of Venezuela (Haug *et al.*, 2001), south and east into central China, where warmer conditions may have persisted into the fifteenth century (Paulsen *et al.*, 2003), as well as in parts of the southern hemisphere (Cook *et al.*, 2002). Within the same time interval there is evidence for climatic anomalies elsewhere, including severe drought in California from AD 900–1100 and 1210–1330 (Stine, 1994). The precise expression and the overall duration of warmer or anomalous conditions varies from region to region. Nevertheless, the number of instances is growing and is based on records from quite diverse archives. In the reconstruction by Mann and Jones (2003) of surface temperature for the last 1800 years (see also Jones and Mann, 2004), most of the period from AD 800 to 1400 emerges as relatively warm in the northern hemisphere, (Figure 7.11) but the situation is much less clear in the southern hemisphere.

The evidence for a subsequent period of distinctively cooler conditions relative to those before and after comes from sites in both hemispheres. Around the North Atlantic, the LIA appears to have begun by the early fourteenth century, if not before (Grove, 1988; 2001). Close examination of the chronologies of glacier readvance show that they are far from perfectly synchronous from region to region or even from valley to valley in the same region. Glaciers react to climate forcing individually and often with significant and varied time lags. Looking to other proxies less prone to lags, such as documentary evidence (Pfister, 1992) or tree rings (Briffa, 2000), the picture is still complex, with different regions experiencing minimum temperatures during different decades, though always within the same broad overall timeframe. When the ensemble of proxy records is combined over the whole of the northern hemisphere, there are two periods of minimum temperature falling within the seventeenth and nineteenth centuries. It is difficult to escape the conclusion that between *c*. AD 1550 and 1850 (Bradley and Jones, 1992), there was, despite spatial and temporal variations in detail, a relatively coherent cooling, certainly at hemispheric scale. Although cooler conditions were widespread, at least some areas in the tropics were probably warmer during this interval (Gagan *et al*., 2004).

There is still some doubt as to the amplitude of change in mean temperature between the MWP, LIA and recent times. Reconstructions based on borehole temperatures (Huang *et al*., 2000) point to a level of warming post AD 1500 some 0.4 °C greater than has been suggested from some multi-proxy reconstructions (Harris and Chapman, 2001). Mann and Schmidt (2003) argue that this is in part the result of snow cover decoupling air and ground temperatures, and thereby introducing errors in the borehole-derived reconstructions. Chapman *et al*. (2004) and Pollack and Smerdon (2004) take the opposite view and present evidence in support of the fidelity of the borehole reconstructions. The latter authors claim that they have overcome the criticisms made by Mann and Scmidt (2003) and Mann *et al*. (2003) and they conclude that northern-hemisphere warming over the last 500 years falls within the range 1.02–1.06 °C. Further doubt arises from the comparisons presented by Esper *et al*. (2004). They point out that whereas there is quite good agreement between the reconstructed decadal scale, high-frequency variability in the series by Jones *et al*. (1998), Mann *et al*. (1999), Briffa (2000) and Esper *et al*. (2002), there is quite a strong divergence in the pattern of low frequency temperature change, especially during the period of transition from the MWP to the LIA (see Figure 7.11) Whereas the amplitude of overall change is low and the differences between mean MWP and LIA temperature are minimal in the reconstructions of Mann *et al*. (1999) and Jones *et al*. (1998), Esper *et al*. (2002) calculate almost twice the total amplitude of variability, and both their series and that of Briffa (2000) show a stronger decline in temperature between MWP and LIA conditions. Esper *et al*. (2004) attribute these differences in calculated amplitude to the use of different methods for de-trending long-term tree-ring data. The latest corrections by Rutherford and Mann (2004) go part way to closing the gap between the estimates by Mann *et al*. (1999; 2003) of post-LIA warming and the higher figures derived from boreholes and alternative interpretations of multi-proxy series. Huang (2004) deals with the divergence between all these series by merging them to generate an integrated temperature series for the period since AD 1500 that compares well with the surface-temperature trends generated by combining the effects of each of the main forcings. In so far as this study resolves discrepancies over the amplitude of low-frequency temperature changes during the last 500 years, it greatly strengthens the value of proxy-based reconstructions as constraints on climate models designed to project future changes (see 13.1.3).

Soon and Baliunas (2003), Soon *et al.* (2003) and McIntyre and McKitrick (2003) have made even more extreme claims for the MWP. On the basis of their revaluations of the estimates of Mann *et al.* (1999), they have proposed a much warmer MWP and thereby have sought to discount the view that late-twentieth-century temperatures were in any way extreme relative to natural variability. These arguments are among several used by sceptics of global warming. They are considered further in Section 16.2.1.

7.10   Signs of earlier synchronous changes from the mid Holocene onwards

Despite the highly differentiated record of climate change from region to region, there are indications that some widespread, quasi-synchronous changes may have occurred at earlier times during the second half of the Holocene. Evidence for a roughly synchronous mid-Holocene change in surface processes comes from the peaks in dust deposition around 4000 BP in ice cores from Peru and Kenya, as well as a marine core from the Gulf of Oman (Thompson *et al.*, 2002) and several lines of evidence from the Mediterranean region. One of the most convincing claims for widespread climate shift is that made by van Geel *et al.* (1996) for the period around 2650 BP. They summarise a range of evidence, mostly from Eurasia , suggesting that this period was one of rapid and dramatic climate change, involving, in western Europe for example, a shift to cool, wet conditions. They point to the fact that these changes coincide with a plateau in the ^{14}C calibration curve and infer from this that external forcing, probably arising from solar variability, is likely to have led to the climate shift that they document.

On the longer timescale of the whole Holocene, Bond *et al.* (1997; 2001), supported by Bianchi and McCave (1999), have claimed that the area around the North Atlantic has experienced a sequence of climate changes that probably reflect the continuation of the periodicity of the Dansgaard/Oeschger oscillations, around 1450 years, but with a much attenuated amplitude and expression during the Holocene. Their evidence comes mainly from mineralogical and rock magnetic indications of southerly extensions of the area of the North Atlantic covered by ice-rafted detritus. The indications, along with stable-isotope evidence for variations in deepwater formation from records in marine sediments from the northeastern Atlantic, point to oceanographic variability in the region of the North Atlantic that in turn probably impacted climate in the surrounding regions (Oppo *et al.*, 2003).

Bradley (2003) reviews the evidence for late-Holocene climate variability on hemispheric and regional scales. His analysis of proxy records from around the

North Atlantic illustrates well the mixture of coherence and diversity characteristic of much of the evidence for recent climate change.

Signs of likely coherence are not confined to extra-tropical latitudes. The
indications in the low-latitude hydrological records from Africa, South America
and Asia of broadly parallel changes from more- to less-moist regimes in the
mid Holocene have already been noted. Haug *et al.* (2001) use their ultra-high-
resolution proxy record of palaeoprecipitation from the Cariaco Basin to infer
that these changes, as well as the ones recorded in their own sequence, may have
been largely due to the southward migration of the intertropical convergence
zone. Indications that there were some periods of widespread climate change
that transgressed the modes of variability considered earlier in the chapter, and
probably reconfigured their interactions, open up the wider questions of forcings
and feedbacks during the later part of the Holocene.

7.11   External forcing

7.11.1   Solar variability

Orbital parameters continued to change smoothly throughtout the Holocene, and
the overall decline in mean temperature at high latitudes in the northern hemisphere may reflect the continuing decrease in summer-insolation receipts. Orbital
changes were also instrumental in modulating monsoon regimes. In both case, as
shown above, the changes in climate and ecosystems were often sudden, and the
trends showed a great deal of detailed temporal and spatial variability. To what
extent may solar variability have contributed to this?

The variations in total solar irradiance (TSI) between the extremes of the 11 year
Schwabe cycle are of the order of 0.08%, with higher variability at the ultraviolet
end of the spectrum. This latter variability leads to significant changes in the heating
of the stratosphere, with effects that may propagate to the lower atmosphere. Over
longer timescales, the 11-year cycle itself varies in amplitude and becomes superimposed on lower-frequency variability (Figure 4.6). Sunspot counts in documentary records show that there were periods of reduced solar activity and variability,
the most marked of which was the Maunder minimum from ~AD 1675 to 1715,
coinciding with one of the coolest periods during the LIA.

Direct evidence for solar variability before the seventeenth century is lacking,
but the proxies discussed in Chapter 4 (4.3.2; Figure 4.6) may well provide, with
the qualifications noted there, a record of changes in solar irradiance reaching
the lower atmosphere throughout the Holocene. What is the strength of the
evidence in favour of the solar variability so recorded playing a major role in
Holocene climate change?

Figure 7.12 shows low-latitude Holocene proxy records alongside reconstructed
solar variability. In each case, the coherence is close and quite convincing.
As already noted above, similar claims have been made for higher-latitude

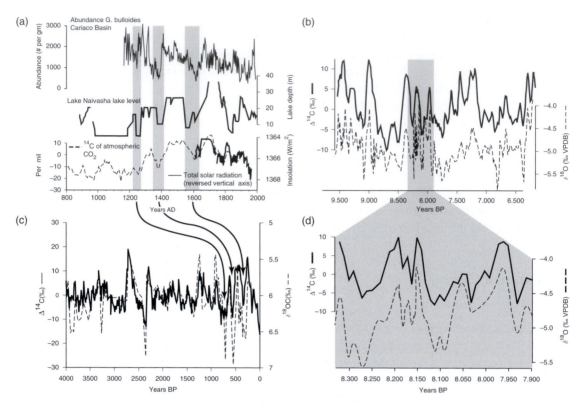

Figure 7.12 Examples of parallels between sequences of proxy-based climate changes and inferred changes in solar radiation during the Holocene. In each case, changes in atmospheric ^{14}C (residual Δ ^{14}C ‰) are used as proxies for solar irradiance (lower values of Δ ^{14}C correspond with higher solar irradiance). (a) Solar irradiance as reconstructed by Lean *et al.*, (1995) back to the early seventeenth century and from $\Delta 14C$ prior to that, compared with changes in the level of Lake Naivasha, Kenya (see Figure 7.10b), and changes in trade-wind strength inferred from the abundance of *Globigerina bulloides* in the Cariaco Basin sediments. (Modified from Bradley *et al.*, 2003.) (b) Changes in δ ^{18}O over the last 4000 years in a stalagmite from northwestern Germany plotted against Δ ^{14}C measured in European tree rings (see Stuiver *et al.*, 1998). (Modified from Niggermann *et al.*, 2003.) (c) Variations in δ ^{18}O of an early Holocene speleothem from Oman, plotted against Δ ^{14}C. (Adapted from Neff *et al.*, 2001.) (d) An expanded high-resolution record from the same sequence around 8000 BP, showing the parallels in detail.

variability around 2650 BP (van Geel *et al.*, 1996). Solar variability has also been invoked to explain the apparent persistence of the *c.* 1450-year variability rhythm in the North Atlantic during the Holocene (Bond *et al.*, 1997; 2001). Links have also been suggested with drought frequency in continental North America (Cook *et al.*, 1997; Yu and Ito, 1999; Dean *et al.*, 2002), changes in peat stratigraphy in northwest Europe (Blaauw, 2003; Blaauw *et al.*, 2004), lake-level variations in central Europe (Magny, 1993), oceanographic variability around Iceland (Andrews *et al.*, 2003) and the variations in oxygen-isotope data from central Greenland (Stuiver *et al.*, 1991). Shindell *et al.* (2001) suggest that links between solar variability and regional climate at high latitudes may be reinforced by the influence of solar variability on the NAO/AO mode (see 7.7). If, as seems increasingly likely, any of these putative links between climate and solar variability stand the test of time, the question arises – what was the total amplitude of Holocene solar variability and how might this have been translated into climate variability at the Earth's surface?

Estimates made by Lean *et al.* (1992) suggest that the total variability between the Maunder minimum and the present-day mean was of the order of 0.24%. Temperature reconstructions indicate that the comparable range of temperature variability over the northern hemisphere was between 0.2 and 0.4 °C. Several simulations (Rind *et al.*, 1999; Crowley, 2000; Shindell *et al.* 2001) suggest that this could be largely accounted for by the inferred solar variability. Moreover, the spatial pattern of simulated solar-forcing effects over the same period shows some similarities to the actual pattern of changes reconstructed from empirical studies. In particular, the period of the Maunder minimum saw strong cooling over mid- to high-latitude continental interiors. This in turn suggests that solar variability may have acted, in part, through modulation of the NAO on decadal or multi-decadal timescales.

Using the reconstructed ^{14}C variability as a basis for scaling solar variability over the Holocene as a whole, the total range may have been around 0.40%. By analogy with the inferred effects over the last four centuries, this could quite credibly have accounted for the links between inferred forcing and reconstructed climate variability summarised in Figure 7.12, though the possible dynamics of the links are still barely explored. Bradley (2003), referring to model simulations, points to the possibility that solar variability had a major effect on the strength of the Hadley-cell circulation, with consequences for climates at higher latitudes through as yet poorly understood teleconnections.

The whole question of how quite small variations in solar irradiance may have exerted a major influence on climate variability has generated a good deal of recent debate. One hypothesis, advanced by Svensmark and Friis-Christensen (1996) and modified by Marsh and Svensmark (2000), proposes a link between the strength of the cosmic-ray flux (which is negatively correlated with solar activity) and the formation of cloud condensation nuclei in the atmosphere. The resulting feedbacks could include a reduction in temperature as a result of the cloudier conditions. Carslaw *et al.* (2003) provide some cautious support for the hypothesis by proposing physical mechanisms that could lead to the observed cosmic-ray–cloudiness link upon which the 'cosmic-ray–cloud hypothesis' largely rests. Sun and Bradley (2002), however, cast serious doubt on the hypothesis by showing that the putative correlation between cloud cover and cosmic-ray flux is strong only for the period 1983–1991 and for the area over the Atlantic Ocean. Extending the time span over which to examine the proposed correlation, and taking more complete and more recent data sets into account led them to discount the correlation for the second half of the twentieth century. The re-examination of the issue by Sun and Bradley has not gone unchallenged

(Marsh and Svensmark, 2004). Much depends on the relative reliability of satellite- versus surface-based observations of cloud cover. Sun and Bradley (2004) reaffirm that only the former show correlations between low cloud cover and cosmic-ray flux and then only for a limited area and time interval.

In his brief review, Rind (2003) makes the important point that solar influences on climate may reflect more than one single mechanism, as well as a complex system of feedbacks. The issue is far from resolved and its implications are considered further in light of the need to estimate the effects of solar variability on the most recent and future climate changes.

7.11.2 Volcanic activity

Both instrumental and annually resolved proxy-climate records (see, e.g., Briffa *et al.*, 1998) confirm that explosive volcanic eruptions have a generally brief, cooling effect on temperatures. The aerosols released reduce the solar radiation received at the Earth's surface. There are also associated circulation changes that may reinforce the cooling effect in some areas and moderate or even reverse it in others. Major eruptions clearly had widespread effects, well beyond the region directly impacted (see, e.g., De Silva and Zielinski, 1998; D'Arrigo *et al.*, 2001). Although individual eruptions have only a short-term transient effect, periods with more numerous, or severe eruptions, may have given rise to longer-lasting climate impacts. Several studies (summarised in Bradley, 2003) suggest that one such period coincided with and reinforced cooling in western Europe and southern Alaska during the later part of the Maunder minimum. Volcanic activity may also interact with the modes of variability already discussed above. As already noted, volcanicity may help to 'prime' the onset of El Niño events (Adams *et al.*, 2003). Also, on the basis of an analysis of the winter-climate anomalies that followed major tropical volcanic eruptions since AD 1600, Shindell *et al.* (2004) suggest that in many cases, eruptions led to changes in stratospheric temperature and wind anomalies that led to a pattern of response in surface temperatures similar to that characteristic of the NAO/AO systems.

7.12 Feedbacks

Several modelling studies suggest that the combination of solar and volcanic forcing can account for almost all the temperature variability at hemispheric or global level over the last 300 to 1000 years, except for the last few decades (Bradley, 2003; Crowley and Kim, 1999; Free and Robock, 1999; Crowley, 2000; Amman *et al.* 2003; Broccoli *et al.* 2003; Jones and Mann, 2004; Figures 12.11 and 12.12). At the same time, it is clear from Sections 7.3 and 7.4 that several of the most dramatic changes in Holocene climate can only be explained when feedbacks from the cryosphere, oceans and terrestrial biosphere are fully taken into

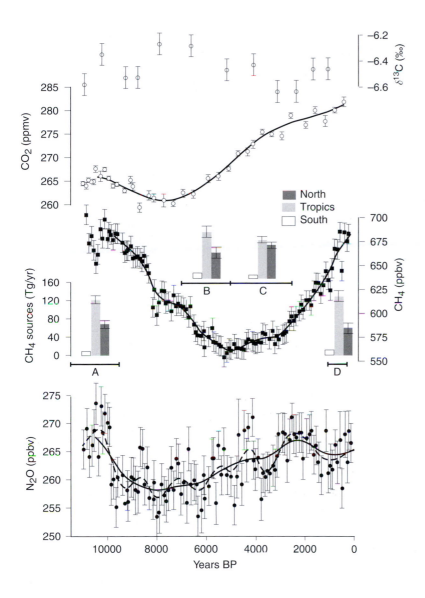

Figure 7.13 The history of atmospheric trace-gas concentrations during the Holocene. The source ascriptions for methane are calculated from a box model and refer to latitudinal bands from 90° N to 30° N (north), 30° N to 30° S (tropics) and 30° S to 90° S (south). The $\delta^{13}C$ record is from Taylor Dome and the N_2O record from Dome C. (Adapted from Raynaud *et al.*, 2003 and Flückiger *et al.*, 2001.)

account. Even in the late Holocene, climate variability, as expressed through the major climate modes, clearly involved changes in ocean circulation, sea-surface temperatures, sea-ice extent and land cover. Until the processes involved are better characterised empirically and can be modelled more effectively with high spatial and temporal resolution, it is difficult to provide any conclusive evaluation of their significance as modulators of the Holocene-climate record.

   One of the most provocative hypotheses advanced concerns the increases in both CO_2 and methane (see Ruddiman and Thompson, 2001) that occurred during the Holocene, from 8000 and 5000 BP, respectively (Figures 7.13 and

Figure 7.14 Part of the basis for the Ruddiman hypothesis of human modulation of Holocene variations in CO_2 and CH_4. The plot shows the apparently unique divergence (shaded) between variations in atmospheric methane during the Holocene and the insolation changes to which they are linked (see Figure 5.5). (Adapted from Ruddiman, 2003b.)

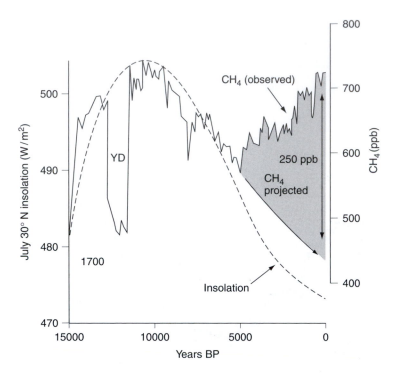

7.14). Ruddiman (2003b) claims that both increases are largely the result of human activities. He marshals several arguments to support this contention:

- Non-anthropogenic explanations for the increases are not in full agreement with the evidence available.
- The orbital configurations to which CO_2 and methane concentrations have been linked in the past would have led to a decline in both during the second half of the Holocene.
- None of the previous interglacials in the Vostok record show similar increases.
- The increase in CO_2 around 8000 years ago coincides with the early development of agriculture, requiring forest clearance and consequent carbon release to the atmosphere.
- The methane increase three millennia later is in good agreement with the development of intensive rice cultivation and flood-irrigated regions of Asia.
- Fluctuations in CO_2 over the last 2000 years are correlated with demographic changes, including outbreaks of plague that he believes are responsible for forest regrowth, increased carbon sequestration and reduced atmospheric concentrations.

Carcaillet *et al.* (2002) show that the global-fire indices compiled from an extensive evaluation of fire-history records throughout the Holocene for each region of the Earth parallel the increase of atmospheric CO_2 recorded in Antarctic ice cores. From this, they hypothesise that that biomass burning may have been a major factor in the increase in atmospheric CO_2 concentrations from

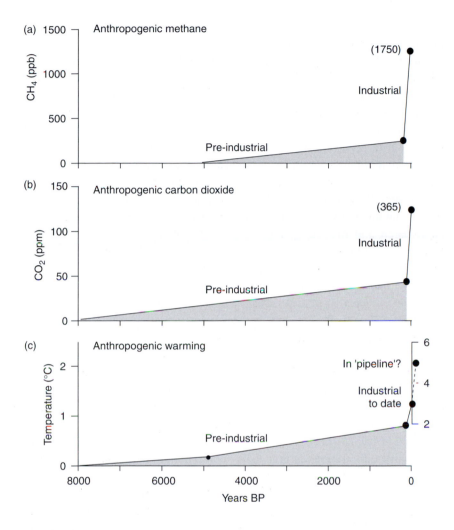

Figure 7.15 Schematic representation of the divergence between insolation-forced and measured atmospheric (a) CH_4 and (b) CO_2 concentrations from 5000 BP and 8000 onwards respectively, and (c) their inferred implications for global temperature. (Adapted from Ruddiman, 2004.)

8000 BP onwards. The analysis by Wu *et al.* (2003) of the loss of storage of organic carbon in soils in China as a result of cultivation lends further credibility to Ruddiman's main contentions.

   Ruddiman presents a detailed evaluation of the historical, archaeological and palaeobotanical evidence in support of his contentions. Having provided support for his hypothesis, he then proceeds to assess the likely effects on climate of the anthropogenic increases in CO_2 and methane, assuming a sensitivity within the range used by the IPCC (Figure 7.15). On the longer timescale, he believes it quite possible that the anthropogenic increases may have forestalled extensive and persistent snow and ice cover in northeastern Canada that could have been the precursor to glaciation. In the shorter term, he considers that the plague-linked dips in atmospheric CO_2 also contributed to climate variability during the last two millennia.

Joos *et al.* (2004) evaluate Ruddiman's hypothesis as it applies to CO_2 by driving the carbon component of a carbon cycle climate model (which links a dynamic global vegetation model to an atmosphere–ocean–sediment model) with climate fields simulated for time slices for the last 21 000 years derived from two global climate models. They conclude that the processes incorporated into the models adequately reproduce the measured changes in CO_2 during the Holocene without recourse to anthropogenic influences of the kind cited by Ruddiman. They further claim that the level of terrestrial carbon release implied by Ruddiman during the pre-industrial Holocene is not compatible with measurements of changing $\delta^{13}C$ in ice cores (see Figure 7.13)

Although Ruddiman's hypotheses have not been universally accepted, they are undoubtedly being taken very seriously and will have a major impact on research and thinking (Mason, 2004). If subsequent research supports them, they will provide one of the strongest lines of evidence so far to emerge for the effects of greenhouse-gas concentrations on climate, but it may be difficult to substantiate the hypotheses without a great deal more work. Two strands of evidence need to be strengthened and linked. On the one hand, it is important to reconcile the wide spread of estimates given for the change in carbon storage in the terrestrial biosphere between the LGM and recent times (see Figure 5.4). Additional evidence must also come from pollen analysis (see 4.5). By emphasising the links between pollen-analytical data and climate so exclusively, recent studies have too often ignored or excluded evidence for human impacts on the extent of forest and cleared land. Even in publications that highlight the importance of human activities in prehistory (Oldfield *et al.*, 2003a; b), the evidence falls short of providing the kind of quantitative information required to develop estimates of carbon sequestration and release as a result of deforestation or land abandonment. Even in areas where the pollen-analytical evidence is relatively unambiguous, this requires quite complex modelling (Broström *et al.*, 1998b; Gaillard *et al.*, 1998; Sugita *et al.*, 1999). In many parts of the world, pollen-analytical data are not readily converted into even subjective estimates of landscape openness. Alongside a more quantitative sense of long-term human impacts on the balance between forest and non-forest, a stronger sense is needed of the net effect of deforestation on carbon budgets in different types of biome, both through direct observations (see, e.g., Achard *et al.*, 2004) and modelling (see, e.g., Matthews *et al.*, 2004). Ruddiman has opened up a range of possibilities that will take a long time to evaluate fully.

7.13   Concluding comments

The ongoing changes in boundary conditions and forcing during and in the wake of deglaciation distinguish the first half of the Holocene from the last 3000 to 4000 years. Some of the most important characteristics of this later period are:

- The persistence of significant variability on annual to millennial timescales, especially in hydrological regimes.
- Strong evidence for both solar and volcanic forcing.
- An emerging debate about the origin and possible climatic consequences of early increases and subsequent variability in atmospheric CO_2 and methane concentrations. Even if the anthropogenic influences posited by Ruddiman turn out to be relatively minor, the possibility remains that the increasing concentrations, howsoever generated, may have had important implications for ice extent and climate.
- Pre-instrumental climatic extremes, beyond those recorded during the last century, are well documented in many archives and regions, especially for droughts, where severity is related to persistence and duration rather than to peak amplitude.
- There is strong evidence that such extremes had dramatic implications for human societies.

Chapter 8

The Anthropocene – a changing atmosphere

8.1 The idea of the Anthropocene

Crutzen and Stoermer (2001) have proposed that the period from the late eighteenth century onwards merits separate designation as the Anthropocene. Their suggestion arises from .an acknowledgement of the increasing role of human activities in the functioning of the Earth system, as witnessed by, for example:

- The accelerated use of fossil-fuel resources, bringing with it steeply rising emissions and an unprecedented gradient of change in atmospheric CO_2 concentrations.
- The impact of human activities on other key global biogeochemical cycles, including nitrogen.
- Major changes in the Earth's surface cover as a result of processes such as deforestation, land reclamation, irrigation and soil degradation induced by human activities.

Irrespective of doubts about the choice of a starting date (Ruddiman, 2003b; Steffen and Crutzen, 2003) and of the extent to which the concept of the Anthropocene will be justified as the long-term future unfolds, the term helps to capture and dramatise the idea that, as Steffen *et al.* (2004) state: 'The Earth is currently operating in a no-analogue state. In terms of key environmental parameters, the Earth System has recently moved well outside the range of the natural variability exhibited over at least the last half million years. The *nature* of changes now occurring *simultaneously* in the Earth System, their *magnitudes* and *rates of change* are unprecedented.'

Figure 8.1 shows graphically some of the key indicators of the current no-analogue state as they have evolved over the last three centuries. Although some of the diagnostic changes begin early in the period, almost all accelerate steeply from around 1950 onwards. Thus most of the distinctive changes that characterise the Anthropocene have come about in less than a human lifetime. All of them are linked, directly or indirectly, to the growth of human populations, along with increased per capita consumption, initially and still dominantly in the developed world. These have led to a massive rise in gross domestic product, with increasing pressure on the Earth's resources, both renewable and non-renewable.

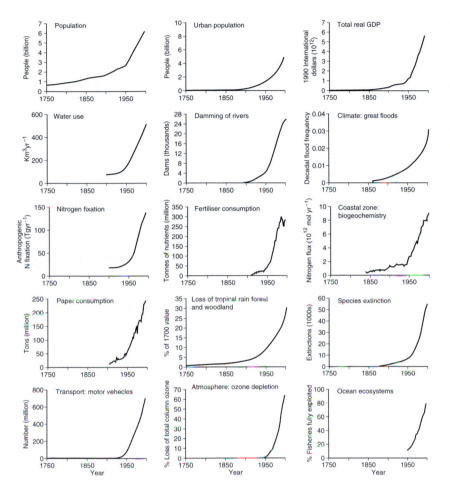

Figure 8.1 Thumbnail graphs of some of the key indicators of the Anthropocene. Note the degree to which all the rising trends steepen from the mid twentieth century onwards. (Compiled and adapted from Steffen *et al.*, 2004.)

Some of the main transformations brought about by human activities are considered in this and the next three chapters. Where appropriate, they are set within the context of human–environment interactions on longer timescales, as, for example, in sections 9.4, 10.2 and 10.4, as well as Figures 9.3 and 10.1.

So far, evidence from the present day and recent past has been used mainly to set out in a preliminary way the nature of current and future problems, and as the basis for calibrating research results aimed at reconstructing the past. From now on, the appraisal of present-day processes and their future implications necessitates new, additional modes of study, as set out in Section 1.4. These have to be integrated with what has been learned from the past to inform our view of the future.

8.2   Increasing concentrations of greenhouse gases

By combining measurements from ice cores with direct measurements from 1958 onwards, it is now possible to reconstruct the history of the main

Figure 8.2 Atmospheric trace-gas and sulphate trends over the last 1000 years compiled from ice-core records and direct measurements. The gas measurements are all derived from ice-core and firn measurements, except for the latest CH_4 (open inverted triangles) and CO_2 values (solid line). (Modified from Raynaud *et al.*, 2003.)

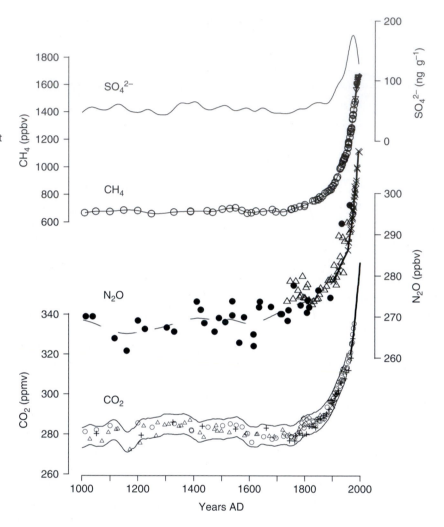

atmospheric greenhouse-gas concentrations (excepting water vapour) continuously up to the present day (Figure 8.2). All increase during the latter part of the eighteenth century. In each case, the trend steepens during the twentieth century and especially since 1950.

8.2.1   Carbon dioxide (CO_2)

Since the mid eighteenth century, atmospheric concentrations have risen from around 280 to over 370 mmol mol^{-1}. Most of this increase has been the result of fossil-fuel burning, cement manufacture, deforestation and biomass burning. Sabine *et al.* (2004) apportion total emissions for the last two centuries to each of the main sources. Fossil fuels and cement manufacture account for some

244 ± 20 Pg (1 Petagram = 1 Gigatonne = g × zom 10^{15}) of carbon of emissions between 1800 and 1994. Over the same period, the terrestrial-biosphere was a net source of 39 ± 28 Pg of carbon, with emissions from land-use change exceeding the terrestrial-biosphere sink. These data are broadly compatible with earlier estimates of industrial emissions at around 7–8 Pg of carbon *per annum* during the 1980s and 1990s, and with Yevich and Logan's (2003) estimate that in the mid 1980s the combustion of biofuel and agricultural waste in developing countries accounted for less than 15% of total anthropogenic emissions. Achard *et al.* (2004) estimate that changes in land cover in the tropics during the 1990s had the net effect of releasing 1.1 ± 0.3 Pg of carbon *per annum*. Their calculations take into account both the short- and longer-term carbon losses linked to deforestation, the impacts of major fires and the effects of regrowth.

A molecule of CO_2 spends, on average, three years in the atmosphere before it is incorporated in the terrestrial biosphere or the oceans. From here, though, most of it will return to the atmosphere within a few years, therefore the effective residence time before long-term sequestration in the deep ocean is of the order of thousands of years (Brasseur, 2003). Out of the 7–8 Gt (Gigatonnes) of carbon released from fossil-fuel combustion and tropical-forest clearance each year during the 1980s and 1990s, less than half (~3 Gt) accumulated in the atmosphere. Atmospheric CO_2 concentrations are therefore growing at less than half the rate to be expected if all the CO_2 released by fossil-fuel combustion and land-use change were to remain in the atmosphere. The atmospheric concentrations reflect the balance between production by a combination of natural and anthropogenic processes, and sequestration in the oceans and the terrestrial biosphere. According to Sabine *et al.* (2004), about two-thirds of the total emissions since 1800 have remained in the atmosphere. It follows from their data that the only true net sink over the last 200 years has been the ocean. They show that the strength of the ocean sink varies greatly between oceans, with the North Atlantic storing some 23% of the global oceanic anthropogenic CO_2 and the Southern Ocean only 9%. Between 1980 and 1999, they estimate that emissions from fossil fuels and cement manufacture were 117 ± 5 Pg of carbon, with the terrestrial biosphere acting as a small net sink of 15 ± 9 Pg of carbon, compared with 37 ± 8 Pg of carbon for the ocean. Over this recent period, the terrestrial-biosphere sink has taken up more carbon than the total emissions from land-use change. A shift in the balance between ocean- and terrestrial-sink strength over the last two decades seems likely, though this is difficult to quantify in view of the uncertainties attached to the estimates of terrestrial-biosphere processes.

Figure 5.3 shows the main carbon reservoirs in the Earth system and the rates of exchange between them. The main concern here is with those fluxes that take place on timescales of a century or less. Figure 8.3 shows estimates of the main sources, sinks and fluxes averaged over the 1980s. It distinguishes between the numbers reflecting natural processes and those resulting from human actions.

Figure 8.3 Main global carbon fluxes and reservoirs for the 1980s. Bold italic numbers refer to anthropogenic fluxes. Bold shaded numbers in the reservoir boxes give the changes in size since pre-industrial times as a result of human activities. The positive number in the 'land' box represents the inferred terrestrial sink; the negative number refers to the decrease as a result of deforestation. NPP – net primary productivity. (Modified from Sarmiento and Gruber, 2002.)

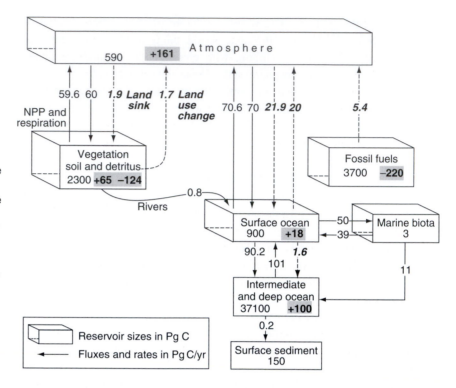

Balancing the budget, even for the present day, proves controversial, for not all the reservoirs and fluxes can be adequately measured on a global basis. Atmospheric concentrations of CO_2 are relatively easy to quantify, as the gas is non-reactive and becomes rapidly well mixed. In the case of the oceanic and terrestrial reservoirs, spatial heterogeneity makes it difficult to calculate global quantities from localised measurements. Inter-annual variability further complicates the problem. In practice, estimates are made by a combination of direct measurements and both forward and inverse modelling, but there remain sinificant doubts as to the relative importance of the oceanic and terrestrial sinks (Adams and Piovesan, 2002). In most attempts at a global budget, there remains a 'missing' carbon sink. Kaufmann and Stock (2003) use a reconstructed time series from 1860 to 1990 to test the alternative hypotheses that have been proposed. They conclude that increases in carbon emissions may generate short-term increases in ocean uptake that are not well simulated in the existing carbon models.

Schimel *et al.* (2001) and Canadell and Pataki (2002) both suggest that the terrestrial biosphere, in the northern hemisphere rather than the tropics, became a significant carbon sink during the 1990s. They apportion the fluxes between biosphere and atmosphere in different regions within the global value of 1.4 ± 0.7 Pg *per annum* used in the IPCC TAR (2001). More recently, Plattner *et al.* (2002) suggest that the rate of uptake by the terrestrial biosphere during the

1990s was less than the mean value used in the IPCC TAR (2001) by a factor of two and propose a figure of 0.7 ± 0.8 Pg *per annum*. Their values are also compatible with those calculated by Bopp *et al.* (2002) and bring estimates into much closer agreement with those based on ocean models and with the calculations made by Sabine *et al.* (2004). Joos *et al.* (2003) highlight the possibility that these revised values reflect the beginning of a reorganisation of large-scale ocean circulation. In these studies, the downward revision of the net rate of terrestrial-carbon uptake is the corollary of an upward revision in oceanic uptake, inferred from measurements of the dissolved-oxygen content of the ocean. By contrast, Liski *et al.* (2003) suggest that temperate and boreal forests have become an increasingly important carbon sink and propose values for this that tend to support those used in the IPCC TAR (Prentice *et al.*, 2001). McNeill *et al.* (2003) use chlorofluorocarbons as a 'tracer' for CO_2 uptake by the oceans and propose values which are compatible with the lower estimates for the terrestrial biosphere quoted above. In an attempt to calculate the total inventory of carbon stored in the oceans over the whole 250 years since early industrialisation, Lee *et al.* (2003), using their own and earlier data, estimate that some 29% of the CO_2 arising from anthropogenic activities has been sequestered in the oceans – a figure close to the estimates by Sabine *et al.* (2004). There are indications that at least one of the controls on carbon uptake in the oceans has changed over the last two decades. Gregg *et al.* (2003) note a more than 6% decline in global ocean primary productivity, mainly in response to increased sea-surface temperatures (SSTs) at high latitudes. From the foregoing estimates, we may conclude both that disagreements persist with regard to the relative importance of marine and terrestrial sinks over the last two decades, and that their relative strengths may be changing.

Figure 8.4 shows that both seasonal and inter-annual variability is superimposed on the rising trend in atmospheric CO_2 concentrations. One of the keys to better quantifying each of the exchanges in the present-day carbon cycle lies in improving our understanding of this variability. Part of the inter-annual variability appears to be linked to climate, notably the incidence of El Niño episodes. Virtually all of these are associated with increases in the accumulation rate in the atmosphere. Dutta (2002), using measurements of ^{14}C in the atmosphere, concludes that two ENSO-linked processes are involved: reduced upwelling in the tropical Pacific and greater release of carbon from tropical forests. One of the most important processes involved is the release of carbon during major wildfires. Schimel and Baker (2002) calculate that the El Niño-linked fires of 1997–98 in Indonesia released 0.81–2.57 Pg of carbon – between 13 and 40% of annual emissions from anthropogenic fossil-fuel combustion. Langenfields *et al.* (2002) base independent estimates of carbon released from these and the previous fires in 1994/5 on measurements of CO_2 and a range of other trace gases, as well as stable-isotope measurements. They propose figures of 0.6–3.5 and 0.8–3.7 Pg of carbon, respectively, for the two events. Despite the

Figure 8.4 Variability in the rate of increase of atmospheric CO_2 concentrations since 1956 compared with the rate of increase in fossil-fuel emissions and ENSO variability. Note the generally positive correlation between El Niño events and increases in the rate of growth of CO_2 concentrations, except during the effects of the Mount Pinatubo eruption (see 8.2.1). (Modified from Steffen *et al.*, 2004.)

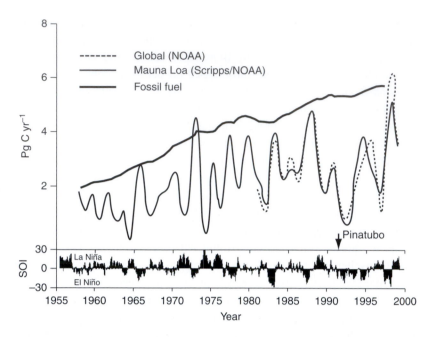

wide range of values proposed by each study, they clearly show that biomass burning can have significant effects on the annual carbon budget. As indicated in the preceding chapter, similar conclusions may be drawn from the longer-term changes in atmospheric CO_2 throughout the Holocene (Carcaillet *et al.*, 2002).

Contributions to inter-annual variations in the carbon budget may arise from ocean processes other than upwelling in the tropical Pacific. Lintner (2002) demonstrates a statistically significant link between the pattern of inter-annual variability in atmospheric CO_2 and the Pacific decadal oscillation (PDO), as well as El Niño southern oscillation (ENSO). Dore *et al.* (2003), using data from the Pacific Ocean near Hawaii provide convincing evidence that changes in salinity, driven by changes in regional P minus E, modulate carbon exchange across the ocean–atmosphere boundary layer. Bates *et al.* (2002) show that changes in the rate of vertical mixing in the sub-tropical region of the North Atlantic may also contribute to inter-annual variability in the marine uptake of CO_2.

In Figure 8.4, the normal link between ENSO and inter-annual CO_2 variability breaks down in the early 1990s. This is thought to be the result of the Mount Pinatubo eruption, though there is no consensus concerning the precise way in which it has affected the carbon cycle. One hypothesis is that an aerosol-generated increase in the diffuse fraction of solar radiation favoured photo-synthesis, hence increased carbon uptake (Gu *et al.*, 2003; Reichenau and Esser, 2003), though Krakauer *et al.* (2003) quote widespread evidence from tree rings to suggest that forest net primary productivity declined in the wake of the eruption (cf. Briffa *et al.*, 1998). Angert *et al.* (2004), using coupled biogeochemical and atmospheric-tracer models claim to disprove the notion that the enhanced post-Pinatubo carbon sink was the result of increased photo-synthesis. Instead, they propose a unique combination of factors, including an enhanced ocean sink, retarded heterotrophic respiration as a result of cooler and drier soils, and reduced biomass burning.

The above account gives some idea of the complex processes involved in balancing the present-day CO_2 budget. Although several of the studies quoted appear to herald a narrowing of the range of uncertainties at the global level, major problems still remain, especially at the level of ascribing appropriate roles to, and quantifying the effects of, the wide range of interacting processes involved, both marine and terrestrial. These problems continue to be com-pounded the difficulties inherent in both explaining past changes in atmospheric concentrations (see 5.2.3, 6.6 and 8.2.1) and predicting future changes for any given emission scenario (see 13.1.3).

8.2.2   Methane (CH$_4$)

Since AD 1700, atmospheric-methane concentrations have more than doubled. Most of this steep increase can be confidently ascribed to human activities. After water vapour and CO_2, methane is the most important greenhouse gas. On a weight for weight or molecule for molecule basis, it is a much more effective greenhouse gas than CO_2. On the other hand, it has a much more rapid atmo-spheric turnover time of around nine years, which means that reductions in emissions could have a relatively rapid effect. In addition to acting as a green-house gas, methane plays an important role in several key reactions in the atmosphere (Wuebbles and Hayhoe, 2002).

Whereas the natural sources of methane are dominated by wetlands, both tropical and boreal, anthropogenic sources of methane are rather varied (Figure 8.5a). Biogenic sources contribute via processes such as anaerobic decom-position and digestion by ruminants and other animals. Wetland sources include natural vegetation as well as rice paddies. Decaying organic matter in landfills, forest fires and gas escapes associated with fossil-fuel extraction are also import-ant sources. Release through biogenic activity is often temperature related, though many other variables can come into play. In boreal peatlands, for example, warming accelerates methane release. Friborg et al. (2003) show that although the extensive wetlands in western Siberia are a net sink for carbon, they are a large net source of methane. Beyond a certain point, however, any warming that leads to increased evaporation will tend to dry out the upper part of the peat column. This, in turn, will lead to conversion of the methane to CO_2 by methanotrophic bacteria. In the case of rice cultivation, the level of methane release is highly dependent on water management, cropping sequences and fertiliser use.

The only major sink of atmospheric methane is through reaction with the hydroxyl radical (OH). This radical also acts as the main oxidant for a wide range of atmospheric pollutants and its capacity to fulfil this role is strongly influenced by methane concentrations. A complex set of rate-dependent reactions summarised by Wuebbles and Hayhoe (2002) lead to a positive feedback between CH_4 and OH: as oxidising capacity is reduced, methane removal slows down, which in turn reduces oxidation capacity, with a consequent further increase in methane concentrations.

Figure 8.5b shows the rate of increase of atmospheric methane from 1984 to 2002. By 1984, the rapid surge in concentrations typical of the late nineteenth and most of the twentieth century had slowed down somewhat, while there has been virtually no increase from 1999 through 2002 (Dlugokencky *et al.*, 2003). The absolute values are higher and the seasonal cycle much wider in the northern compared with the southern hemisphere and the tropics reflecting the extent to which methane sources are terrestrial. Seasonal variations in concentration are the result of seasonal changes in the photochemical sink for methane, which is strongest during summer. The post-1999 stabilisation of atmospheric-methane concentrations is largely a function of declining northern-hemisphere emissions. Although the reasons for the decline are not fully understood, they are thought to reflect a reduction in one or more sources rather than an increase in sink strength. The most likely candidate is the former Soviet Union where emissions arising from fossil-fuel combustion declined steeply by some $10 \, \mathrm{Tg \, yr^{-1}}$ between 1990 and 1995. The recent levelling off of atmospheric methane concentrations has been seen by some (see, e.g., Co2science, 2003) as evidence for approaching equilibrium between atmospheric CH_4 and OH. Given the varied methane sources, the difficulties involved in estimating their strengths, and the subtle nature of the main sink, interpreting the temporal changes in atmospheric concentrations is still fraught with large uncertainties. In particular, it will be important to develop well-validated process models of wetland emissions under changing climatic conditions. Until these issues are resolved, it would be premature to see the most recent trend as evidence for stabilisation, though if this were the case, it would be important not only for future methane projections but for those of tropospheric ozone also (see 8.4). Only the most 'benign' of the IPCC TAR scenarios (B1 – see 13.1.2 and Figure 13.2) has methane concentrations peaking around the current values. All the other scenarios incorporate increases significantly beyond this level.

Consideration of the changes in atmospheric methane during transitions from glacial to interglacial conditions widens rather than narrows the uncertainty about future methane concentrations, for it highlights a further possibility, that of clathrate release from methane hydrates stored in coastal sediments (see 6.5.4 and Somoza *et al.*, 2002) and below permafrost (Dallimore *et al.*, 2002).

8.2.3   Nitrous oxide (N_2O)

Nitrous oxide, with a lifetime in the atmosphere of some 120 years, is an important greenhouse gas that together with methane accounts for around 25%

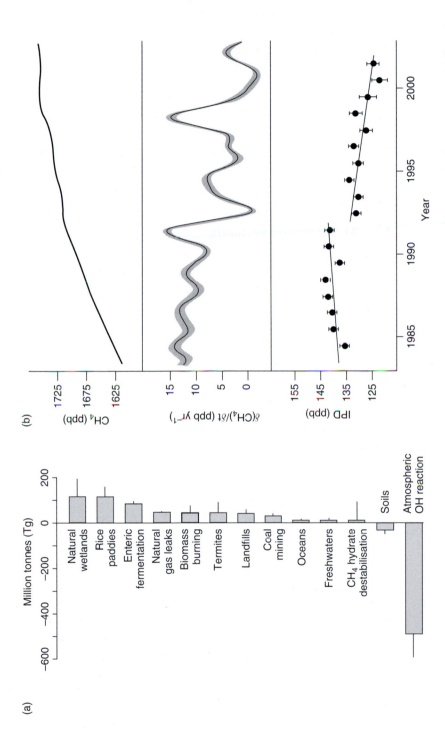

Figure 8.5 (a) Estimated sources and sinks of atmospheric methane, with uncertainties, for the 1990s (Modified from Steffen *et al.*, 2004.) (b) Changing atmospheric concentrations of methane, 1986–2002. The upper graph shows the trend of the annual global average measurements. The middle graph plots the changing growth rate of methane concentrations. The lower graph plots the difference between northern (53° N –90° N) and southern (53° S–90° S) atmospheric concentrations as a function of time. Solid lines are linear least-squares fits to the values for the periods 1984–1991 and 1992–2001. IPD – inter-polar difference. (Modified from Dlugokencky *et al.*, 2003.)

of the total radiative forcing by greenhouse gases since the onset of the industrial revolution (Scholes *et al.*, 2003). Most N_2O arises as a result of microbial processes in soil and aquatic ecosystems, though decomposition of animal wastes, biomass burning and oxidation of ammonia (NH_4) are also significant. Atmospheric concentrations have increased from a pre-industrial level of $270\,nmol^{-1}$ to $314\,nmol^{-1}$ in 2001 largely as a result of both the expansion and intensification of food production and modifications to the global nitrogen cycle considered in more detail in Section 9.1.1. The sequence of changes in N_2O concentrations during the Holocene parallel those for CO_2 rather than CH_4, with minimum values around 8000 BP and a slight subsequent increase, though the reasons for the variations are not well understood (Flückiger *et al.*, 2001).

8.3   Atmospheric aerosols

The increasing importance of anthropogenic aerosols in the radiative balance of the atmosphere must first be set in the context of those that are produced naturally. These comprise primary aerosols (mainly sea salt and soil dust), and secondary aerosols. Sea salt reaches the atmosphere through the evaporation of spray from breaking waves. Dust is mainly the result of entrainment by wind in arid regions, though volcanic ash may also be important following major eruptions. Secondary aerosols arise from chemical reactions and the condensation of atmospheric gases and vapour, largely through the operation of the sulphur cycle. Two processes dominate the production of natural secondary aerosols: marine-biological activity and volcanic eruptions. In the former case, primary production in the oceans leads to the release of dimethylsulphide (DMS), which oxidises to sulphuric acid, mainly in the form of fine aerosol droplets. Secondary volcanic aerosols result from the release of SO_2 during major eruptions.

Anthropogenic sulphur aerosols produced by fossil-fuel combustion, biomass burning and other industrial processes now greatly exceed those arising from natural processes. They peaked at $110\,ng\,g^{-1}$ at the end of the 1960s, compared with $26\,ng\,g^{-1}$ in pre-industrial times (Figure 8.2). Overall, anthropogenic activities are responsible for around 20% of the atmospheric aerosol burden, but this underestimates their importance in the radiative balance of the atmosphere, as they contribute up to 50% to the global mean aerosol optical depth/ thickness (Raynaud *et al.*, 2003). This determines the degree of extinction of incoming solar radiation. An increase in optical depth implies the probability of an overall negative forcing at the Earth's surface. As a generalisation of aerosol effects, this turns out to be an unacceptable simplification, for atmospheric aerosols modify climate in quite complex ways. The direct effect involves the absorption and scattering of solar energy. Indirect effects include the formation of cloud condensation nuclei, along with a complex and still controversial series of effects on radiative balance, precipitation and evaporation.

Progress has been made towards a better understanding of the role of aerosol forcing as a result of studies linked to specific regions, events and processes. Penner *et al.* (2004) attempt to quantify the effects of indirect aerosol forcing on radiative fluxes at sites in Oklahoma and Alaska, using a combination of direct observations and modelling. They conclude that these effects are responsible for a significant negative forcing. Podgorny *et al.* (2003) calculate the impact of the changes in optical depth over part of the equatorial region as a result of the Indonesian forest fires of 1997. They conclude that over the Indian Ocean there was a strong decrease in solar flux at the sea surface and a significant increase in the solar heating of the atmosphere. They infer likely consequences for regional climate, including increased precipitation in the smoke-covered area, possibly leading to a redistribution of tropical rainfall and subsequent feedbacks into the El Niño climate regime itself. Xu (2001) documents similar regional climatic effects from pollution haze in parts of China. Major reductions in solar irradiance at the Earth's surface have also been observed as a result of biomass burning in Amazonia (Procopio *et al.*, 2004) and Zambia (Schafer *et al.*, 2002).

The mean size of the aerosols produced by biomass burning is crucial in determining the indirect effects on climate. In studies from both Amazonia and Africa, the small size of the aerosol generated by land clearance involving biomass burning tends to delay rainfall production. In Amazonia, it can also lead to the entrainment of moisture and pollutants to higher levels, cause intense thunderstorms and modify atmospheric circulation (Artaxo, 2003; Andreae *et al.*, 2004). The effects of smoke on the radiative balance at the Earth's surface may not always be negative. Koren *et al.* (2004) show that by suppressing cloud formation during the dry season in Amazonia, formation of cumulus clouds was inhibited to the point where surface warming was actually enhanced. All these studies point to a range of important but quite complex feedbacks through which changes in land cover lead to the modification of regional climate.

Kaiser and Qian (2002), present evidence from observations and meteorological data in support of the view that the recorded decline in sunshine duration in much of China over the last 50 years has been the result of increased aerosol loading. Observations carried out during the Indian Ocean Experiment campaign show that the combination of biomass burning, industrial development and dust entrainment in the Indian subcontinent generates heavy aerosol loadings over coastal areas and adjacent seas north of the equator – the 'asian brown cloud', within which around 80% of the particles are believed to be anthropogenic in origin. The resulting aerosol forcing at the surface in coastal India is estimated at $-27\,\mathrm{W\,m^{-2}}$ (Jayaraman, 1999). During the dry season, the negative effect of aerosol forcing on radiation at the Earth's surface in southern Asia can greatly exceed the positive effect of greenhouse gases (Lelieveld *et al.*, 2001; Ramanathan *et al.* (2001), though the radiative balance is complicated by the role of soot particles, which absorb radiation and lead to atmospheric warming (see below), as well as by the extent to which the aerosol layer is above or below any existing clouds (Jayaraman and Mitra, 2004). Whereas fossil-fuel combustion is the main source of sulphate aerosols, high levels of carbonaceous particulates are generated through biomass burning. In this region, the effects of burning fuel wood, dung-cakes and crop waste greatly exceed those of forest fires (Reddy and Venkataraman, 2002). As these brief accounts of aerosol effects illustrate, the processes involved are complex and there are major differences from region to region. For these reasons, a 'bottom-up' approach to estimating global aerosol effects, based on aggregating direct observations, is hardly feasible in the present state of knowledge.

Two mutually independent approaches have been used to estimate anthropogenic-aerosol forcing on a global basis: forward modelling based on the physics and chemistry of the aerosols, and inverse modelling whereby the total forcing is brought into line with what is required to match simulations of past-climate change with the temperature changes actually observed (cf. 2.2.13) The forward modelling approach gives a mean negative forcing of $-1.5\,\mathrm{W\,m^{-2}}$ over the industrial era, with a wide range of uncertainty extending beyond $-3\,\mathrm{W\,m^{-2}}$. The inverse modelling approach indicates a negative forcing of $-1.0\,\mathrm{W\,m^{-2}}$, with a narrower range of uncertainty (Anderson et al., 2003). Most simulations of future climate change use values within the second range, but if, in future, values within the wider range arising from forward modelling proved to be more accurate, this would have a significant effect on modelling future temperature changes. As Anderson et al. point out, the choice of a value for aerosol forcing affects the inferred sensitivity of climate to total forcing as well as the degree to which recent warming can be attributed to anthropogenic greenhouse gases (see 13.1.3). Recent attempts to constrain aerosol forcing have concentrated on the indirect effects. An independent study using satellite observations tends to reinforce a value within the narrower range inferred from inverse modelling (Lohmann and Lesins, 2002), as does the detailed evaluation of indirect aerosol forcing outlined by Anderson et al. (2003).

Many of the sources quoted above infer or accept that aerosols, including those arising from biomass burning, have an overall negative effect on temperatures at the Earth's surface. The main exception is black carbon or soot, generated mainly by fossil-fuel combustion, and with a modal size an order of magnitude less than biomass smoke (Heintzenberg et al., 2003). Black carbon is distinguished from organic carbon on the grounds of optical properties rather than chemistry and is thought to have an overall positive effect on temperature. Brasseur's (2003) rough estimate suggests that man-made sources have led to a twenty-fold increase in the atmospheric burden since pre-industrial times. Hansen et al. (2000) and Hansen (2002) claim that the net positive forcing resulting from black carbon on a global scale is an order of magnitude greater than that estimated by the IPCC, with important implications both for future projections of global warming and for possible mitigation strategies. Jacobson (2002) also claims a strong positive forcing for fossil-fuel generated black carbon and associated organic matter, with the corollary that controlling emissions holds out the promise of slowing future global warming. His analysis has, however, attracted a whole sequence of critical comments and counter claims (Chock *et al.*, 2003; Penner, 2003; Feichter *et al.*, 2003) followed by replies from the original author. Wang (2004) infers, from a detailed modelling study, that the effects of black carbon on climate are dominantly regional rather than global. Penner *et al.* (2003) claim that warming by black soot is not inevitable, even at a regional scale, since it appears to depend on the altitude of injection of the aerosol. Menon *et al.* (2002) contend that increased levels of black carbon in the atmosphere may actually be responsible for cooling in the areas directly affected as well as for changes in rainfall regimes. Their study once more emphasises the fact that aerosol-related impacts on climate are pre-dominantly regional, but with the potential for generating changes well beyond the area directly affected. Both model simulations and empirical observations suggest that the likely combined impacts of high-aerosol and tropospheric ozone concen-trations include a significant reduction in crop yields in parts of east Asia as a result of the greatly reduced solar radiation reaching the Earth's surface (see, e.g., Chameides *et al.*, 1994).

It is clear from the above account that considerable doubt still surrounds the precise net effect of aerosols on the radiative budget of the atmosphere, both on a regional and a global basis. Although inclusion of estimates of their effect have significantly improved attempts to model the course of climate change during the twentieth century, the persisting uncertainties necessarily widen the total range of temperature projections derived from future climate scenarios (see 13.1.3). Moreover, the impact of aerosols on ecosystems may be more varied and subtle than the above account suggests. One possibility is that by increasing diffuse relative to direct radiation, aerosols may generate higher levels of carbon uptake through the photosynthetic advantage that diffuse light gives to below-canopy leaves, which would otherwise be shaded by the leaves above (Roderick *et al.*, 2001). Finally, aerosol effects extend beyond climate to include signifi-cant fertilisation of otherwise nutrient-poor areas, for example the Okovango delta, in Botswana. Here, in areas which are not enriched seasonally by flowing water, aerosols supply some 52% of the phosphates and 30% of the nitrates received by the vegetation (Garstang *et al.*, 1998).

8.4 Chlorofluorocarbons and Ozone

Ninety percent of the ozone (O_3) in the atmosphere is within the stratosphere at an altitude above 15 km (Figure 8.6). Whereas the ozone in the stratosphere acts as a UV-radiation shield and has a beneficial role, that in the troposphere generally has a harmful effect. Moreover, it acts as a powerful, long-lived greenhouse gas.

We consider first *stratospheric ozone*, which is formed by photolysis of O_2 and subsequent recombination of the oxygen atoms. Under unperturbed

Figure 8.6 The altitudinal distribution of ozone in the atmosphere. (Modified from Brasseur *et al.* 2003.)

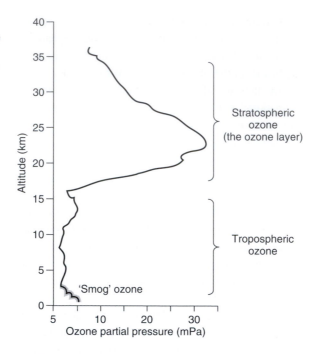

Figure 8.6 The altitudinal distribution of ozone in the atmosphere. (Modified from Brasseur *et al.* 2003.)

conditions, this process is balanced by a series of reactions that lead to the destruction of ozone. Over the last few decades, chlorine compounds, notably chlorofluorocarbon (CFC) gases ($CFCl_3$ and CF_2Cl_2) have been added to the stratosphere such that the concentrations of chlorine are now six times higher than might be expected from natural processes. Serious depletion of stratospheric ozone was first revealed by measurements over Antarctica in 1985, but much subsequent research was needed to work out exactly how the resulting 'ozone hole' had arisen, through a chain of catalytic reactions (Crutzen, 2003). The build up of chlorine concentrations since the first production of CFCs in the 1930s has had a disproportionate effect on the rate of ozone depletion, for the rate of ozone breakdown is proportional to the square of the ClO concentration. As a result, the processes of ozone depletion are now 36 times stronger than they were prior to the production of CFCs. Even in the wake of the Montreal protocal, designed to phase out the manufacture of ozone-depleting chemicals, and subsequent amendments, it is unlikely that stratospheric-ozone levels will reach their 1980 levels until around the middle of this century (Wuebbles *et al.*, 1999). Despite the fact that the production of CFCs had fallen from 1.1 million to 150 000 tons between 1986 and 1999, the 2003 ozone hole over the Antarctic was still around 28 million km^2 at its peak – larger than the previous year and only slightly smaller than the peak extent of September 2000 (Showstack, 2003). Rex *et al.* (2004) show that increasingly cold temperatures in the stratosphere over Antarctica over the last four decades have led to conditions increasingly favourable to ozone loss. Stratospheric cooling may well be the result of increased

greenhouse gas concentrations, in which case the interaction between two unrelated anthropogenic perturbations to the atmosphere has led to mutual reinforce-
ment and delayed recovery. Solomon (2004) succinctly summarises the current
state of the ozone hole, stressing that cold temperatures enhance the impact of
elevated chlorine concentrations. She also points out that, given the long (50–100
year) lifetime of CFCs in the atmosphere, it is quite unrealistic to expect rapid
improvements. It is also unrealistic to read any great significance in the inter-
annual fluctuations, as these mainly reflect variations in atmospheric conditions.

The story of the ozone hole carries with it a double warning. Not only did it
take the scientific community completely by surprise and involve a long series of
complex chemical reactions that would have been well nigh impossible to
predict, it could have been orders of magnitude worse. Bromine is some 100
times more effective than chlorine in destroying ozone. If bromofluorocarbons
rather than CFCs had been chosen for the industrial processes that began in the
1930s, loss of ozone could have been complete. Choice of chlorine- rather than
bromine-based compounds seems to have been a matter of luck rather than
informed scientific judgement (Crutzen, 1995).

Whereas in the stratosphere, ozone depletion is the main cause of concern,
problems have arisen from an increase in ozone concentrations in the troposphere by a factor of two or more. Tropospheric ozone can have a deleterious
effect on plants, animals and human health. The main increases have taken place
in Europe and east Asia, as a result of the interaction between reactive oxides of
nitrogen, hydrocarbons, CO and sunlight. Tropospheric ozone is a greenhouse
gas and it is estimated that it has led to a positive radiative forcing of up to
$0.49\,\mathrm{W\,m^{-2}}$, implying a global mean surface warming of $c.$ 0.28 °C since pre-
industrial times (Mickley et $al.$, 2004). It also breaks down to a primary source of
the hydroxyl radical (OH), the oxidation capacity of which forms a major sink
for many chemical species in the atmosphere. The role of ozone is therefore
complicated and intimately bound up with other reactive atmospheric
gases, especially methane, which is now known to play a significant role in
tropospheric-ozone formation (Fiore et $al.$, 2002).

8.5   Other atmospheric contaminants

Many other compounds have been released to the atmosphere as a result of
human activities. Some, like *polychlorinated biphenols* (PCBs), *polyaromatic
hydrocarbons* (PAHs) and many heavy metals, are known to damage human
health. The histories of production generally fall into three categories. Several of
the heavy metals such as lead and copper were first exploited in prehistory and
the long record of their changing atmospheric concentrations can be reconstructed from ice cores and peat bogs. Other metals, such as cadmium and
chromium, are more closely linked to industrial processes that began during
the Industrial Revolution and signs of the first significant increases in their

atmospheric concentrations come during the nineteenth and early twentieth centuries. Some of the more damaging compounds from a human-health stand-point have an even more recent history. These include pesticides and PAHs, which latter have mainly arisen as a result of the high-temperature combustion of urban waste.

Global levels of radioactivity in the atmosphere peaked in 1963–4 and have declined thereafter as a result of the nuclear-test ban treaty and the radioactive decay of many of the radioisotopes produced. For example, the ^{137}Cs (half life of 30 years) released in the 1960s has now decayed to around 40% of its original activity. Superimposed on this decline, but only in parts of Europe and the former Soviet Union, is the peak in activity arising from the Chernobyl accident. Only in areas close to the latter has there been incontrovertible evidence for damage to living organisms, including human populations.

8.6   Summary and conclusions

The foregoing sections outline some of the main changes in atmospheric composition that have important systemic effects within the Earth system. This account is far from exhaustive as many other inter-related chemical species and reactions have been ignored. For a full account of these, refer to Brasseur *et al.* (2003).

Several essential messages emerge:

- Complex reactions link the chemical species in the atmosphere and give many of them a high degree of interdependence. Accounts dealing with them in succession inevitably fail to capture this fully, but it must always be born in mind when proposals to deal with any single atmospheric constituent are considered.
- Many of these reactions closely link atmospheric chemistry to the climate system at both a regional and a global level. Ignoring or misunderstanding them seriously impairs insight into climate dynamics.
- The production, fluxes, budgets, burdens, breakdown and impacts of many of the atmospheric constituents considered above hinge on exchanges with other components of the Earth system – the oceans and the terrestrial biosphere. They can only be understood and evaluated within the context of the Earth-system as a whole.
- Despite an immense and ever increasing research effort, many uncertainties still surround attempts to quantify the role played by many of the components considered above. Climate and Earth-system models must always seek to convey these uncertainties.
- Atmospheric chemistry is still capable of providing surprises in the form of unanticipated effects and it is unlikely that all the consequences of current trends can be confidently predicted.

Chapter 9
The Anthropocene – changing land

9.1 Changed global nutrient cycles

9.1.1 Nitrogen

We have already seen in the previous chapter that atmospheric concentrations of N_2O have risen over the last two centuries largely as a result of fossil fuel combustion, but this is not the only, or indeed the most significant disruption of the nitrogen cycle as a result of human activities. One of the most remarkable trends during the course of the twentieth century has been the relentless increase in the extent to which anthropogenic processes have begun to dominate the conversion of non-reactive nitrogen to reactive forms, i.e. those that are biologically, photochemically and radiatively active in the biosphere and atmosphere (Galloway, 2004).

Prior to the opening of the twentieth century, reactive nitrogen was produced mainly through nitrogen-fixing organisms – bacteria, both free living and symbiotic, and blue-green algae. Human activities such as the cultivation of rice and leguminous crops contributed around 5% of the total. Anthropogenic contributions since then have increased in several ways. The growing human population has generated an increasing demand for food, leading to higher levels of nitrogen fixation through the cultivation of rice and legumes. By the late twentieth century, annual nitrogen production linked to cultivation was around 33 Terragrams ($1 Tg = 10^{12}$ g). More importantly, the demand for nitrogenous fertilisers quickly outstripped the supplies from guano and nitrate mining, and the development of the Haber–Bosch process led to a massive increase in the quantity of anthropogenically produced reactive nitrogen. It is estimated that the production of nitrogen as an artificial fertiliser had risen to around 78 Tg per year by 1990, and 125 Tg by 2002 (Mosier *et al.*, 2002). In addition, the massive rise in fossil-fuel combustion led to an increase in the output of reactive nitrogen to some 25 Tg. By 1990, these three processes together were generating over 130 Tg of reactive nitrogen annually – over 90% of the total anthropogenic output. By 2002, the total anthropogenic output had reached *c.* 150 Tg. At the same time, natural terrestrial-nitrogen fixation had declined to 85 Tg per year. From around 1970 onwards, anthropogenically produced reactive nitrogen has exceeded that fixed naturally by the terrestrial biosphere.

Of the anthropogenic nitrogen, that generated by fossil-fuel combustion goes into the atmosphere, that produced by nitrogen fixation in rice paddies, pasture and other croplands goes into the food chain, and that used as chemical fertiliser is mainly released into the environment in a variety of ways. Some, in the form of dissolved nitrogen, is carried by rivers to the oceans, where it has promoted extensive eutrophication in the coastal zone. Much is dissipated into the atmosphere and deposited over land and sea. Ultimately, the extra nitrogen load is either denitrified or stored. So far, there is no consensus as to the balance between the two. This impairs any attempt to assess the impacts of future changes in the nitrogen cycle at either regional or global levels.

One major cause for concern about the changed nitrogen cycle arises from the multiple effects that reactive nitrogen can have. Galloway *et al.* (2003) characterise these as a nitrogen cascade and illustrate this by showing how NO released into the atmosphere from fossil-fuel combustion can increase ozone concentrations in the troposphere, reduce atmospheric visibility, increase the concentration of fine particulates in the atmosphere, increase the acidity of precipitation, alter forest productivity, promote surface-water acidification, accelerate eutrophication and eventual hypoxia in the costal zone, increase greenhouse warming and decrease stratospheric ozone. There is also concern about the long-term effects of enhanced nitrogen deposition on forests, since there is growing evidence that damage from nitrogen deposition is beginning to outweigh the positive effects of fertilisation in many regions (Nosengo, 2003). Already, elevated levels of inorganic-nitrogen deposition have been shown to reduce species richness in areas of temperate grassland growing on acid, normally nutrient-poor soils (Stevens *et al.*, 2004).

Set against these deleterious effects are enormous benefits in terms of increased global food production over the last 50 years, part of which can be ascribed to the use of nitrogen-rich fertilisers. Moreover, nitrogen deposition may have contributed to higher levels of storage of CO_2 in forest biomass and soils than would otherwise have been possible.

9.1.2   Phosphorus

Unlike nitrogen, the phosphorus cycle does not have a significant atmospheric component, save through the transport of dust particles. It therefore does not generate the type of complex interactions within the Earth system that have been outlined above. It is, however, the major limiting nutrient in many aquatic environments, both freshwater and marine, and in some terrestrial ecosystems, including mature, undisturbed forests (Wardle *et al.*, 2004). Changes in the flux of biologically available phosphorus, mostly in the form of phosphates, are therefore environmentally and ecologically highly significant. The increase in the flux of phosphates in the environment to a current 12.5 Tg yr^{-1} (Jahnke, 2000) from the natural levels of around 2.2 Tg yr^{-1} (Reeburgh, 1997) is the result of the mining of phosphate and its conversion to fertilisers and detergents. Although the former is used largely for application to farmed land, it, like detergent phosphate, often results in the enrichment of lakes, rivers and the coastal zone (see 10.2.1 and 10.4).

9.2 Deforestation

The clearance of forests is but one aspect of land-use and land-cover change, albeit a dominant one spatially, historically, and from the standpoint of Earth-system functioning and the provision of ecosystem goods and services. It is also often a precursor to land degradation, the theme of the next section.

In all but the most remote, lightly settled or inhospitable environments, the earliest unambiguous evidence for deforestation by human populations lies thousands of years in the past. The earliest sites lie close to the regions of origin for the main cereal crops such as rice, wheat and maize. Rice cultivation certainly started no later than 8500 BP, wheat even earlier (Yasuda, 2002; Wright and Thorpe, 2003). The global trend over the last 300 years has been towards the expansion of pasture and cropland at the expense of forest (Goldewijk and Battjes, 1997; Goldewijk 2003; Ramankutty and Foley, 1999 and Figure 9.1). The estimates, summarised by Lambin *et al.* (2003) suggest a four- to five-fold increase in cropland area, from 300–400 million ha. in AD 1700, to 1500–1800 million ha. in 1990, with a net increase of around 50% during the twentieth century alone. This, coupled with an even more dramatic, though less well quantified increase in the extent of pasture lands, has led to a decrease in the area of forest cover from 5000–6200 million ha., to 4300–5300 million ha. over the same period. Over much of Europe and parts of eastern North America, the period of maximum deforestation now lies in the past, and the current trend is towards reforestation either through commercial and amenity planting, or as a result of the invasion of abandoned farmland. By contrast, in many parts of the tropics, deforestation has accelerated in the second half of the twentieth century and it is this trend that causes most concern at the present day.

Figure 9.1 Changes in land use since AD 1700 as estimated by Goldewijk and Battjes (1997). One of the main questions surrounding such a graph is the extent to which the 'other' category, which reached almost 50% in AD 1700, relates to vegetation cover unmodified by early human activity, or landscapes previously impacted by human activity, but no longer managed. (Modified from Steffen *et al.*, 2004.)

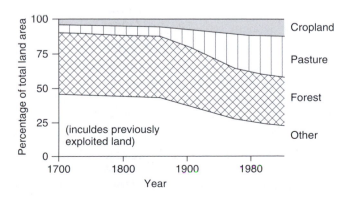

The rate of tropical deforestation since 1980 has proved difficult to assess. The figure derived from aggregated ground-based observations tends to exceed that based on satellite images. Achard *et al.* (2002), for example, estimate that the calculated ground-based deforestation rates for the humid tropics as a whole between 1990 and 1997 were overestimated by some 23%. Similar conclusions have been reached by DeFries *et al.* (2002) for the 1980s. DeFries (2004) summarises the different estimates in terms of mean annual changes from 1980 to 1990 and 1990 to 2000. Despite the differences in absolute estimates, there is agreement that the rates of deforestation have declined in Latin America (at least until 2000) and probably Africa, but satellite-based observations suggest that they have increased in southeast Asia. In both Latin America and southeast Asia, rates of reforestation since 1990 are generally less than 80% of the deforestation rates. Moreover, reforestation does not imply re-establishment of primary forest, but rather the early stage of a secondary succession, highly dependent on factors such as lithology and previous land-cover history, with no guarantee that the original forest will ever be re-established. Earlier studies of tropical-forest ecosystems point to the probability of rapid soil degradation in the wake of deforestation in tropical rainforests, in view of the extent to which the store of nutrients and organic matter is largely retained in the living biomass, rapidly recycled and, with the loss of living biomass, rapidly depleted.

The causes and mechanisms of tropical deforestation have been studied exhaustively by Lambin *et al.* (2001), Geist and Lambin (2002), Lambin and Geist (2003) and Lambin *et al.* (2003). They conclude that there is no single global explanation for tropical deforestation and that simple generalisations linking the process to overpopulation or poverty are misleading. Both the underlying and proximal causes vary greatly from region to region, with the only common factor being interaction between local and wider-scale pressures at regional, national or even global level. In their attempt to synthesise the results of numerous and diverse regional studies, Lambin *et al.* (2003) propose a typology of forces driving land-use and land-cover change and identify a restricted range of dominant pathways linked to fundamental, generic causes such as resource scarcity, changed market opportunities, policy intervention, loss of adaptive capacity leading to increased vulnerability, as well as changes in social organisation, in access to resources, and in attitudes. They also highlight the need to look at the changes, both cultural and biophysical, on a timescale much longer than that spanned by the era of remote sensing.

Studies undertaken within the framework of the LBA (Large-Scale Biosphere-Atmosphere Experiment in Amazonia) are now attempting to quantify the carbon, moisture, energy, trace-gas and nutrient cycles at many sites throughout the region (Nobre *et al.*, 2001). The main increase in deforestation rates in Amazonia has occurred since 1970 and by now, in the Brazilian part of Amazonia, around 15% of the forest has been cut down. In this same region, the annual rate of deforestation has been estimated at 15 000–20 000 km^2 (Nobre,

2004). Alongside the changes in nutrient cycling, deforestation also brings about significant changes in the exchange of energy, water vapour, greenhouse gases and volatile organic compounds (VOCs) between the biosphere and the atmosphere. Whereas the undisturbed forests in Amazonia appear to be a significant carbon sink, fragmentation and clearance through deforestation and burning lead to significant greenhouse-gas emissions. Moreover, logging increases fire risk. The VOCs released into the atmosphere by the Amazonian rainforests act as condensation nuclei, so changes in the VOC flux arising from land-cover changes have additional impacts on cloud formation and rainfall. Many other studies strengthen the view that in the tropics recent land-cover change and associated changes in soil characteristics are having widespread effects on climate and hydrology (see, e.g., Osborne *et al.*, 2004; Voldoire and Royer, 2004), as well as global atmospheric CO_2 (Achard *et al.*, 2004). Land and atmosphere are linked in ways that transcend the organisation of themes adopted here.

It is debatable whether much of the current research on tropical deforestation is set within a sufficiently extended time perspective to characterise fully the longer-term impacts. For example, Fu (2003), by comparing actual vegetation as revealed by satellite images, with model-simulated potential vegetation, claims that human activities in east Asia over the last 3000 years have resulted in the transformation of the land cover of more than 60% of the region, with consequent changes in albedo, surface roughness, leaf-area index and vegetation cover. Model simulations suggest that these changes may have weakened the summer monsoon and enhanced the winter monsoon, with, as one possible consequence, the trend towards aridification in parts of the region, notably northern China. Similar considerations for other parts of the world should be born in mind when interpreting Figure 9.1 from Goldewijk and Battjes (1997). The category 'other' that ascribes almost 50% of the land surface to non-domesticated land could include many areas previously deforested or degraded by human activities now no longer current – a key issue in relation to Ruddiman's (2003b) hypothesis (see 7.12).

9.3   Land degradation and desertification

Two processes are responsible for most land degradation – deforestation to create cropland in humid regions, and overgrazing of rangeland in drylands (Glenn *et al.*, 1998). Rangelands in semi-arid areas are especially vulnerable, with around 73% being degraded at the present day. Almost half the un-irrigated croplands in areas of marginal rainfall are degraded, mainly as a result of soil erosion. Around 30% of irrigated cropland is also being degraded, largely through salinisation – deposition at the soil surface as salts are drawn upwards by capillary action, driven by high rates of evaporation and waterlogging (Kassas, 1995). Land-degradation problems are particularly extreme in sub-Saharan Africa where 20–50% of the land and around 200 million people are affected (Nachtergaele, 2002), with some 135 million people deemed vulnerable to the potential collapse of their traditional land-use systems. The driving forces of land degradation are both socio-economic and biophysical, with authorities differing markedly in the extent to which they invoke one or the other as the dominant factor in any given region. One way in which the interactions between climate variability and societal responses has been studied is through the concept of 'syndromes'. In this case, what has been termed the Sahel syndrome reflects the interaction of biophysical and societal processes that typifies this particular type of degraded land system (Petschel-Held, 2001).

Land degradation generally implies deterioration in key properties such as soil nutrients, organic matter and moisture status, to the point where productivity is reduced and, over wide areas, recovery on a timescale of years or even decades is unlikely. There are huge areas where the combination of marginality, abusive exploitation and environmental variability has led to the high levels of degradation noted above. Erosion of topsoil removes the nutrients stored in the surface layers of the soil, especially those retained within the organic matter generated by the decomposition of the vegetation. This, in turn, reduces moisture-holding capacity. Any form of exploitation that leads to depletion of the soil nutrients at a rate exceeding the rate of their renewal eventually leads to soil degradation as the ecosystem shifts to one based on a lower level of nutrient recycling and availability.

9.4   The long-term palaeo-perspective

Although the most dramatic instances of land degradation may have come to light only in recent years, there is clearly a need for a long-term perspective since any 'cycle' of degradation and subsequent recovery, where this is possible, involves a high level of hysteresis, with the timescales of regeneration often orders of magnitude longer than the timescales of degradation. One of the processes involved in land degradation – erosion from the land surface – has been reconstructed from many sites and on a range of timescales. In Figures 9.2 and 9.3 changes in sediment yields rather than on-site erosion measurements are plotted. This is because long-term records of the latter are generally lacking. It is important to realise that in any given catchment sediment-yield records provide a minimum figure for erosion, and one that is subject to lags, the importance of which tends to increase with the size of the catchment and the degree of sediment storage taking place within it. Figure 9.2 records several estimates of changing sediment yield over the last 250 years. Figure 9.3 considers rather more diverse evidence for changing sediment yields spanning all or most of the Holocene.

Case studies taking the long-term view may help to guide us towards answers to some vital questions: which ecosystems have been irreparably damaged, when

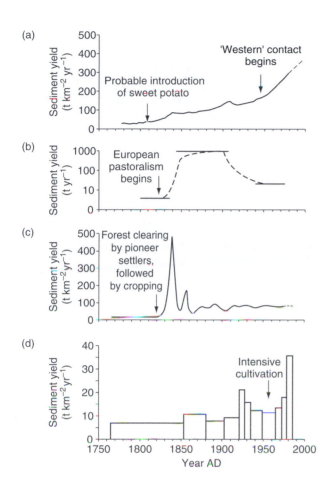

Figure 9.2 Changing sediment yields over the last 250 years in small lake and stream catchments: (a) Lake Egari, Papua New Guinea. (b) Southeastern Australia. (c) Frain's Lake, Michigan, USA. (d) Seeswood Pool, English midlands. Note how the initial responses and subsequent trajectories vary between catchments. (Modified from Oldfield and Dearing, 2003.)

and how? What were the degradation trajectories? How have human impacts interacted with climate variability and ecosystem processes to generate degradation trajectories of different kinds? What were the pre-intervention characteristics and histories of areas that have become entirely dependent for sustained productivity on major inputs of finite, or increasingly scarce resources such as artificial fertilisers, or irrigation water? Which types of ecosystem can be either restored or acceptably transformed and maintained with sustainable levels of material and energy inputs? And arising from the last question – where 'restoration' or 'conservation' are the main goals, what state is to be restored, or conserved, and what are the key processes that should ensure this in the long term? Lack of convincing answers to most of these questions highlights the gap between 'contemporary' and 'historical' studies in an area of global change where the unresolved questions are of increasing importance.

Of particular interest in this context are those cases where the combination of environmental and societal processes has led to major changes in ecosystem

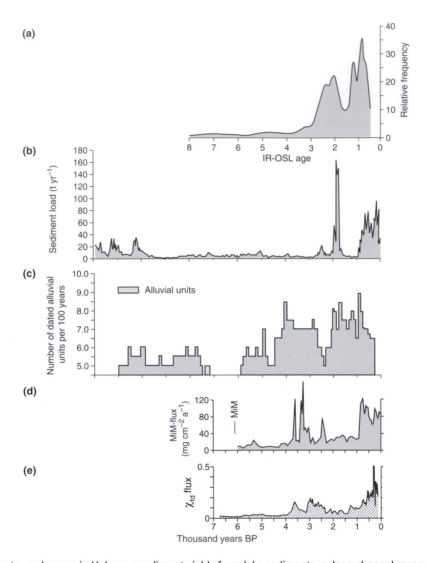

Figure 9.3 Long-term changes in Holocene sediment yields from lake sediments, palaeo-channel measurements and alluvial sequences: (a) Frequency of optically stimulated luminescene (OSL)-dated soil-erosion-derived sediments from sites in the loess hills of southern Germany (Lang, 2003). (b) Lake-sediment accumulation rates at Holzmaar, western Germany (Zolitschka, 1998). (c) Frequency of dated alluvial units in British rivers (Macklin, 1999). (d) Changes in the depositional flux of inorganic-mineral matter at Lago di Mezzano, central Italy (Ramrath *et al.*, 2000). (e) Depositional flux of magnetic minerals derived from surface-soil erosion into the mid Adriatic core RF 93–30 (Oldfield *et al.*, 2003a). (f) Alluvial-accumulation rates in the Yellow River Basin, China (Xu, 1999). (g) Flood reconstructions derived from palaeo-channel cross-sections in southwest Wisconsin, USA (Knox, 2000). Note the strong parallels between (a) and (b), and between (d) and (e). In each case the curves are based on quite independent lines of evidence. Also, whereas (b) and (d) refer to single sites with small drainage basins, (a) and (e) integrate evidence from a much larger region.

The balance between forcing by climate and human activities varies through time and between regions. Most of the post-4000 BP increases in the European sequences correlate with episodes of greater human impact. The sequence from China moves from base-line rates in the early Holocene, through climatically induced increases, to anthropogenically driven changes during the last 2000 to 3000 years. The North American example is dominated by climatic forcing throughout. (Based on Oldfield and Dearing, 2003.)

(f)

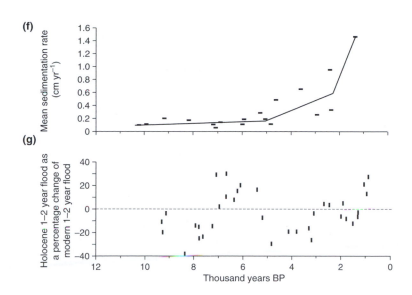

Figure 9.3 (Cont.)

functions, surface processes, resources and social organisation – even, in some cases, to the 'collapse' of previously successful cultures. This whole field is bedevilled by conflicting perspectives. Many social scientists reject or downplay the role climate variability and especially the occurrence of severe drought may have played in triggering social change (e.g., Redman, 1999). By contrast, some palaeo-scientists may have oversimplified the links between environmental change and human catastrophe, leaving their analyses open to accusations, whether justifiable or not, of old-fashioned environmental determinism (deMenocal, 2001; Hodell *et al.*, 1995; Weiss *et al.* 1993; Weiss, 1997; Weiss and Bradley, 2001). All too few studies present a balanced view in which the interactions between social and environmental processes are carefully explored. Shennan (2003) makes the crucial point that the impact of climatic change on human societies 'has to be conceived in terms of perceived costs and benefits in the context of relevant constraints'. Until more research emerges that respects and integrates both biophysical and cultural perspectives, many of the classic examples of the 'collapse of civilisations' will remain unnecessarily contentious (Oldfield and Dearing, 2003; Oldfield, in press and section 14.7).

One example where the evidence for a dramatic and dominantly human-induced decline in population and economy emerges from all the available evidence is that of Easter Island. Bahn and Flenley's (1992) account opens up the possibility that the island and its history may serve as a microcosm of the increasingly over-exploited and vulnerable Earth system as a whole. If so, the story carries with it a dire warning. Their cautious summary of the evidence concludes that forest clearance led to its eventual total removal, along with a resulting decline in soil fertility and increased erosion. This occurred alongside the depletion and disappearance of other key resources. In combination, the

Figure 9.4 A conceptual model of the sequence of interactions between human immigration and subsistence systems that could have led to population decline on Easter Island. The shaded boxes identify those processes and interactions that are less specific to the particular Easter Island environment. (Modified from Bahn and Flenley, 1992.)

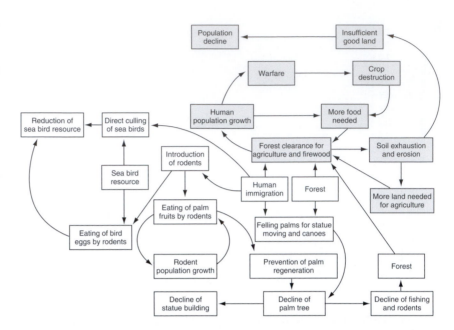

processes led to irreversible environmental degradation, which brought about the rapid collapse of what had previously been a remarkably stable and successful culture. Figure 9.4 is a summary of the hypothesised links they draw between the several interacting processes that eventually led to cultural collapse. Although the contrast in scale between a tiny, isolated island and the Earth system totally confounds any attempt to see Easter Island as a microcosm for the future fate of the world, the authors identify some disturbing parallels linked to the human drive for growth, competitive advantage and short-term gain.

Chapter 10
The Anthropocene: changing aquatic environments and ecosystems

10.1 Introduction

Water in the form of lakes, rivers or the ocean, and the sediments that accumulate below water, are inevitably impacted, either directly or indirectly, by many of the changes that have affected the land and the atmosphere. In some cases, water simply provides one of the major links in biogeochemical cycling. Often, as in the case of carbon, it provides, along with accumulating sediments, one of the major sinks. Irrespective of the role it fulfils with respect to any given process within the Earth system, its quality, distribution and availability are altered by many of the changes already described in the two previous chapters. Since water is one of the essentials for the maintenance of life on Earth, as well as a vital ingredient in a vast range of industrial and domestic processes, the changes to lakes, rivers and the oceans brought about by human activity are a key part of global change. Anthropogenic transformations of both quantitative and qualitative aspects of the hydrological cycle are of vital concern.

10.2 Changes in lake-water chemistry

10.2.1 Cultural eutrophication

Eutrophication can be the outcome of natural processes and many lakes are naturally eutrophic. *Cultural* eutrophication is the chemical enrichment of the water body through human impacts on nutrient supplies. It generally leads to increased productivity, hence higher quantities of organic carbon, which can result in major changes in the structure and functioning of the lake ecosystem, especially where the decomposition of the additional carbon involves high levels of oxygen demand by decomposer organisms. Where water bodies are well mixed, oxygen may be renewed through contact with the atmosphere, but where the water column is strongly stratified, as a result of thermal or chemical gradients, the oxygen used in respiration by the decomposer organisms will not be readily replaced for as long as stratification persists. The end result may be anoxia in all or part of the water column below the level to which exchange with the atmosphere is possible. This may be a seasonal phenomenon, as is often the case in lakes where high spring and summer biological productivity coincides with well-developed stratification linked to a strong thermal gradient (thermocline) in the water. In the most extreme cases, bottom-water anoxia may persist throughout the year.

The key chemical nutrients responsible for eutrophication are phosphorus and nitrogen. Since phosphorus is often the main limiting nutrient in aquatic ecosystems, and nitrogen can be fixed by several commonly occurring aquatic organisms, enrichment by bio-available phosphates has proved to be the key factor in many cases. Piecing together the origin and history of eutrophication involved monitoring the annual nutrient, productivity and redox cycles in the affected lakes, as well as reconstructing the onset and development of eutrophication using chemical and biological tracers in the sediment record.

Although several studies have shown that small lakes close to prehistoric settlements became temporarily and mildly eutrophic during early periods of nearby habitation and agriculture (e.g., Renberg, 1990; Gaillard et al., 1991), the quasi-world-wide trend to eutrophication is a much more recent phenomenon. There are many examples of lakes in western Europe where the recent history of cultural eutrophication has been well documented. One of the earliest studies was on Lough Neagh in Northern Ireland. Despite its shallow and generally well-mixed nature, algal blooms became a serious problem in 1964, with damaging consequences for water supplies, amenity and sewage disposal. By analysing the changing frequencies of diatoms (Battarbee, 1978) preserved in dated sediments, it was possible to show that the trend to highly eutrophic conditions had begun in the mid nineteenth century and accelerated from the 1950s onwards. Similar historical records of eutrophication have been reconstructed for other lakes in Europe and North America. From these, it has been possible to establish the causes and trajectories of eutrophication. Urbanisation in the nineteenth and twentieth centuries increased pressure on waste-disposal systems. This, combined with the introduction of integrated sewerage networks and the growth of industries discharging organic waste led to increased nutrient inputs to many lakes, especially in the industrialised world. The increased use of phosphate-rich detergents during the 1950s and 1960s created a further major boost to nutrient input. The combination of these two processes made urban areas increasingly important point sources of chemical 'pollution' through enrichment. Only with the introduction of phosphate-free detergents and tertiary sewage treatment plants was the trend reversed. Non-point sources also increased greatly over the same period as livestock densities increased and as the addition of artificial fertilisers to catchment soils became more and more prevalent, often in injudicious quantities, applied at times of the year when runoff into lakes was more likely than retention in the soil. Billen and Garnier (1999) estimate that in the most highly fertilised catchments, dissolved inorganic-nitrogen fluxes have increased ten-fold over the last century.

At the present day, eutrophication linked to high-population densities, poverty and lack of infrastructure, seriously affects water resources in many developing countries where the cost of mitigation lies beyond the reach of ever-growing populations. There are serious health implications, as the same water is often used for waste disposal, bathing and drinking.

10.2.2 Surface-water acidification

The likelihood that industrial processes such as coal- and oil-based power generation might be seriously damaging aquatic ecosystems came to public attention during the 1980s. In both Europe and North America, declining pH in lakes over wide areas prompted several ingenious and well-coordinated studies, notably the surface-water acidification project, SWAP, (Battarbee *et al.*, 1990), and palaeoecological investigations of recent lake acidification PIRLA studies (Charles *et al.*, 1990). They provide a fascinating insight into the role that 'palaeo' research can play in addressing contemporary issues and in *post hoc* hypothesis testing. They also confirm the value of long-term perspectives in evaluating environmental processes that began before systematic, direct observations were initiated.

Realisation that the pH of many freshwater bodies in Scandinavia and western Britain, as well as in northeastern North America, had declined during the decades leading up to the 1980s gave rise to serious concerns at national and international levels. Several possible causes were proposed. One was that the changes were the late stages of a natural process of acidification reflecting the evolution of water bodies surrounded by soils from which nutrients had been gradually leached on the timescale of the whole Holocene. An alternative explanation was that acidification had arisen as a result of the commercial afforestation of catchments using tree species the leaf litter of which tends to accelerate soil acidification. It was also suggested that in some parts of Scandinavia especially, acidification could have arisen through the abandonment of upland farming. All these hypotheses were proposed as alternatives to the explanation favoured by environmentalists, namely that the acidification had arisen as a result of industrialisation and especially fossil-fuel based power generation.

The most successful approach to resolving the problem was essentially one of *post hoc* hypothesis testing. The pathways of acidification in paired lakes, respectively with and without agricultural land-use changes, or commercial afforestation in their catchments, were compared. Past lake pH was reconstructed using carefully calibrated records (cf. 10.2.2) of changing lake biota, mainly diatoms, preserved in the recent sediments of each lake. The changes were dated using the radioactive decay of lead-210 (see 3.4.7 and Figure 3.7). The evidence for progressive acidification inferred from the diatom record was compared with the history of accumulation of chemical contaminants and particulate matter directly

Figure 10.1 Records of changing surface water pH over different timescales: (a) and (b) Daily and weekly (respectively) measurements of water pH at Loch Dee, Galloway, southwest Scotland. (c) Diatom-based reconstruction of lake-water pH changes at Loch Dee since the early nineteenth century, from sediment analyses, dated by ^{210}Pb. (d) Results of detailed diatom analyses from Lilla Oresjøn, southern Sweden. The evidence from the sediments of Loch Dee shows how sediment-based reconstructions can reveal long-term trends, despite the high amplitude of short-term variability. The Lilla Oresjøn record confirms that the steep decline in lake-water pH that occurred there during the twentieth century was unprecedented throughout the whole of the Holocene. Loch Dee data from Battarbee (1998); Lilla Oresjøn data from Renberg *et al.* (1990). (Modified from Oldfield and Dearing, 2003.)

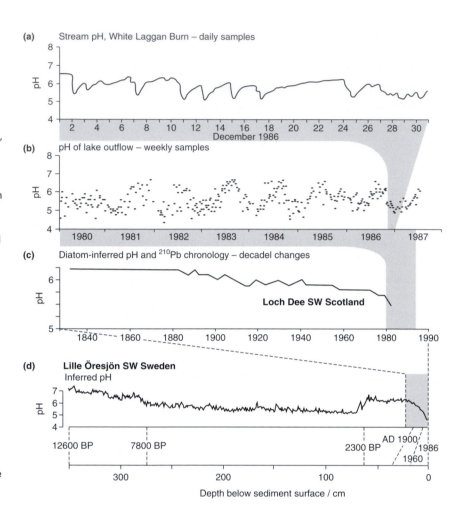

associated with fossil-fuel combustion. The results showed that acidification occurred irrespective of land-use change, or conifer plantation. Comparison between the decline in diatom-inferred pH and the deposition of heavy metals and combustion-associated particulates confirmed consistent parallels in virtually every acidified lake studied on both sides of the Atlantic.

The compelling nature of all the lake-sediment based evidence was crucial in highlighting the need to develop technology for reducing sulphur emissions from power stations using fossil fuel. Only by tracing the history of acidification in the sedimentary record was it possible to separate the nineteenth and twentieth century trend from the daily, weekly and annual 'noise' arising from individual weather events and the changing seasons. Equally, only the lake-sediment record provided evidence to show that the changes linked to industrial processes were unique in the context of the whole Holocene record (Figure 10.1). By now, the same kind of approach can be used to track and evaluate the progress made in reversing previous acidification through reducing emissions (Smol *et al.* (1998). As with so many other environmental processes, acidification does not act in isolation. By promoting declines in dissolved organic carbon in unproductive lake waters, acidification, which may actually be exacerbated by climate warming (Schindler, 2001), increases the vulnerability of aquatic organisms to damage by increased exposure to ultraviolet (uv) radiation (Schindler and Curtis, 1997).

The extent to which acid precipitation damages terrestrial as well as aquatic ecosystems depends very much on the buffering capacity of the soils and water bodies affected. On a global scale, the regions with the lowest pH of precipitation are in the United States, Europe and China. Where this is not neutralised by cation-rich dust deposition, acidification remains a threat to both aquatic and terrestrial ecosystems. Rodhe *et al.* (2002) also take into account the potential interactions between acid deposition and soil-nitrogen saturation, and point out that this combination is also likely to leave ecosystems vulnerable to future acidification in southern, southeastern and eastern Asia, as well as parts of central South America. Lapenis *et al.* (2004) confirm that forest soils in Russia have experienced acidification over the last century, especially in areas with the highest atmospheric deposition.

10.3 Other hydrological changes – rivers and groundwater

During the period from 1940 to 1990, there was a more than four-fold increase in water abstraction for human use. By now, some 40% of total global run-off to the oceans is intercepted by large dams (Vörösmarty *et al.*, 2003). The major changes in flow regime that have resulted from hydrological engineering are a consequence of the following activities (Meybeck and Vörösmarty, 2004):

- Damming for power and/or irrigation, (e.g., the Aswan Dam and Lake Nasser, on the River Nile).
- High levels of water extraction, mainly for irrigation, as in the case of the Syr-Darya River, one of the inputs to the now seriously reduced Aral Sea.
- The redirection of flows as a result of interbasin transfers such as those designed to optimise hydroelectricity production by rivers flowing into Hudson Bay.

The downstream impacts of these activities include total disruption of the flow regime, loss of silt and nutrient supply, habitat disruption and salinisation. One of several extreme examples of these effects is the Aral Sea which has shrunk to less than half its size since the 1960s largely as a result of upstream irrigation. It is also strongly affected by salinisation and pollution, including heavy pesticide enrichment (Löffler, 2004).

One additional consequence of artificial impoundments has been a 700% increase in the volume of standing water in river channels worldwide (Vörösmarty *et al.*, 1997), which in turn has increased the global mean-residence time of water within river systems from around two weeks to four months.

The changes are even more severe in several large river basins such as the Nile, Colorado and Rio Grande (Vörösmarty and Sahagian, 2000). A further consequence has been a huge increase in the volume of sediment retained in river systems, to the point where, as estimated by Vörösmarty *et al.* (2003) almost 30% of the global sediment flux from the continents is now retained in the world's 45 000 largest reservoirs. Small, artificial water bodies, of which there are an estimated 2.6 million in the USA alone, may have, proportionally, an even greater impact especially on sediment retention (Smith *et al.*, 2002). Along with these essentially quantitative changes have come massive changes in river-water chemistry. Once again this can be illustrated by the case of nitrogen. Meybeck and Ragu (1997) estimate that dissolved inorganic-nitrogen concentrations in the world's rivers has increased from 0.13 to 0.33 mg l^{-1} since the turn of the twentieth century.

One dramatic regional effect of human activities on hydrology has been the result of land reclamation in the middle Yangtze Plain. Geographical Information System-based surveys coupled with longer-term palaeolimnological and geomorphological studies in and around Hong Hu, one of the major lakes (352 km^2) that stores flood waters from the Yangtze during high flow, has quantified the loss of storage capacity between 1953 and 1976 as a result of the reclamation of open water and lake-marginal land for some form of cultivation. Between 30 and 60% of the flood-storage capacity of the lake was lost over this period (Oldfield, unpublished data, 1998; Yu, personal communication, 2000 – Figure 10.2). The surrounding region was severely affected by flooding in 1991 and 1998. Loss of storage capacity was a major contributing factor.

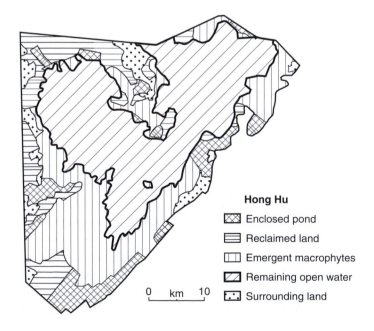

Figure 10.2 The extent of land reclamation around Hong Hu, a large shallow lake in the middle Yangtze Plain between 1953 and 1976, (Yu, personal communication, 2000). The decline in surface area led to a serious reduction in storage capacity during Yangtze floods, with serious consequences in 1991 and 1998.

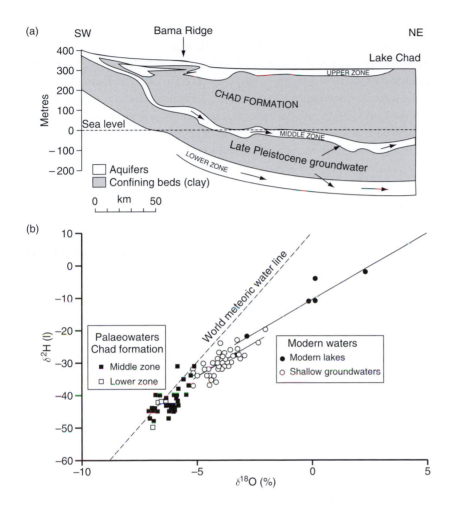

Figure 10.3 Mining of palaeogroundwaters in the Chad Basin, Nigeria (Edmunds *et al.*, 1999). The upper diagram (a) shows a cross section through the aquifers and intervening clay beds. The lower plot shows the differences in the $\delta^{18}O$ to δ^2H (deuterium) ratio in the different zones of groundwater, relative to that in modern lakes in the region. Only the recent waters diverge from the world meteoric water line through evaporative enrichment. Over much of the region, water dated to between 18 000 and 25 000 BP is being extracted, with only local, limited recharge. (Modified from Steffen *et al.*, 2004.)

Raising the banks of the Yangtze, which in this part of its course flows in a channel often above the surrounding land, protects the region against moderate floods, but greatly increases the hazard when extreme events threaten to overtop the embankments (Zong and Chen, 2000).

Of major concern in some parts of the world is the depletion and contamination of groundwater resources. Depletion arises when extraction from a given aquifer exceeds the current rate of recharge. This can result in land subsidence, salt-water intrusion in coastal regions and an increasingly costly and ever deeper search for water, with the likelihood of eventual depletion beyond the point at which extraction becomes feasible. In many sub-arid tropical regions, the groundwater currently being extracted is the result of more favourable recharge conditions during the last glacial maximum (LGM) and early Holocene. The water being extracted is therefore many thousands of years old (Edmunds *et al.*, 1999). Figure 10.3 illustrates this with respect to the Chad Basin in northern

Nigeria. In the most extreme cases, groundwater extraction under these conditions constitutes the mining of a non-renewable resource.

There are all too many instances of falling water tables. The effect is to diminish the capacity of a region to respond to future drought. Examples range from the impoverished region of Gujarat in northwest India to the Great Plains area of the United States, where there is over-extraction from the huge Ogallala aquifer (Woods *et al.*, 2000). Especially in developing countries, groundwater contamination is also an increasingly serious problem, going hand in hand with the contamination of surface waters and falling water tables.

10.4   Coastal and marine impacts

Many of the processes described above in this and the previous chapter have important implications for coastal and marine ecosystems. One consequence of the changes to the nitrogen and phosphorus cycles has been a massive increase in the flux of these nutrients to many coastal regions and near-shore marine environments. This process has been reinforced by the massive increase in the number of people living in coastal regions and generating organic waste, much of it for ultimate disposal in the sea. As a result of these trends, there are now many examples of marine eutrophication. Elsewhere, damming and disruption of river flows have cut off the sediment supply to some coastal regions, with serious consequences for coastal protection and marine habitats.

Marine eutrophication has been detected in a wide range of near-shore environments. A remarkable pre-Anthropocene example of the early effects of terrigenous inputs to a marine environment comes from the central Adriatic, close to regions of strong human impact from Bronze-age times onwards (Oldfield *et al.*, 2003). Pollen analyses point to the beginning of a period of extensive deforestation and cultivation from around 4000 BP onwards which led to an increase in the flux of soil-derived sediment at the site. This was accompanied by changes in the foraminifera present in the sediments. These changes were most probably a response to a more stressed and oxygen-depleted benthic environment as a result of the increased deposition of eroded, soil-derived particulates. Even more dramatic changes occur in sediments dating from *c*. 700 BP onwards during the medieval period. In addition to increased sedimentation, the sediments also record an increase in organic carbon content resulting from higher marine productivity (Asioli, 1996). Further changes in foraminera assemblages occurred, suggesting more severe oxygen depletion in the benthic environment. Clearly, in the seas adjacent to long-settled areas, human impact on benthic ecosystems has a long history. In most regions, though, discernable impacts begin during the nineteenth or twentieth centuries.

Andren *et al.* (1999 and 2000) in their studies of the impact of climate change and human activities on marine ecosystems in the southern Baltic, interpret a shift in the balance of diatom productivity from benthic to planktonic communities, detected in the fossil record from recent sediments as a response to reduced light penetration as a result of cultural eutrophication. In cores from the area closest to the densely populated regions around the southern shores of the Baltic, the change begins in the mid nineteenth century. The timing parallels the record of cultural eutrophication in many north European lakes. In the Gotland Basin to the north, the first clear evidence of an ecosystem response to increased nutrient supply is in sediments dated to the mid twentieth century. This coincides with the rapid increase in the use of artificial fertilisers and phosphate-rich detergents.

Similar evidence spanning the same period comes from Chesapeake Bay, where increases in organic-carbon inputs and sedimentation rates, as well as changes in benthic foraminifera assemblages, diatom-species distributions, dinoflagellate assemblages (Willard *et al.*, 2003) and sediment chemistry point to a suite of changes in the Bay synchronous with and resulting from the sequence of land clearance and erosion from the seventeenth century onwards (Karlsen *et al.*, 2004). Bottom-water anoxia began later, during the twentieth century, accompanied by increased turbidity, mainly as a result of increased nutrient and sediment inputs (Cooper and Brush, 1991; 1993; Cooper, 1995, Adelson and Helz, 2001).

One of the most dramatic examples of marine eutrophication is of even more recent origin. Anoxia has recently been observed on a much wider scale in the northern Gulf of Mexico (Sanderson, 2004). Here, the oxygen demand placed on the system by the decomposition of algal blooms has given rise to a 'dead zone', the size of which doubled from around $6400 \, km^2$ to over $13000 \, km^2$ between 1993 and 1999. The 'dead zone' is characterised by episodes of die-off and out-migration of marine organisms, as well as persistent reductions in productivity and in the abundance of a wide range of species. The strong thermal and chemical stratification of the waters within the northern part of the gulf during summer months is an important factor in the process of oxygen depletion. Warm, fresh water caps underlying salty water and inhibits vertical mixing. This isolates the bottom water for long periods of the year and creates the perfect environment for severe oxygen depletion. This would not occur, however, without the development of massive algal blooms. These can be ascribed to the input of nitrates from the drainage basin of the Mississippi and Atchafalaya rivers. High levels of nutrient enrichment began in the 1960s and have led to increasingly severe eutrophication from that time onwards (Goolsby, 2000), though the trend to increased-nitrogen flux probably began some 150 years ago (CENR, 2000). Since then, nitrogen export has increased by between 2.5 and 7.5 times (Howarth *et al.*, 1996), two thirds of the increase being ascribed to the application of nitrogen fertilisers to the farmlands of the mid west of the United States. Although most authors have emphasised the role of nitrogen in generating the dead zone, concomitant increases in phosphate loading may also have been critically important.

All the above examples point to a strong link between human-induced changes on land and changes in coastal ecosystems. Increased rates of erosion alone have been of major significance, since many of the reactive chemical compounds carried by rivers to the coastal zone are quickly adsorbed onto fine particles and hence transported as part of the particulate load within the water. Although the earliest evidence comes from prehistory, the processes involved have, like all the others considered in this and the preceding two chapters, accelerated and become ever more widespread during the Anthropocene.

In addition to the increased input of nutrients to the coastal zone have come many by-products of industrial processes, including heavy metals, persistent organic pollutants, radioactive materials and effectively non-degradable materials like many plastics. Along coastlines, human impacts have included almost ubiquitous construction wherever coastal populations have required protection from the sea, can wrest land from the sea, need to modify the shoreline to exploit its resources, or have decided to develop port and industrial facilities. One rapidly expanding impact in tropical regions has been the conversion of mangrove forests to prawn and shrimp farms. This and other forms of economic exploitation have reduced the area of mangroves by around 50% (World Resources Institute, 2000). Jackson *et al.* (2001) document other, less direct impacts on marine ecosystems. Both in the Gulf of Maine and along the west coast from Alaska to California, exploitation of marine-mammal and fish populations has, by reducing the predation pressure on sea urchins, led to an expansion of their population to the point where increased grazing has led to a significant decline in kelp populations. In Chesapeake Bay, the ecosystem response to eutrophication already referred to above may have been reduced had there not been over-exploitation of the coastal oyster beds. Oysters, by filter feeding on phytoplankton, remove large quantities or organic matter that, in their absence, becomes subject to bacterial degradation, with a consequent increase in the rate of oxygen depletion.

The impact of human activities on marine ecosystems stretches well beyond the coastal zone, largely as a repppsult of the ever-increasing demand for animal protein in the form of fish. The Food and Agriculture Organization (2000) estimates that 47–50% of the main marine-fish stocks for which information is available are fully exploited, 15–18% over-exploited and 9–10% depleted or recovering from depletion. The impact on fish populations extends beyond the target species since the same source quotes a figure of 25% for the percentage of the annual marine-fisheries production discarded as 'bycatch'. Most of the commercial fisheries are focused on continental shelf and upwelling areas. In some key areas of heavy demand and exploitation the combined effects of over-exploitation and climate-induced changes in food-webs and species distributions threaten the long-term viability of traditional fisheries.

The last example raises again the difficulties that arise when trying to disentangle the impacts of human activities and climate change. What is clear is that there are well-documented instances in marine as well as terrestrial environments where their effects are mutually reinforcing. The increasing range of stresses on coral reefs is a case in point. In addition to the climate-linked stresses, increased nutrient and sediment loading, intensive fishing and the rapid growth of tourism in reef areas are having cumulative effects in many regions. Changing sea-surface temperatures and carbonate chemistry along with a range of human activities combine to change disturbance regimes and reduce the ability of coral reefs to withstand perturbations without experiencing major shifts in ecosystem structure and function (see 15.2.5).

Chapter 11
Changing biodiversity

11.1 Extinctions

Some 1.7 million species have been identified, but these probably comprise less than 15% of the total number of species thought to exist on Earth (Hammond, 1995). Of the major taxonomic groups, only in the case of plants and vertebrates have more than 80% of all the species been described and it is only for these groups that any direct assessment of rates of extinction can be calculated on a percentage basis. One of the few comparisons between regional extinction rates in birds, plants and insects – in this case, butterflies in Britain – suggests that extinction rates calculated for birds and plants may well be equally indicative of trends in insect populations (Thomas *et al.*, 2004). Although the comparisons made are for a small area in global terms, and the populations and drivers of change involved may not be representative of those in the wider world, the results still lend some support to the view that groups of organisms for which data are sparse may be just as threatened by extinction as those studied more comprehensively.

Figure 11.1 shows estimates of the percentage of species of birds, mammals, fish and plants regarded as currently under threat of extinction (Pimm *et al.*, 1995). McCann (2000) states that one third of the plant and animal species in the United States are at risk of extinction. Current estimates suggest that rates of extinction in vertebrates and vascular plants have already increased by between 50 and 100 times as a result of human activities (Pimm *et al.*, 1995; Lawton and May, 1995). The increase in the absolute rate of species loss due to human activity may be orders of magnitude greater in tropical rainforests (Wilson, 1988; Vitousek *et al.*, 1997). Some authorities see this as evidence for regarding the currents trends as part of the Earth's sixth major extinction event, especially when the effects of human activities are combined with those of projected climate change (Thomas *et al.*, 2004).

Globally, the existing large-scale variations in biodiversity can be linked statistically to relatively few environmental variables that include latitude and available energy (Gaston, 2000). Refining the perspective, we see that high biodiversity is concentrated in specific types of ecosystems and locations. Around 44% of the known biodiversity of plants and 35% of all non-fish vertebrates are endemic to 25 biodiversity 'hotspots' covering no more than

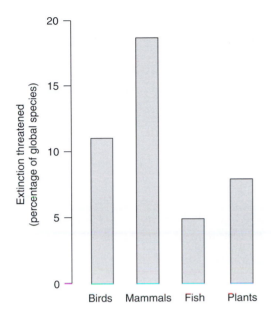

Figure 11.1 Percentage of bird, mammal, fish and plant species currently threatened by global extinction. (Modified from Pimm *et al.*, 1995.)

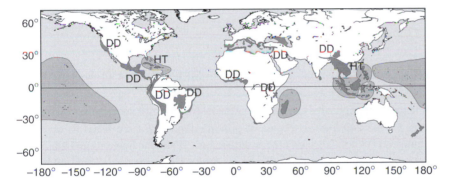

Figure 11.2 Annotated map of biodiversity hotspots (Myers *et al.*, 2000). DD denotes hotspots known to have experienced prolonged and severe droughts in the past; HT denotes hotspots prone to shifts in tropical storm regimes. (Modified from Overpeck *et al.*, 2003.)

12% of the Earth's surface (Kitching, 2000; Myers *et al.*, 2000; Figure 11.2). Virtually all of these areas are seriously threatened through forest clearance and other types of land-cover change. These are indeed the main drivers of biodiversity loss, especially in lower latitudes, though the growing impact of climate change on high-latitude ecosystems (see 12.5) should also be borne in mind as an increasingly important factor. Some of the consequences for global change of the resulting loss of biodiversity may be inferred from Figure 11.3.

Loss of biodiversity through the extinction of species and landscape change is a complex problem with many ramifications. The Swiss Biodiversity Forum (2003) identifies a range of values associated with biodiversity. These include ethical (right to exist), aesthetic (beauty), cultural (systems of high-conservation value created and maintained by human activities), socio-economic (health products for example) and ecological (ecosystem functioning and sustainability) values. Within

Figure 11.3 The role of
biodiversity within the
wider context of global
change. (Modified from
Chapin *et al.*, 2000a.)

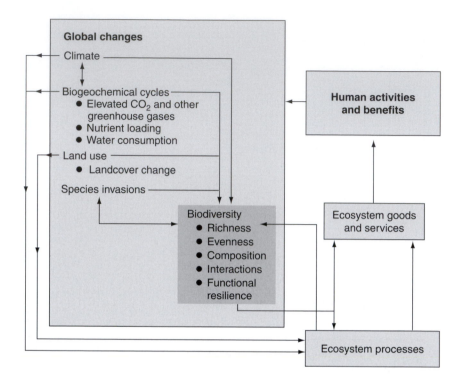

the context of this book, the last will receive most attention, but all should be borne in mind as they enrich human life in a wide variety of ways. The ethical issue strikes home most forcibly in the case of primates close to our own species in evolutionary terms. Of the 240 known species of primate almost half are listed as critically endangered (19), endangered (46) or vulnerable (51) by the International Union for the Conservation of Nature (Steffen *et al.*, 2004). The aesthetic and cultural values are extremely difficult to quantify but easily appreciated and never undervalued by anyone whose life has been enhanced by the beauty and challenge of wild environments, the harmony of many landscapes created and maintained over the long term as sustainable embodiments of human effort, the order and symmetry of plants at all scales from the tiniest flower to the living architecture of great forest trees, and the wonderful and diverse functionality of animal movement. The socio-economic value extends beyond the familiar link between many plant products and both medicines and cosmetics. It includes the need to preserve the genetic diversity that allows breeding for disease resistance in important commercial and subsistence crops (Chapin *et al.*, 2000a; 2000b).

11.2   Species diversity

Although it is becoming increasingly clear that declining biodiversity is an important feature of aquatic ecosystems, both freshwater (e.g., Revenga *et al.*,

2000) and marine, most of the research on the causes and effects of the loss of biodiversity has been focused on terrestrial ecosystems. The review by Irigolen *et al.* (2004) of diversity in marine-plankton populations suggests that at least some of the general features of biodiversity patterns are common to both terrestrial and marine ecosystems. For example, in both cases biodiversity shows a unimodal distribution in relation to biological productivity, peaking at intermediate levels and declining towards both high and low extremes.

Most authors concentrate on *species* biodiversity since it is the easiest to quantify and the best documented, but it is important to realise that biodiversity exists at a range of levels in the hierarchy of life, from the gene upwards. Genetic diversity, already noted in an example above, refers to the total gene pool of a particular region, ecosystem, species or group of species; techniques for studying it effectively over large areas are rather new and few results are available. Petit *et al.* (2003) use the diversity within the chloroplast DNA of each species at a range of locations within Europe to explore within-species diversity for several widely dispersed European trees and shrubs. They find that the areas of maximum diversity lie not in the regions around the Mediterranean that are known to have served as refugia during glacial times, but in more northerly regions where populations from different refugia were able to interbreed. The primary aim of their paper is to test the possible link between the location of glacial refugia and the survival of genetic diversity, but this type of research could also have important implications for conservation in so far as it may point the way to retaining maximum genetic diversity for the future within key species. Maintaining genetic diversity within the surviving populations of any species threatened with extinction is of great significance for long-term survival (see, e.g., Harte *et al.*, 2004).

Species diversity has been studied at a wide range of spatial scales. These have been characterised in different ways by different authors. They are differentiated as local, landscape and macro-scales by Whittaker *et al.* (2001). In some studies the focus is on the *distinctiveness* of a particular flora or fauna, especially in those situations where the main interest arises from trying to understand or conserve species endemic to a particular locality or ecosystem. In many other studies, the focus is on species *richness* – at its simplest, the number of species in a given area. Diversity at the landscape scale is intimately linked with species biodiversity, since it is only through the maintenance or establishment of a mosaic of suitable habitats on appropriate scales and with necessary linkages, both spatial and functional, that species diversity can be retained (Waldhardt, 2003).

11.3   Consequences for ecosystem function

The nature of any links that may exist between species diversity and ecosystem function has been the subject of controversy for several decades (see, e.g., Bolger, 2001). The idea that increasing the number of species in an ecosystem increases its stability was first stated formally by MacArthur (1995), though the basic idea goes back to Darwin (1859) and was also strongly favoured by Elton (1958) as a result of his research on plant and animal invasions. From the 1970s until recently the alternative view has held sway, largely as a result of model simulations showing no clear correlation between the complexity of ecosystems and their stability (May, 1973). There is now a growing literature dealing with the role diversity may play in ecosystem functioning and resilience (the ability of an ecosystem to withstand perturbations without undergoing major shifts in the processes and structures required to maintain it). By now there are several models of how species diversity may influence the stability of ecosystem functioning (Peterson *et al.*, 1998). The differences between models hinge largely on the extent to which ecosystem functions are seen as determined purely by historical contingency, by the way in which species may replace each other functionally as species composition changes, or by a limited number of crucial ('driver' or' keystone') species that control the functional relationships within the whole ecosystem. These somewhat theoretical notions become highly relevant when we examine the functional consequences of the loss of a given species or group of species within a particular ecosystem. Are all equally essential? Is there redundancy that may lead to the replacement of the functions performed by one species by another with a similar role in the ecosystem? Are some species both irreplaceable and vital to ecosystem resilience and eventual survival?

In the above debates, increasing emphasis has been given to the concept of *functional* diversity rather than species diversity as such. Diaz and Cabindo (2001) consider these issues in some detail.They show how the effects of diversity on ecosystem processes depend not so much on species numbers but on the functional characteristics of each species, the interactions between them and the ways in which they affect each other's environment. Species richness and functional richness are not necessarily correlated. In order to explain the notion of functional richness, the authors summarise the concept of plant-functional types. These are 'groups of plant species showing similar responses to the abiotic and biotic environment and similar effects on ecosystem functioning'. Functional types may be defined in terms of their *response* to the environment (similar degrees of fire, or drought resistance for example), or in terms of their *effects* on dominant ecosystem processes (nitrogen fixers or major primary producers for example). Diaz and Cabindo (2001) show that functional diversity is an important factor in ecosystem resilience. This arises in part through redundancy, whereby the existence of several species within the same functional type allows for mutual compensation under conditions of varying stress. The notion of 'insurance' is also involved, for greater functional richness increases the potential repertoire of responses to external perturbations. There seems to be a growing consensus that diversity is important for ecosystem functioning and resilience (McCann, 2000), but the theoretical basis for this is far from clear. It harks back to fundamental questions about the origin and persistence of diversity *per se*. Tilman (2000) still poses the question: 'why is the world so diverse?'. Although most of the research cited above refers to terrestrial ecosystems, similar links between functional diversity and resilience can be made for coral reefs (Bellwood *et al.*, 2004).

Chapin *et al.* (2000b) note some of the types of effect that changes in species composition may have, especially through the spread of introduced species. They may totally modify the energy and material fluxes within an ecosystem, as with the case of *Myrica faya*, a nitrogen-fixing tree introduced into Hawaii. Its spread led to a five-fold increase in nitrogen availability in previously nitrogen-limited ecosystems. The deep-rooted *Tamarix* species introduced into the arid southwest of North America increased the availability of water and soil solutes, with a resulting increase in productivity, but a decline in biodiversity as regeneration by native species was inhibited. Introduced species can also have a major effect on disturbance regimes, notably fire. One extreme example is the spread of the grass *Bromus tectorum* in western North America, where its introduction has led to a greater than ten-fold increase in fire frequency over more than 40 million hectares. The costs attached to responding to introduced species with major impacts on processes ranging from river flow and sediment storage to human health are huge and rapidly rising. Furthermore, it seems inevitable that the costs will increase non-linearly in some cases as feedbacks into the climate system become more apparent. This is especially so where deforestation leads to the replacement of forest by persistent grassland communities. Ozanne *et al.* (2003) see this kind of transformation as a major factor in global change. They point out that forest canopies form the functional interface between the biosphere and atmosphere over more than 25% of the Earth's land surface. They play a key role in the carbon cycle, hydrological patterns and processes, atmospheric chemistry, cloud formation and climate. This interplay becomes even more complex and unpredictable as ecosystem responses to rising CO_2 concentrations are taken into account (Diaz, 2001). Forest canopies, as well as being among the most threatened habitats in the world, are also among the most important from the perspective of biodiversity, since 22 of the 25 global biodiversity hotspots include forest habitats that 'combine high levels of endemism with the imminent threat of degradation' (Ozanne *et al.*, 2003).

Biodiversity is a multifaceted aspect of global change, interwoven with virtually every other theme considered in the second half of this book. There is no simple and universally applicable measure (Purvis and Hector, 2000). Attempting to quantify biodiversity loss highlights key areas of ignorance, not least those of taxonomy, where uncertainty about the total number of species in many groups of organisms limits our understanding of the scale of the problem, and phylogeny, which lies at the heart of the development of diversity and could help to set priorities for its preservation (Mace *et al.*, 2003). Some aspects of the biodiversity theme are also relatively new foci for scientific enquiry. For example, Tilman (2000) points to a publication as recent as that by Schulze and

Mooney (1993) as the starting point for the current interest in the effects of changing biodiversity on ecosystem processes.

One of the gaps in our knowledge of biodiversity arises from the partial and biased nature of the fossil record. For most species, even within the geologically recent temporal framework of the last four glacial cycles, it is not possible to reconstruct their history and changing distribution with the degree of detail and completeness required to understand all the factors that have contributed to their survival. In the case of some aquatic taxa (diatoms for example) the fossil remains represent the whole organism and provide the basis for identification to species, in some cases even sub-species level. In this way, long histories of speciation and survival are available from the sediment record in major lakes like those of the African Rift Valley, or Lake Baikal (Cohen, 2003). In the case of most terrestrial plants, the closest to a continuous fossil record is that reconstructed from pollen and spore counts. From these, all too few species are unambiguously identifiable and at all but a few sites world-wide the record is temporally discontinuous and open to much controversy, especially in critical regions like the Amazon Basin (Colinvaux et al., 2000;). Macrofossil remains (seeds and fruit for example) allow recognition of species more readily in many cases, but the record for these is even more discontinuous. At one time it seemed likely that by identifying, from their pollen, for example, assemblages of species or genera currently characteristic of particular plant communities or ecosystems, it would be possible to trace the history of the biotic assemblages themselves. This now seems a vain hope, as many studies have shown that despite the strong interaction between species within an ecosystem at any one time, many behave in the long term and under changing environmental stresses, in a rather individualistic way, with, as a result, communities developing and transforming themselves through time (Bennett, 1997).

One thing that can be gleaned from the fossil record is the importance of particular habitats and niches. For temperate species, surviving the rigours of the glacial periods depended on the ability of plants and animals to find favoured refugia in lower latitudes, from which they were able to spread during interglacial intervals (Taberlet and Cheddadi, 2002), though refugia were not necessarily the areas within which biodiversity evolved (Knapp and Mallet, 2003; Petit et al., 2003). For many temperate, montane species, surviving the maximum forest extent in the early–mid Holocene depended on the existence of high-altitude habitats where continuous tree cover could not develop. During periods of deforestation by human activity, one of the early and often enduring corollaries was a higher level of biodiversity at landscape level, as new habitats were created without entirely destroying earlier ones, and a rich mosaic of varied ecological niches was created (Lotter, 1999). As an ongoing and virtually ubiquitous process of considerable complexity, biodiversity loss not only raises many questions about the past, it places an enormous burden on monitoring at all scales now and in the future.

Chapter 12
Detection and attribution

12.1 Introduction

Of all the changes resulting from human activities over the past two to three centuries, the possible effects on climate have rightly received most attention both in scientific literature and in the media. This chapter seeks to outline some of the accumulating body of evidence for changes in climate that have occurred over the last few decades. It also considers the question – to what extent can these be attributed to human activities?

The analyses of northern-hemisphere temperatures over the last thousand years summarised in Section 7.9 (Figure 7.11) have set the scene. They highlight the outstandingly warm conditions during the 1990s, culminating in 1998, the warmest year of the millennium, even when the uncertainties associated with the proxy-based temperature reconstructions are taken into account. In all of the reconstructions, as well as in the globally integrated observations (Figure 12.1), the strong warming trend begins in the mid nineteenth century, slows down, even reverses, between *c.* 1946 and 1970, then resumes, continuing into the 1990s. As it stands, all that this tells us is that unusually warm conditions have prevailed in recent times. It does not answer the key question – does this reflect natural, low-frequency variability (recovery from the 'Little Ice Age' perhaps) or is it, in whole or part, a consequence of increased atmospheric concentrations of greenhouse gases?

12.2 Detection and attribution – context and distinctions

The authors of the IPCC TAR (2001) make the distinction between *detection* and *attribution* in the following terms: '*Detection* (of climate change) is the process of demonstrating that an observed change is significantly different (in a statistical sense) than can be explained by natural internal variability. *Attribution* of change to human activity requires showing that the observed change cannot be explained by natural causes'. Detection is often based on statistical analysis of long-term temperature reconstructions and long climate-model integrations designed to characterise natural-climate variability. Attribution requires calculation or estimation of the effects of human inputs to the climate system, the

Figure 12.1 Changes in
global air temperature
from the mid nineteenth
century to 2002 (from
www.cru.uea.ac.uk/).

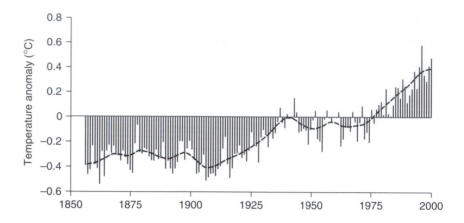

incorporation of these in climate, or Earth-system models, and comparison
between the resulting simulations and observed behaviour. Confirming the
importance of anthropogenic forcing as an underlying cause of recent climate
change hinges on demonstrating that observed changes are unlikely to be the
result of internal variability alone, that they are consistent with responses
expected from combinations of forcings that include anthropogenic factors,
and that they are inconsistent with combinations that do not include anthropogenic factors. This approach relies on using the observed changes to test the
extent to which the estimates of each type of forcing used in models actually
generates simulations that match observations. For those simulations that match
observations within prescribed statistical limits, the attributions arising from the
combination of forcings used serve as unfalsified hypotheses. They are also an
essential platform from which to launch simulations of future climate scenarios.
Given the urgency with which estimates of future climate change are required,
the patterns of attribution arising from the 'successful' model simulations of past
climate change become more than hypotheses; they form part of the basis from
which future policy responses must evolve.

Although from a purely logical standpoint detection and attribution can be
clearly separated, in practice they tend to merge. Given the short period of direct
observations, the current uncertainties in both long-term reconstructions and in
parameterisation, and the possible complicating effects of poorly quantified
feedbacks arising from processes such as land-cover change (Chase *et al.*,
2002), totally unambiguous attribution in any individual study is presently
unattainable. In consequence, evidence in support of attribution to human
causation has been accumulated through a wide variety of studies in which
consistency with anthropogenic forcing and/or inconsistency with naturally
forced variability has been claimed with varying degrees of confidence. In the
sections below, 12.3 considers some of the evidence that has been put forward in
support of the view that recent changes in the Earth system lie beyond those that
can be explained by natural variability alone. Section 12.4 examines the extent to

which recent changes in global mean sea-level may be linked to climate change; 12.5 summarises some of the more notable ecosystem responses to recent climate change; and 12.6 outlines some of the evidence for attributing recent climate changes to anthropogenic forcing.

12.3 Detection – a warming world

12.3.1 High-latitude warming and 'polar amplification'

One of the most robust outcomes of climate-model simulations forced by increased atmospheric greenhouse gases is disproportionate warming at high latitudes. Given the way in which the glacial to interglacial shifts in temperature at high latitudes greatly exceed those in the tropics (see 5.5), this is hardly surprising. Overpeck *et al.* (1997) were the first to carry out a systematic survey of recent climate change at circum-Arctic sites, using a range of calibrated climate proxies including summer-ice melt layers, tree rings, stable isotopes, lake-sediment varve thickness and foraminiferal changes. They show that the Arctic has warmed rapidly from the mid nineteenth century onwards. Taking the circum-Arctic as a whole, the mid twentieth century saw peak temperatures, since when the trend has been less consistent as a result of the spatial variability superimposed on any global trend. The few instrumental records available for the region suggest that during the twentieth century the overall pace and magnitude of warming (some 0.6 °C) exceeded that for the northern hemisphere as a whole.

During the last two decades of the twentieth century, surface temperatures rose largely as a result of warming in spring and summer (Wang and Key, 2003). Stone *et al.* (2002) note that the date of final snow melt in northern Alaska has advanced by around eight days since the mid 1960s as a result of reduced winter snowfall and warmer springs. This change, with its implied positive feedback to any warming trend as a result of the change in albedo as the period of snow cover declines, was partly offset by increased cloudiness during the same seasons as well as by declining winter temperatures. Both the temperature changes and the changes in cloud cover appear to be strongly correlated with the Arctic oscillation (AO) (see 7.7). These two features – a strongly seasonal character to the temperature trends and a strong link to a dominant natural mode of variability – are reminders that mean annual trends do not tell the whole story of climate change and that responses to anthropogenic forcing interact with and may even be, to a large degree, expressed through changes in pre-existing modes of variability.

By now, there are many studies documenting the reality and the impact of increasing temperatures in the circum-Arctic region. Peterson *et al.* (2002) show that river discharge from the major Eurasian rivers flowing into the Arctic Ocean increased by 7% between 1936 and 1999. The rate of increase averaged around 2.0 cubic kilometres per year, leading to a rate of discharge at the end of the century some 128 cubic kilometres per year greater than at the beginning of the

Figure 12.2 The trend in
annual minimum ice cover
over the whole Arctic
1979 –2000, plotted
against summer
temperatures over the ice-
covered areas. (Modified
from Comiso, 2002.)

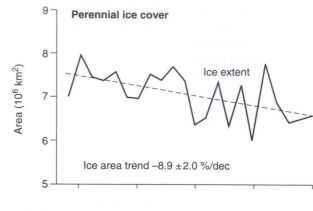

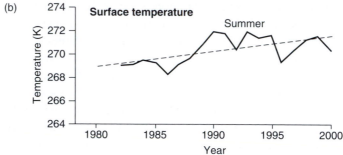

period of measurement. Deviations from the overall trend in discharge (i.e., variations from the mean, long-term trend) appear closely linked to the North Atlantic oscillation (NAO) (see 7.7), as well as to global-temperature changes. Although the overall increase is still an order of magnitude less than that required in model simulations to trigger a break down in North Atlantic deep water (NADW) formation (see 13.3), a continuation and especially any acceleration in the trend, set alongside any concomitant increases in precipitation and ice melting, could clearly have important implications for ocean circulation, hence global climate.

By using satellite data to reconstruct the minimum extent and areal coverage of perennial sea-ice in the Arctic, Comiso (2002) shows that during the period 1978 to 2000, the area has declined at the rate of 9 % per decade (Figure 12.2a). If this trend were maintained during the present century, it could lead to the end of perennial sea-ice cover. Along with evidence for a decrease in the spatial extent of ice, Comiso also quotes evidence pointing to a thinning of sea ice, an increase in the period of annual melting and an overall reduction in ice volume. One of the processes involved, as in Antarctica, is fragmentation.

Extensive changes in sea-ice cover have a major impact on Arctic climate, since a decline in ice cover reduces surface albedo – which in turn exerts a positive feedback by accelerating the warming trend at the Earth's surface.

The associated changes in energy fluxes across the air–ocean boundary and on the stratification of the Arctic Ocean could be dramatic, leading to major changes in Arctic climate and marine productivity. The observed changes in sea-ice cover are broadly consistent with those simulated by the Hadley Centre HadCM3 atmosphere-ocean global climate model (AOGCM) (Gregory *et al.*, 2002), lending some credence to the conclusions based on the model, namely that natural forcings and modes of variability alone are unlikely to have been sufficient to generate the decrease, and that an ice-free Arctic Ocean in late summer towards the end of this century is not out of the question.

There is a broad consensus that high-latitude warming is taking place at the Earth's surface and that this warming exceeds that at lower latitudes – the 'polar amplification' consistent with model simulations. It is not yet certain to what extent this polar amplification is unique to recent decades. Moreover, the precise extent to which the observed changes have been the result of an enhanced greenhouse effect, natural variations in external forcing, or changing internal dynamics, remains an open question. As indicated above, model simulations, despite their limitations (see Shindell, 2003), suggest that the first of these is playing a significant role.

Although some of the strongest evidence for recent warming comes from high latitudes, there is evidence for perceptible trends from many other parts of the world. For example, Kunkel *et al.* (2004) show that in the west of the United States, growing season length has increased in two major steps by some 25 days since 1910 (Figure 12.8a); half the increase has occurred since 1980. Saaroni *et al.* (2003), using NCEP/NCAR reanalysis data for the period 1928–2002 from the east Mediterranean region, show that there has been a long-term warming trend of just over 0.01 °C per year. Superimposed on this are short-term fluctuations, with the warmest period postdating the mid 1990s. Alongside the overall trend they detect a significant increase in the range of extreme values, largely as a result of an increase in maximum temperatures.

12.3.2   Thawing circum-Arctic permafrost

Many accounts highlight recent changes in surface processes in the circum-Arctic regions and link them to rising temperatures. Among the most damaging are the effects of thawing permafrost. Similar changes in the extent and persistence of permafrost are occurring in temperate mountain regions. Schiermeier (2003) quotes evidence for a temperature increase of 0.5 to 2.0 °C over the last 80 years in the regions of permafrost soils ranging from the mountains of the Sierra Nevada in Spain to the arctic islands of Spitzbergen. In the Arctic region of Alaska, the region of permafrost may have warmed by as much as 2–4 °C (Lachenbruch and Marshall, 1986). Whereas complete thawing of the permafrost world-wide at present rates would take many centuries, perhaps even millennia, negative consequences occur from the beginning of the process, including not only damage to infrastructure, but also hydrological disruption and release of methane, as the ice in the uppermost layers begins to thaw (Romanovsky *et al.*, 2002).

12.3.3   Warming mountains; retreating glaciers

Using the NCEP/NCAR reanalysis data (see 4.1.1) and snow-cover data set for the northern hemisphere, Diaz *et al.* (2003) present an analysis of changes in the height of the freezing level and the extent of snow cover over the last 30 to 50 years. They find that all the major mountain chains have experienced an upward shift in the height of the freezing level during this period and that the pattern recorded compares well with spatial and temporal changes in snow cover. The effects of this trend reached the headlines during 2003 as a result of the extreme warmth at high altitude in the European Alps. Whereas ice above 3000 m has normally persisted throughout the year, with only a brief period of melting, during 2003 sustained melting occurred up to 4600 m, with dramatic conse-quence for the stability of rock faces.

The decreases in tropical and temperate glacier ice from the mid nineteenth century to the present day are virtually global, with few areas registering glaciers with a positive mass balance. This theme is further considered below in relation to sea-level change (12.4.1). An increasing number of remote sensing studies have been able to quantify the current rate of glacier retreat for many parts of the world. Paul's (2002) analysis of glacier retreat and disappearance in the Tyrolean Alps (Austria) is one example of this type of study. Percentage loss is negatively correlated with area and ranges from 10 to 100% (see Figure 12.3). Arendt *et al.* (2002) report similar changes, in this case of ice volume in Alaska.

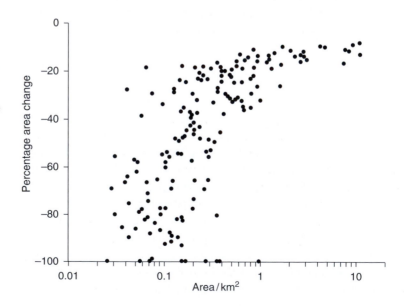

Figure 12.3 Shrinking alpine glaciers: glacier changes in the Tyrolean Alps, 1985–99. The graph plots the percentage area change against the total surface area (log scale) of each of 170 glaciers. (Modified from Paul, 2002.)

Their study calculates volume loss since the mid 1950s for 67 glaciers. They show that the absolute decrease in thickness increases steeply towards the snout of each glacier and that the mean values over all the glaciers studied is a net loss of 0.52 m per year. Since 1980 glacier retreat in the temperate glaciers in China, many of them monsoon-fed, has become virtually universal (He *et al.*, (2003). In east Africa the rate of ice retreat of the ice cap on Kilimanjaro points to its likely disappearance within the next few decades.

12.3.4   The changing mass balance of polar ice sheets

Assessing the current mass balance (ice accumulation minus ice loss) of major *polar* ice sheets is a much more complex issue. Part of the difficulty arises from the realisation that the ice sheets in Antarctica are not in equilibrium with present-day conditions (Ackert, 2003). The response of the east-Antarctic ice sheet, for example, is thought to be lagged by up to 35 000 years and some parts of the west-Antarctic ice sheet may still be melting in response to the warming at the transition to the Holocene (Stone *et al.*, 2003). This means that any observed, contemporary mass wasting may reflect the continuation of long-term responses to earlier events, rather than to the current environment. The key question then becomes whether changes that we see at the present day are significantly different from the long-term responses. To complicate matters even further, ice-core studies from Greenland and Antarctica show that rapid warming during deglaciation was accompanied by *increases* in ice accumulation at the sites sampled (see, e.g., Alley *et al.*, 1993) as precipitation increased. Gildor (2003) infers from this and other lines of evidence that we cannot assume that global warming will necessarily lead to a sustained, negative mass balance, and there are certainly well-documented increases in snow accumulation linked to warming over the middle of the nineteenth century in some locations. The study by Moore *et al.* (2001) of Mount Logan in the Yukon region of northwest Canada shows that the accelerating increase in snow accumulation over this period parallels a warming trend over northwestern North America as a whole and is also strongly coherent with decadal modes of variability such as the Pacific decadal oscillation (see 7.7).

Despite uncertainties there is a growing consensus that, at least at the margins, there is significant loss of ice mass in both Greenland and western Antarctica. Rignot and Thomas (2002) use mainly mass-budget calculations and measurements of elevation change to estimate changes in Greenland and Antarctica since 1978. They find evidence for a thinning of the Greenland ice in virtually all the coastal areas studied, whether or not the glaciers have floating ice shelves. They infer that surface warming may have increased lubrication of the basal layers of the ice as more water has reached the bed of the glacier through crevasses and moulins. Their analysis of Antarctica is less complete, but nevertheless includes 33 glaciers, including 25 of the 30 largest ice producers. Whereas the east Antarctic ice sheet shows a positive balance overall, the west Antarctic ice sheet shows an even greater negative balance. This is most extreme in the case of the ice flowing into the Amundsen Sea Embaymnent where ice loss is estimated as $72 \pm 12\,km^3$ per year. The margins of the west Antarctic ice sheet have proved to be capable of rapid changes in ice thickness, flow velocities and rates of melting and disintegration. One of the most dramatic events in recent years was the detachment of an ice berg some 48 km by 17 km from the floating tongue of the Pine Island glacier in this region during 2001 (Bindschalder *et al.*, 2002), after the period analysed by Rignot and Thomas (2002). The longer-term implications of these processes remain a subject of conjecture. Long *et al.* (2002) suggest that the apparent increase in the number of Antarctic icebergs (including the largest ever, B15, released from the Ross ice shelf in 2000 and measuring 295 km by 37 km) is probably more a function of better observations than a climatically linked trend. Periods of increased ice calving occur naturally. On the other hand, ice shelf collapse can lead to subsequent glacier surging, as was the case when the Larsen ice shelf collapsed in 1995 (de Angelis and Skvarca, 2003). These authors point out that the disintegration of the Larsen B ice shelf in 2002 may well trigger surging behaviour in the glaciers that formerly nourished it. Loss of ice shelves may thus herald greater instability in the future. Whereas total mass balance in Antarctica may be changing little and warming may even give rise to a more positive balance through enhanced precipitation, peripheral instability seems to be occurring now and may have serious consequences for global sea-level (Raymond, 2002). There are similar indications from the Arctic. Since 2000, a major ice shelf in arctic Canada has also broken up (Mueller *et al.*, 2003). Smith *et al.* (2003) show that melting of Arctic ice caps is occurring, with a 20 % increase in the intensity of summer surface melting during the 1990s.

12.3.5   Warming oceans

As with the atmosphere, so with the oceans: any trends observed over the last few decades have to be evaluated in the context of natural modes of variability. Lau and Weng (1999) identified a slow warming of global mean sea-surface temperatures (SSTs) between 1955 and 1997 of about 0.1 °C per decade, with a steeper increase of some 0.2 °C to 0.3 °C during the second half of the period. They regard the acceleration as reflecting the warm phase of a decadal–interdecadal–oscillation. Their reconstructed trends in SSTs vary from region to region and respond strongly to El Niño southern oscillation (ENSO) variability. Despite the inevitable interaction between natural variability and any anthropogenically forced trends in recent times, Banks and Wood (2002), using a coupled climate model, claim that SSTs, especially in the Arctic, Atlantic, North Pacific and Southern oceans, are likely to provide robust early indications of anthropogenic climate change. Levitus *et al.* (2000) provide a general summary of the evidence for world ocean warming and Barnett *et al.* (2001) note that the spatial and temporal characteristics of the increased heat content of the oceans compares well with model simulations incorporating anthropogenic forcing. Their results give only a 5% probability that the observed changes could have been produced by natural, internal forcing alone. These results both broaden the support for attribution to human activities and, by generating realistic values for ocean-heat uptake, they provide tighter constraints on future models (see, e.g., Knutti *et al.*, 2003). Reichert *et al.* (2002) give independent confirmation of the strong accord between observed ocean warming, both at the surface and at depth, over the last five decades, and the results of global climate model (GCM) simulations that include not only increased greenhouse-gas concentrations, but also the effects of changes in aerosols and ozone.

Alongside observations of ocean warming, the melting of Arctic sea ice and of increasing freshwater discharges from rivers draining into the Arctic Ocean (see 12.3.1) have come parallel observations of a rapid freshening of the North Atlantic (Dickson *et al.*, 2002). These authors point to observations showing that the flow of dense, cold water into the North Atlantic from the Arctic Ocean, via the Denmark Strait and the Faroe Islands channel, has been slowing down. This dense, descending water is a key element in the global thermohaline circulation. Their results suggest that the slowdown has been accompanied by an enhanced freshwater input into the North Atlantic that has already modified the descending limb of the meridional overturning circulation in the Atlantic. High-latitude climate change may already be translating into changes in the deep ocean. The recorded changes are, however, strongly linked to the recent changes in the NAO. Similar freshening over the past four decades has been recorded in the Ross Sea, on the margins of Antarctica (Jacobs *et al.*, 2002) and it is still not clear to what extent this is the result of anthropogenic forcing. The authors attribute it to the combined effects of increased precipitation, reduced sea-ice formation and the accelerated melting of the west Antarctic ice sheet. Curry *et al.* (2003) compare salinities along a transect from 50° S and 60° S in the Atlantic over the period from the 1950s to the 1990s. They find the already noted decrease in salinities at high latitudes, but at low latitudes salinities have increased over the past four decades. This confirms that freshwater is being lost at low latitudes and gained at high latitudes at a rate faster than can be compensated by ocean circulation. Curry *et al.* (2003) infer that these changes are indicative of a global ocean response that they tentatively link to global warming and associated changes in the hydrological cycle. Nevertheless, some of the changes in ocean-water properties at lower latitudes may be more strongly linked to persistent oscillations than to any recent trend (Bryden *et al.*, 2003).

12.4 Rising sea-level

12.4.1 The last 100 years

The last two sections above deal with processes that affect sea-level. The waxing and waning of ice sheets have provided the main control on what is termed eustatic

Figure 12.4 Time series of relative sea-level change over the last 300 years from tide-gauge measurements at sites in Europe (from IPCC TAR, 2001). All series show a rising trend despite discontinuities and short-term oscillations.

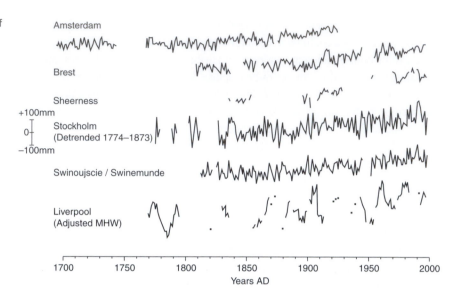

sea-level, on Milankovitch timescales (see 3.4.1, 4.4.1 and 6.4.2). In addition, changes in the temperature and, to a lesser extent, the salinity, hence density of the upper layers of the ocean, lead to expansion and contraction (so-called volume (steric) effects) that also contribute to changes in global sea-level. Measuring the change in sea-level over the last two to three centuries has posed a major challenge. Long tide-gauge records are sparse and some stations lack continuity. In many cases, the range of variability within some decades is almost as great as the total smoothed change over the last three centuries. Moreover, the most any given station can record is relative sea-level, that is to say sea-level as measured against a fixed and immediately adjacent land datum. It is therefore primarily a local, not a global record. Each stretch of coast is subject to regional and local influences reflecting tectonic processes as well as changes in coastal configuration, wind-driven ocean circulation, changing atmospheric pressure, tidal range and sediment supply. Tectonic effects include the response, in any given region, to global isostatic adjustment, a complex process of recovery from the changing distribution of loading on the Earth's crust during the last glaciation. They may also reflect more localised Earth movements. Despite the limitations attached to any single tide-gauge record, careful evaluation and analysis of tide-gauge records from around the world has led to the conclusion that global mean sea-level has been rising for at least the last 100 years and perhaps longer (Figure 12.4).

Munk (2003) suggests a figure of 1.4 mm yr^{-1} for the eustatic contribution and 0.55 mm yr^{-1} (of which over 90% is attributable to temperature) for the steric effect, giving a total of around 1.9 mm yr^{-1} for the second half of the twentieth century. Both the total figure and the relative importance of eustatic and steric effects are matters of debate, since the tide-gauge measurements that provide the framework for the above estimates give a much higher figure for twentieth century

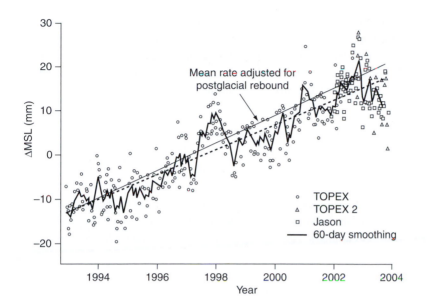

Figure 12.5 Global mean sea-level variations from 1993 to 2003, from Topex–Poseidon and Jason satellite altimetry. The mean unadjusted rate (heavy dashed line) is $2.8 \pm 0.4\,\mathrm{mm\,yr^{-1}}$. (Modified from Cazenave and Nerem, 2004.)

sea-level rise than do calculations derived from estimating the two effects (Miller and Douglas, 2004). These authors conclude, from their own analysis of the data, that the tide-gauge-based estimates are likely to be reliable, that the mass (eustatic) effect exceeded the volume (steric) effect and that the twentieth century rate of sea-level rise was probably between 1.5 and 2.0 mm yr^{-1}.

Leatherman *et al.* (2003) calculate the rise in mean sea-level directly from Topex–Poseidon satellite data for the period 1993 to 2003 to be 2.39 ± 0.19 mm yr^{-1}. Cazenave and Nerem (2004), using both Topex–Poseidon and Jason altimeter measurements and correcting for postglacial rebound, arrive at a figure of 3.1 mm yr^{-1} for the same period (Figure 12.5). One of the restraints on sea-level rise, especially between 1960 and 1980, was the damming of rivers for water storage on land (Sahagian, 2000). Although not fully quantified, this effect may have exceeded 0.5 mm yr^{-1}. This possible restraint on sea-level rise would be reduced in the event of a decline in the rate of dam construction in the future, though, as Cazenave and Nerem show, it is one of several continental processes that influences mean sea-level, and most are rather poorly quantified.

Arendt *et al.* (2002), using laser-altimetry-based measurements of changing ice volume calculate that for Alaskan glaciers alone, the average contribution to rising sea-level was 0.14 ± 0.04 mm yr^{-1} between 1950 and the mid 1990s. Thereafter, the contribution may have been as high as 0.27 ± 0.10 mm yr^{-1}. Mitrovica *et al.* (2001) estimate that melting of Greenland ice contributed around 0.06 mm yr^{-1} over the last century whereas Rignot and Thomas (2002) estimate that the current rate of loss is sufficient to raise sea level by 0.13 mm yr^{-1}. Overall, Antarctica may have played a neutral role (Goodwin, 2003), with increased ice accumulation as a result of higher precipitation over the east Antarctic ice sheet balanced by continued loss of ice over the west Antarctic ice sheet, which Rignot and Thomas (2002) suggest may be contributing a net 0.2 mm yr^{-1} to rising sea-level. Where the latter is unstable and prone to glacier surges (De Angelis and Skvarca, 2003), the potential exists for additional contributions.

In summary:

- The rate of sea-level rise over the last decade appears to have exceeded the mean rate for previous decades during the twentieth century; by how much will remain in doubt until the issue of the representivity of tide-gauge records is fully resolved.
- There are growing signs that both enhanced ocean warming and ice melting over the last decade have contributed to this acceleration in sea-level rise, though a longer period of direct observations will be needed to confirm this.
- The possibility that global warming is responsible for these changes cannot be excluded.

12.4.2   Longer-term perspectives

Since the period of documented sea-level rise spans most of the period during which global temperatures have been rising, it would seem reasonable to link the rise in sea-level to changing climate. However, the period for which tide-gauge and temperature observation overlap is short, therefore, inferring a causal relationship, howsoever plausible it may seem, on the basis of recent trends, is risky. Before doing so, it is important to look more closely at the longer-term relationship between climate and sea-level.

The link between climate, ice volume and sea-level on Milankovitch timescales is well established, but in Greenland and Antarctica it is mediated in part through processes that have long response times. For smaller glaciers there are often delays of several decades between changes in local climate and the response of the glacier. Even this is significant on the timescales under consideration here (Goodwin, 2003). Is it possible, despite complex and varied lag effects, to confirm climate–sea-level linkages on sub-millennial timescales and if so, what are the main mechanisms?

Siddall *et al*. (2003) present a reconstruction of relative sea-level over the last 470 000 years using sediment records from the Red Sea. The authors combine the record of oxygen-isotope ratio changes in foraminifera preserved in sediment cores, with a model of the relevant geochemical and physical processes, to calculate, for each sample, the sill depth at the point where the Red Sea exchanges waters with the open ocean. From this, changes in sea-level can be directly inferred. This study is of particular relevance as it is the first one to provide a record of sea-level variability sufficiently well resolved to permit close comparison with the record of isotopically inferred temperature variation in polar ice cores. As well as showing the familiar sea-level maxima and minima during MIS 5e and 2, respectively, it traces a sequence of changes between

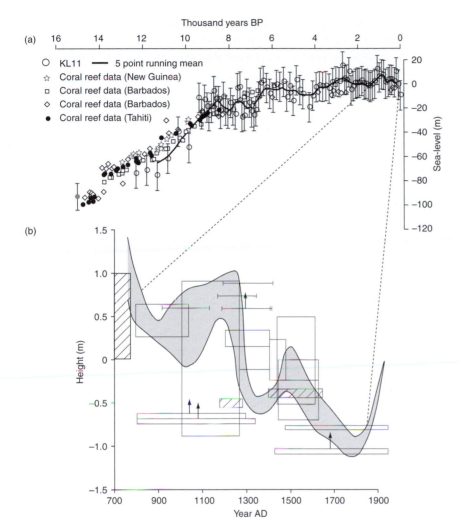

Figure 12.6 (a) Reconstructed global sea-level changes since the LGM. (Modified from Siddall et al., 2003.) The solid line showing the reconstruction derived from the Red Sea, beginning during the Younger Dryas, includes high-resolution measurements for the last 2000 years. (b) The relative sea-level envelope for the equatorial and southwest Pacific (shaded) from AD 700 (Nunn, 1998). Unshaded boxes show the modifications made by Gehrels (2001), taking into account uncertainties in radiocarbon dates and height ascriptions. Cross-hatched boxes show sea-level reconstructions for the Langebaan Lagoon in South Africa (Compton, 2001). Despite uncertainties, both series show a decline over the period of measurement, as does the Red Sea sequence plotted in (a). (Modified from Goodwin, 2003.)

20 000 and 70 000 years BP that track rather closely the temperature record in Antarctica. The indications are that even on centennial to millennial timescales, and despite the complication of lagged responses, higher sea-levels and Antarctic warming may be correlated, at least during the last glacial period.

Siddall *et al.* (2003) also provide a more detailed reconstruction of sea-level changes since the last glacial maximum (LGM) (Figure 12.6a). Their data, spanning the period of sea-level rise resulting from melting of the major ice sheets, parallel those from coral-reef studies, but whereas the latter provide only a very discontinuous record for the last 6000 years, the Red Sea record is detailed and continuous. Of especial interest is the record spanning the last thousand years, for it points to peak levels until around 700 years ago, with a decline thereafter. To what extent may this trace be regarded as a record of sea-level change that can be compared directly with reconstructions of global temperature over the same period? Clearly a record of relative sea-level from a single region and one with large, but consistent errors on the calculated absolute heights is not, on its own, sufficient.

Some broad parallel between sea-level and global temperature during the second half of the Holocene, once sea-levels had fully recovered from glacial levels, around 6000 BP, may be tentatively inferred from the evidence for sea-levels 1–2 m higher than now in the equatorial Pacific during the mid–late Holocene (Grossman *et al.*, 1998); however, it is only over the last thousand years that changes in global climate have been reconstructed with sufficient confidence to permit close comparison with sea-level data. Unfortunately, detailed records of relative sea-level variations between the mid Holocene and the beginning of tide-gauge records are very sparse indeed. For example, of the 1200-plus radiocarbon dates used in Shennan and Horton's (2002) summary of evidence for Holocene land- and sea-level variations in Great Britain, virtually none lie within the last thousand years.

Goodwin (2003) presents a detailed evaluation of the evidence for links between climate and sea-level over the last 2000 years, relying largely on dated sequences from micro-atolls, and biological indicators of sea-level on rocky coasts and saltmarshes. Figure 12.6b from Goodwin's account, summarises Nunn's (1998) reconstruction of relative sea-level for much of the southwest Pacific over the last 1300 years. Even when additional data are added to accommodate statistical uncertainties, radiocarbon dates are calibrated (Gehrels, 2001) and evidence from South Africa is taken into account (Compton, 2001), the sequence points to changes in relative sea-level of up to 2 m during the period, with a maximum before AD 1300, and a minimum around AD 1800. A similar, but more detailed reconstruction is that by van de Plassche *et al.* (1998), based on a salt-marsh sequence from the Connecticut coast. The parallels with northern-hemisphere temperature changes are quite striking. Other data from the east coast of North America reinforce the indication that sea-level has risen by around 0.2–0.3 m over the last 200 years, before which the full range of evidence is more ambiguous. In particular, the reconstruction by Gehrels *et al.* (2002), based on foraminiferal analysis of dated salt-marsh sequences from the Gulf of Maine, also shows high sea-levels before *c.* AD 1300, followed by lower levels until *c.* AD 1800 and a steep rise thereafter (Figure 12.7). van de Plassche (2000) and van de Plassche *et al.* (2003) extend the record back to *c.* AD 500 and claim correlations between at least some of the reconstructed fluctuations and the isotopically inferred temperature record from central Greenland, as well as a response to solar forcing. They suggest that the correlations proposed imply time lags of up to a century between Greenland temperature and sea-level change, and of around 125 years between solar forcing and sea-level. Goodwin (2003) concludes that the evidence from both hemispheres favours the view that sea-level between *c.* AD 1400 and 1850 was some 0.2–0.5 m lower than during the

(a)

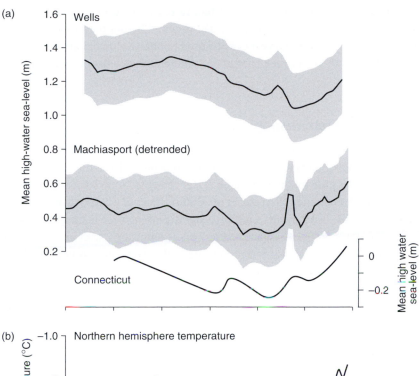

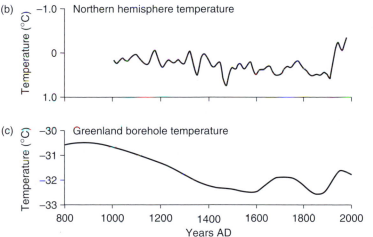

Figure 12.7 Reconstructed sea-level changes over the last 1200 years compared to evidence for northern hemisphere temperature changes. The sea-level curves for Wells and Machiasport are from sites in the Gulf of Maine (Gehrels *et al.*, 2002). The Connecticut sequence is from van de Plassche *et al.* (1998). All three series are based on analyses of saltmarsh stratigraphy. The northern hemisphere temperature graph is from Mann *et al.* (1999) and the Greenland borehole temperature record from Beltrami *et al.* (2000).

late twentieth century. Evidence for higher sea-levels (+0.5–1.0 m) between AD 1000 and 1300, and for detailed variability in parallel with temperature changes, is less strong, though there are clear similarities between the trends in Nunn's (1998) reconstruction, those from the east coast of North America for that period, and the record over most of the same time interval from the Red Sea (Siddall *et al.*, 2003 and Figures 12.6 and 12.7).

The date at which the ongoing increase in sea-level began is still not precisely determined, but it must lie within the last 200 years. The possible parallels with climate are tantalising and come from quite diverse sources, archives and areas, but a great deal more well-dated and tightly constrained records of sea-level from many more coastal environments are required before the full record from the last one to two thousand years can be used to improve quantitative estimates of the complex links between current and future climate change, and sea-level (see 13.2).

12.5   Ecosystem responses

The ecological consequences of the circum-Arctic warming noted above include increased shrub abundance and a northern movement of the tree-line in many areas. It is also probable that the tundra has changed from a net sink to a net source of carbon dioxide (Serreze, et al., 2000). Jia et al. (2003), using satellite data covering the last 21 years, confirm a significant increase in the greenness of vegetation cover in the Alaskan tundra. This is associated with an increase in above-ground plant biomass in at least some of the Arctic ecosystems involved.

Among the most commonly cited responses to recent global warming are changes in phenology – the timing of seasonal processes in plants and animals. In areas such as western Europe and much of North America, the onset of plant growth in spring is highly correlated with temperature. Consequently, temporal trends in its timing have become, for many observers, dramatic indicators of the response of the terrestrial biosphere to climate change. Most studies show that the dates of flowering and leaf unfolding in Europe and North America have advanced over the last four to five decades, generally by between 1.2 and 3.8 days per decade on average (Menzel and Estrella, 2001; Menzel, 2002). At the other end of the growing season, leaf colouring and leaf fall have, on average, been delayed by between 0.3 and 1.6 days per decade, with a good deal of spatial variability. In the IPCC report (2001), analysis of the results of over 40 studies led the authors to conclude that a climatically induced response from a wide range of biota could be confirmed with a high level of confidence, though for some non-biologists the conclusion remains controversial (Jensen, 2003).

In a study covering more the 1700 species, Parmesan and Yohe (2003) reanalysed the data upon which the IPCC based their conclusions. The responses of 484 species were analysed in detail. This was done by sorting species into four categories – those that responded in accordance with global-warming predic-tions, those that did the opposite, those that failed to show any response and those for which the response was apparently unrelated to global warming. They concluded that 87% of the species behaved as predicted by models of global warming. The changes they record include changes in biological communities, in phenology and in spatial distributions. Their analyses indicate that range shifts for species responsive to global warming have averaged 6.1 km per decade polewards and roughly the same value upwards in metres. Their figure for the recent advancement of spring events is some 2.3 days per decade – in quite good agreement with the range given by Menzel (2002). They further claim that for

(a)

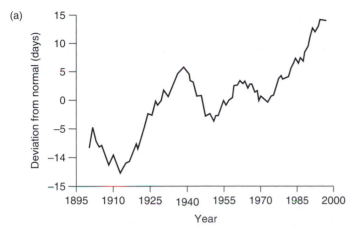

(b)

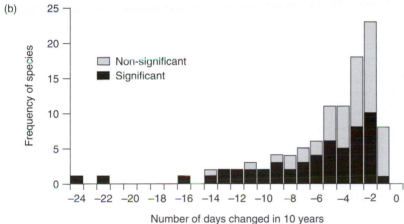

Figure 12.8 Changes in growing season and phenology: (a) Changes in the number of frost-free days during the twentieth century, relative to the mean value, for the area to the west of 100 °C W in the United States. (Modified from Kunkel *et al.*, 2004.) (b) Frequency distribution of species and groups with a temperature-related change in phenology, ordered by number of days change per decade. (Modified from Root *et al.*, 2003.)

279 of the species studied, the response to climate change was so diagnostic as to constitute a fingerprint for twentieth-century warming.

These conclusions are further reinforced by Root *et al.* (2003) who analyse the results of 143 studies, selected from a much larger set, using rigorous criteria that exclude, for example, those with a time span shorter than ten years and those showing links to climate oscillations such as the NAO or ENSO, rather than to sustained trends. Their study focuses on temperature effects while recognising that other aspects of climate change, notably moisture availability, have a major effect on physiology. They note four types of possible biotic response to rising temperatures – changes in species density at a given location, in range, in phenology, and in the genetic composition of populations. Their studies shows that over 80 % of the species analysed responded to global warming in the expected way. Figure 12.8b (Root *et al.*, 2003) shows the results of their analysis of changing phenology in terms of rates of change per decade for all taxa and for groups of organisms. Not only are different groups of organisms responding at different rates, the mean rates of response at high latitudes exceed those at lower latitudes with a statistically significant difference.

Walther *et al.* (2002) point out that freeze-free periods in mid to high latitudes are lengthening, partly because diurnal temperature ranges are decreasing – diurnal minima are increasing about twice as fast as diurnal maxima. Moreover, snow cover and ice extent in the northern hemisphere have decreased by some 10 % since the late 1960s. The same authors provide further support for a coherent response to global warming over the last century. Their evaluation includes the effects of both temperature and precipitation changes, as well as the impacts that modes of climate variability impose on longer-term trends. Nevertheless, they conclude that many of the processes and spatial patterns of ecological change can be linked to climate change over a wide range of species and biomes. Additional detailed and diverse indications of phenological and ecotone responses to climate change emerge from a growing number of studies at more local scales (e.g., Ye *et al.*, 2003; Krajick, 2004; Penuelas and Boada 2003).

Ecological responses to global warming are not confined to terrestrial biota. Aquatic ecosystems record some of the clearest responses to high-latitude warming since they record changes continuously as sediment accumulates, and many of their included biological remains are of species highly sensitive to changes in water temperature, seasonal ice cover and lake stratification. Using diatom remains from five lakes in Finnish Lapland, Sorvari *et al.* (2002) show that the first principle component of the overall diatom changes in each lake over the last 200 years closely parallels the increase in spring temperatures in the region.

As Livingstone (2003) notes, the recent warming of freshwater lakes is not confined to high latitudes. O'Reilly *et al.* (2003) summarise evidence for warming in Lake Tanganyika, which they link to climate change, despite doubts raised by Eschenbach (2004). Lying between 3° S and 9° S, it is a large Rift Valley lake in east Africa, over 1300 m deep in both northern and southern basins. Clear warming trends have been recorded in the upper part of the water column since 1913, as well as, to a significantly lesser extent, in the deep water, around 600 m since 1938. Similar trends have been recorded in lakes Victoria, Albert and Malawi. These trends are linked to well documented increases in surface-air temperature in east Africa since 1910, and especially post-1970. The changes in Lake Tanganyika have had a major impact on the physical limnology. The lake, like many deep lakes in the tropics, is permanently stratified. Aquatic productivity depends largely on the extent to which wind-driven mixing can generate upwelling at the southern end of the lake to provide a supply of nutrients from deep in the water column, nutrients that would otherwise be lost to the aquatic organisms living in the surface layers. Surface heating has sharpened the density gradient within the lake (Verburg *et al.*, 2003). This, together with an overall decline in wind speed, has increased the stability of the water column, thereby reducing the effectiveness of upwelling. The resulting decline in nutrient supply has seriously reduced aquatic productivity and fish yields. The ensemble of evidence produced by these studies is summarised by Verschuren (2003). The yields of sardine, the main commercial catch, have declined by 30–50% since the late 1970s. Although some local over-fishing may have occurred, there is a compelling case for ascribing the general decline in productivity to climate change. Since fish from the lake form a major source of protein for people over a wide area, the decline in productivity has serious human consequences.

There are growing numbers of marine organisms and ecosystems recording major impacts, among the most dramatic being the effects of elevated sea-surface temperatures on coral reefs. Hoegh-Guldberg (1999) cite six periods of mass coral bleaching since 1979, each linked to periods when summer temperatures in the region have exceeded their long-term average values by more than 1 °C for several weeks. It is estimated that in 1998 some 16% of the world's reef-building corals died (Wilkinson, 2000). The most severe bleaching events can wipe out all corals present; less-severe events are selective in their impact, affecting thin-tissued species more than the massive ones.

Many other impacts of climate change on marine ecosystems are being recognised, either as long-term trends or as fluctuations closely linked to changing modes of climate variability. Among the former are the effects of reduced sea-ice formation on krill, hence on the food webs linked to their distribution and abundance. Among the latter are the close links between fish populations and modes of climate variability in both the North Atlantic and Pacific (Mantua et al., 1997; Finney et al., 2000).

By now, the number of instances of inferred biological responses to climate change could be multiplied almost endlessly. Several key points emerge from this mix of global surveys, local studies, phenological analyses, ecosystem shifts and growth trends:

- Despite the use of terms like 'fingerprints', and no matter how well the changes map onto measured climate change or onto anthropogenically forced simulations, the data on ecological responses alone cannot be considered as conclusive evidence for ascribing the warming trends and associated changes to anthropogenic forcing.
- That said, the consistency with which ecological responses to a global mean warming of around 0.6 °C over the last century can be unambiguously identified gives some indication of the likely scales and rates of response to the predicted magnitude of future global warming.
- Given that in many environments recent ecological changes have already had significant resource implications for human populations, the potential resource implications of future ecological responses give cause for serious concern in many parts of the world (cf. the IPCC TAR, 2001).
- Many of the ecological responses already noted are also strongly affected by the range of human impacts and environmental manipulations considered in Chapters 8 to 11.

- Most of the changes noted above fall a long way short of identifying the functional implications of the ecosystem changes that have been recorded. Even where changes appear to be linearly related to the modest forcing so far observed, it is worth remembering that past ecosystem responses to environmental change show a high degree of non-linearity, with thresholds and hysteresis more the rule than the exception.
- At least some of the ecosystem changes driven by the combination of human and natural forcing will have serious future implications for biodiversity (Chapter 11), which in turn will often lead to further modifications to ecosystem functioning.
- Changes in the structure and function of ecosystems, both terrestrial and aquatic, have major implications for the exchange of energy and materials across the boundary layer with the atmosphere. As the palaeo-record illustrates, the feedbacks resulting from changes in properties such as land cover, soil moisture and albedo interact with external forcing to generate disproportionate responses in the climate system.

Among the many interactions between changing climate and changing biomes, the issue of carbon sequestration and release calls for special attention. Nemani *et al*. (2003) suggest that over the 18 years from 1982 to 1999, global net terrestrial primary production increased by around 6 %, with over 40 % of the increase accounted for by Amazonian rainforests. However, this is only part of the terrestrial biosphere–atmosphere carbon-exchange equation. One indirect inference from their results is that simultaneous changes in respiration are also of considerable significance. Oechel *et al*. (2000) have attempted to summarise ecosystem responses to warming in the Alaskan Arctic. Until recently the ecosystems of the region were long-term sinks for atmospheric CO_2 as a result of the slow rates of decomposition in the perennially cold, wet soils. The authors show that the previously persistent long-term sequestration of carbon in the tundra areas was reversed in the early 1980s as a result of the warming and drying of the climate. As soils dried out, rates of plant and soil respiration increased more rapidly than net primary productivity. This led to an initial major loss of terrestrial carbon, only partly offset by recent adjustments including the development of a net carbon sink in summer. Overall, the Arctic ecosystems of the region remain a net source due to the release of CO_2 in winter. There is an urgent need for more studies of this kind in other major world biomes. Chapin *et al*. (2000b) consider both Arctic and boreal ecosystems. They highlight the importance of shifts in albedo as conifer forests replace tundra and as deciduous-forest trees replace conifers (cf. 7.3). One of the key factors in changing the proportions of coniferous and deciduous species is fire frequency, which has already begun to increase in boreal forests in North America. The analysis by Chapin *et al*., of the current CO_2 budget in boreal and Arctic ecosystems fails to resolve with certainty whether or not high-latitude ecosystems form a net source or sink at the present day since measurements indicate the former, modelling studies the latter. It is important to remember that short-term measurements of current carbon budgets can be very misleading, especially in ecosystems where the main carbon stores are linked to the dynamics of long-lived species, the population structure of which has been affected by land-use changes and other past human impacts. This is well illustrated by the case of the forests of the eastern United States where the current carbon sink largely reflects the age structure of the stands, not their long-term behaviour (Foster *et al.*, 2002a; 2002b; Hall *et al.*, 2002).

12.6   Attribution

12.6.1   Trends and forcings

The studies cited above, taken together, constitute strong, varied and often mutually independent lines of evidence for global warming and other associated climate changes that are often exceptional in the context of natural variability during the late Holocene. They nevertheless fall short of conclusively attributing the changes to anthropogenic forcing. This is well illustrated by the analysis of Overland *et al.* (2004) of some 86 time-series of physical and biotic change scattered across the whole circum-Arctic region. Their multivariate statistical analysis reveals that over half of the full range of changes can be resolved into three dominant temporal patterns, the most important of which appears as a major regime shift *c.* 1989. In the view of the authors, this could be the result of greenhouse-gas forcing, internal processes, or the coincidence of decadal and longer-term cycles of variability. Only longer-term observations can resolve the question. The second most significant pattern is one of inter-decadal variability and the third is a linear trend over the past 30 years. In the authors' view, the case for coherent change is strong, but the evidence for conclusive attribution is still lacking.

   This and the next section consider some of the main lines of evidence that have been used in the *attribution* of recent warming to specific causes. To begin with, we consider evidence that hinges on apportioning changes in climate, mainly temperature, over the last few centuries, to currently recognised forcing and feedback mechanisms. This type of research serves an additional purpose, for not only does it make a major contribution to resolving the issue of attribution, it forms a vital stepping-stone to modelling future climate change. If a model using a particular combination of external forcing, climate sensitivity and atmosphere–biosphere–ocean interactions can generate simulations of recent climate change that match observations, it is much more likely to be a realistic guide to future climate change than one that fails this essential test. One of the assumptions upon which this approach to attribution relies is that the responses of temperature to each type of anthropogenic forcing (greenhouse gases and sulphate aerosols, for example) are linearly additive once the forcings are combined. In view of all that has been said earlier about non-linear changes in the Earth system, it is perhaps rather surprising that this assumption appears to be justified with regard to the way in which the responses combine in simulations using the HadCM2 model (Gillett *et al.*, 2004).

In a precursor to their 1999 paper, Mann *et al.* (1998) considered only the period since AD 1400. They broached the issue of attribution by subdividing the last six centuries into 50-year sliding windows. For each of these, they assessed the degree to which the temporal changes in northern-hemisphere temperatures correlated with each of the three main forcing factors – solar variability, explosive volcanicity and atmospheric greenhouse-gas concentrations. Although this was a test of correlation and temporal coherence rather than attribution, their results were nevertheless indicative. They concluded that the strongest correlations for the changes in temperature before *c.* 1920 were with solar and volcanic variability, whereas after that date there was an increasing tendency for the changes to correlate with rising greenhouse-gas concentrations.

Subsequent studies have gone beyond this type of approach and compared hemispheric or global temperature series with model simulations forced by the main natural and anthropogenic processes (Figure 12.9). In most cases, the comparisons between simulations and observational data are focused on the last 150 years, for which detailed instrumental series are available. Tett *et al.* (1999) compare the output from the UK Hadley Centre HadCM2 coupled AOGCM with 50-year sequences of decadal near-surface mean temperatures from 1906 onwards. Their model includes forcing by solar and volcanic variability, greenhouse gases and sulphate aerosols. Ensemble averages were computed using four simulations, each with different initial conditions. They conclude that the sequence of changes in twentieth-century temperatures cannot be explained by any combination of natural forcing processes. For the period of twentieth-century warming prior to 1946, their results suggest that the trend is best explained by a combination of greenhouse-gas forcing and natural variability, with some significant uncertainty regarding the degree to which changes in solar irradiation may have been responsible for the observed changes. The lack of any significant warming between 1946 and 1970 they ascribe to the countervailing effects of greenhouse gases and aerosols. From 1970 onwards, their results agree with other studies in ascribing a dominant role to greenhouse-gas forcing, partly attenuated by aerosol influences.

In a later study, Stott *et al.* (2001) use a technique of 'optimal detection' applied to both annual and seasonal near-surface-temperature variations. This involves first using the HadCM2 model to estimate the response of the climate systems to each of the four main forcing processes considered by Tett *et al.* (1999), as well as to characterise natural variability. This in turn allows an analysis of the effects of any given combination of forcing, in terms of the ability of the combination to generate changes consistent with the observational record. As a further refinement, Stott *et al.* (2001) reject combinations where the comparison between simulation and observations yields residuals that that are

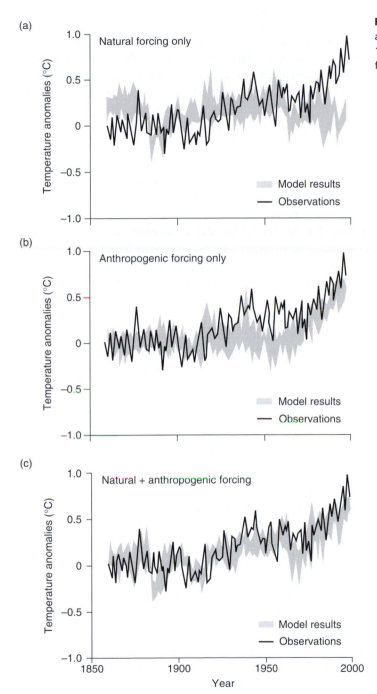

(a)

Natural forcing only

(b)

Anthropogenic forcing only

(c)

Natural + anthropogenic forcing

Figure 12.9 Comparison between observed and modelled temperature changes since 1860, using different combinations of forcing (from IPCC TAR, 2001).

inconsistent with natural variability. By using two alternative reconstructions of solar variability, by Lean *et al.* (1995) and by Hoyt and Schattén (1993), they reinforce the earlier conclusions to the effect that post-1946 warming cannot be explained by natural forcing alone, irrespective of the series used. The extent to which solar variability can be shown to have contributed to changes in temperature during the first half of the twentieth century depends on the choice of reconstruction and on the way in which the effects of greenhouse-gas and aerosol forcing are prescribed in the model. While stressing the robust nature of these conclusions, they also point to the limitations of their study. The indirect effects on climate of sulphate aerosols via cloud albedo are not included, nor are the possible effects of other aerosols. Likewise, the possible effects of changes in land cover and ozone have not been considered. All these, they claim, are likely to have been much less significant than the forcing factors they consider. Gillett *et al.* (2004) show that the consistency between the response of models forced by greenhouse gases and sulphate aerosols, and instrumental observations is improved by using an ensemble of five current climate models; once again, the results strongly reinforce the main conclusions reached by Tett *et al.* (1999) regarding anthropogenic forcing in the second half of the twentieth century.

Tett *et al.* (2002) also follow up the earlier study, this time using the HadCM3 AOGCM, an improved model that requires no flux adjustments and has a more realistic representation of secondary aerosol effects. They conclude that prior to 1957 anthropogenic effects tended to be mutually compensating. They therefore made only a negligible contribution to warming during that period. In ascribing the warming trend during the first half of the twentieth century largely to natural forcings, Tett *et al.* (2002) draw attention to its unusually rapid pace. The late twentieth century warming over the last 30 to 50 years they ascribe to the positive effects of greenhouse gases ($+0.9 \pm 0.24$ °C per century), offset in part by volcanic and anthropogenic aerosols, which achieve a negative forcing of -0.4 ± 0.26 °C per century. They state that natural forcing made no net contribution to warming during this period. Stott *et al.* (2003), in a study using an improved method of 'optimal' fingerprinting, reinforce the main conclusions of Tett *et al.* (2002) and others, but with the caveat that the models have probably underestimated both the strength of solar forcing in the early part of the twentieth century and that of anthropogenic effects, both greenhouse warming and sulphate-aerosol cooling, during the later part.

The results of efforts to detect anthropogenic forcing in recent changes in precipitation have proved less conclusive. Lambert *et al.* (2004), apply 'optimal fingerprinting' to their comparison between simulated land precipitation using HadCM3 and the observed changes during the twentieth century. They are able to show that the observed changes are inconsistent with modelled natural variability, but consistent with the combined effects of natural and anthropogenic forcing. They are, however, not able to detect a statistically significant response to anthropogenic forcing in isolation.

Returning to temperature, the only significant inconsistency in the analysis by Tett *et al.* (2002) arises from the fact that the simulations overestimate the increase in tropospheric temperatures by some 50% when compared with actual observations. This issue is address by both Thorne *et al.* (2003) and Jones *et al.* (2003). The first of these reinforces the view that the temperature changes in the troposphere from 1960–1994 are probably the result of anthropogenic forcing. The tendency for the model simulations to provide estimates of tropospheric warming higher than the rates recorded by direct observation is not explained, though the authors point out that uncertainties in the observations may be sufficiently significant to allow overlap between the observed temperature changes and those simulated with anthropogenic forcing. Jones *et al.* (2003) present an attribution study in which both near-surface and free-atmosphere temperature changes are used in combination for the period 1960 to 1999. Their study allows confident detection of both a well-mixed greenhouse-gas signal and a combined sulphate-aerosol and ozone signal. Use of the near-surface and free-atmospheric change also allows them, for the first time, to detect both a stratospheric volcanic-aerosol effect and, less securely, the subsidiary effects of solar forcing over this period.

The tendency for models apparently to overestimate tropospheric warming over the last two to three decades, relative to observations, has become one of the areas of criticism used by sceptics of global warming. Lindzen and Giannitsis (2002), for example, use the discrepancy to estimate a low value of less than 1 °C for climate sensitivity to CO_2 doubling. Vinnikov and Grody (2003) reconstruct tropospheric warming from satellites for the period 1978–2002 and propose a rate of from $+0.22$ to $+0.26$ °C per decade. They regard this as entirely consistent with the global warming trend of 0.17 °C per decade derived from surface observations. The question of tropospheric warming, which is linked to an increase in the height of the tropopause, remains controversial (Santer *et al.*, 2003; 2004; Pielke and Chase, 2004), though the latest study by Fu *et al.* (2004) is viewed by some (Kerr, 2004; Schiermeier. 2004b) as resolving the issue. Fu *et al.* show that by correcting satellite-measured tropospheric temperatures for the effects of cooling in the stratosphere, the measured rate of warming becomes entirely consistent with the surface trends. From this study, Kerr (2004) concludes that 'for now, satellite temperatures can no longer be used to portray a feeble greenhouse effect'. This conclusion is consistent with that reached by Santer *et al.* (2003), who consider the increase in the height of the tropopause – the boundary between troposphere and stratosphere – that has been observed since 1979 an increase in line with reanalysis data and climate-model simulations. They conclude that around 80% of the change can be ascribed to anthropogenic influences. Two processes are involved – a cooling of the stratosphere and a warming of the troposphere as a result of increased greenhouse gases. Their study reinforces the view that recent warming of the lower atmosphere is real and can be attributed largely to anthropogenic effects. De Laat and Maurellis (2004) claim, however, that their analysis of spatial variations in lower-tropospheric warming correlates well with the spatial pattern of CO_2 emissions. From this, and other lines of evidence, they infer that *local* surface heating processes may have made a significant contribution to recent warming. The debate about the disparity between tropospheric and surface warming continues. Douglass *et al.* (2004) claim that the disparity is real, but that it is mainly confined to tropical, oceanic regions. Their results tend to downplay, though not entirely exclude the possible effects of land-surface processes such as those invoked by Kalnay and Cai (2003) and de Laat and Maurellis (2004). Instead, they propose that the disparity arises from processes at the ocean–atmosphere interface.

All the above studies are concerned with temperature alone. Brocolli *et al.* (2003), using the GFDL R30-coupled model, present simulations for the period 1865 to 1997 for both temperature and precipitation in which greenhouse gases, anthropogenic sulphate aerosols, solar variability and volcanic aerosols in the stratosphere are progressively added. Addition of the last two natural forcings improves the comparison with observations, thus reinforcing the conclusions reached by Jones *et al.* (2003). They point out that since the net effect of natural forcing in the latter part of the twentieth century has been negative, the fact that inclusion of natural forcing in the simulations improves the match with observations for both temperature and precipitation (Fig 12.10), strengthens rather than weakens the case for anthropogenically induced warming.

One of the challenges arising from the availability of millennium-long proxy-based hemispheric and global temperature reconstructions (see 7.9) has been to use these both as a test of model skill and as a contribution to attribution studies, by placing the simulated effects of each forcing variable alongside evidence for longer-term variability. Hegerl *et al.* (2003) do this by comparing the simulations carried out using a energy balance model (see 2.3) with a range of proxy-based reconstructions, including a modified version of the Crowley and Lowery (2000) series and that of Briffa *et al.* (2001). Hegerl *et al.* find excellent agreement between the first of these and the model simulations for the northern hemisphere (Figure 12.11) and rather less good agreement between model simulations and the second, which refers to northern-hemisphere growing-season temperature. In the first comparison, anthropogenic, volcanic and solar effects are all detected with high statistical significance; in the latter only the first two. Analysis of the divergences between the simulations and observations as well as between some of the observed series are especially interesting. On the basis of these, the authors tentatively infer the influence of a fall in CO_2 concentrations on cooling during the Maunder minimum as well as a possible response to land-use change in the reconstruction by Mann *et al.* (1999). Their results point to the detection of a CO_2 response in most of the series used by the middle of the twentieth century and in all by 1980. They also suggest a rather modest and intermittent impact of solar forcing on multi-decadal changes in hemispherically averaged temperatures.

(a)

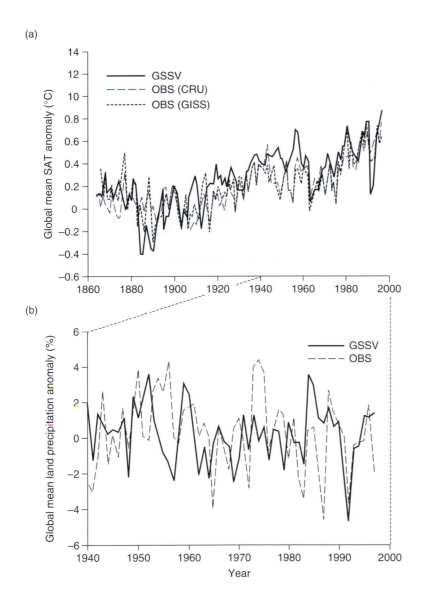

Figure 12.10 Simulations of past global temperature and land precipitation changes, compared with observations. (a) Comparison between observed (OBS) and modelled temperature changes since 1860, using greenhouse-gas, solar, tropospheric sulphate-aerosol and volcanic-aerosol (GSSV) forcing (modified from Broccoli *et al.*, 2003). The model ensemble is compared with two observed surface-temperature series. CRU – Climate Research Unit, University of East Anglia; GISS – Goddard Institute of Space Science. (b) Comparison between observed and modelled land precipitation changes since 1940, using greenhouse gas, solar, tropospheric sulphate-aerosol and volcanic-aerosol forcing (modified from Broccoli *et al.*, 2003). Note the consistently strong agreement between modelled and observed temperature changes and the much weaker agreement for parts of the precipitation record.

The effects on climate of land-cover changes have generally been neglected in the simulations referred to so far. Bauer *et al.* (2003) are at pains to include this in their assessment of climate forcing over the last thousand years, using the CLIMBER intermediate-complexity model. In general, their results reinforce those in the studies already summarised above, but they infer, in addition, a cooling effect from deforestation. To this they attribute a significant part of the cooling during the second half of the nineteenth century. Matthews *et al.* (2004), using a dynamic vegetation model coupled to an intermediate-complexity Earth-system climate model, also accept that the biogeophysical effects of land clearance

Figure 12.11 Comparisons between modelled and proxy-based changes in northern hemisphere temperature, plus the attribution of components of the series to different forcing mechanisms: volcanism; solar radiation; greenhouse gases + sulphate aerosols. (a) The upper graph shows the Crowley and Lowery (2000) reconstruction from AD 1000 onwards, together with the model-forced series and the shorter instrumental record. Below this is shown the sequence of effects for each forcing mechanism with 5–95% confidence limits. Asterisks denote a response to the forcing that is detected at the 5% significance level. (b) The upper graph shows the reconstruction by Briffa *et al.* (2001) from AD 1400 onwards, together with the model-forced series and the shorter instrumental record. Below this is shown the sequence of effects for each forcing mechanism with 5–95% confidence limits. Asterisks denote a response to the forcing that is detected at the 5% significance level. In both figures (modified from Hegerl *et al.*, 2003) the residual variability, plotted in the original and ascribed to internal climate variability and errors in the reconstructions and forced responses, is omitted.

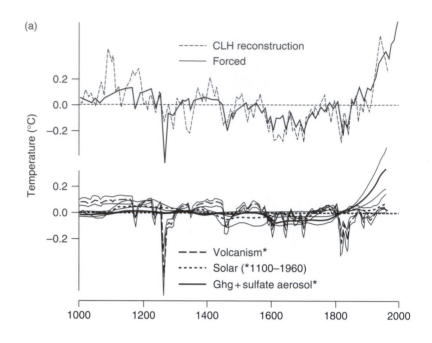

(a)

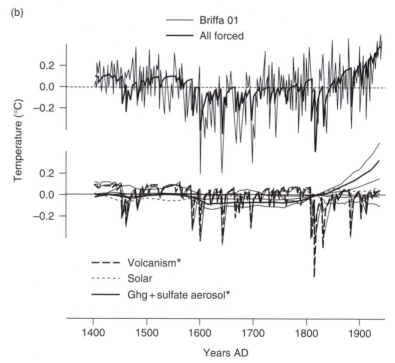

(b)

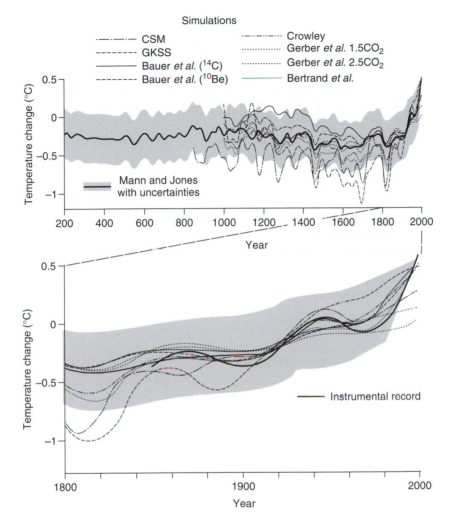

Figure 12.12 Comparisons between several modelled and proxy-based changes in global temperature. The proxy series is that published by Mann and Jones (2003), with the shaded area representing uncertainties. Eight individual simulations are compared with this. The lower graph shows an expansion of the simulated trends for the last two centuries, along with the smoothed instrumental record. (Modified from Jones and Mann, 2004.)

since AD 1700 have led to a net cooling, but they further claim that this has been more than offset by the effects of land-cover change on increases in atmospheric carbon dioxide. Overall, they claim that the net effect of land-cover change has been a small *increase* in global temperature, of around 0.15 °C. Mann and Jones (2003) present a synthesis of several of studies in which model-based temperature series are compared with observations and with their own reconstructed values for the last 1800 years (Figure 12.12). The level of agreement between most of the modelled and reconstructed temperature changes is impressive.

One of the problems in using the above approach to attribution arises from the huge computing power needed to provide ensembles of simulations that include all the likely forcing terms and their associated uncertainties (see 2.3). Knutti *et al.* (2003) have partly overcome this problem by using neural networks. This approach, they claim, increases the efficiency of large climate model ensembles by at least an order of magnitude. The main purpose of their paper is to improve climate-change projections; it is therefore considered in more detail in Sections 13.1.3 and 13.2. Here, it suffices to confirm that their exhaustive treatment of uncertainties in every aspect of model development, together with a rigorous comparison between their model simulations and the observed temperature changes between 1750 and 2000, lead them to propose that the dominant forcing responsible for the difference is increased atmospheric greenhouse gases – CO_2, CH_4, N_2O, SF_6 and halocarbons. According to their best estimates, these, together with tropospheric ozone, contribute a positive forcing over eight times greater than that ascribable to differences in solar forcing between the two dates.

12.6.2   Attribution – towards fingerprinting anthropogenic influences

In this case, attribution relies on identifying characteristics in the *current* processes and patterns that point to a particular combination of forcing – essentially a diagnostic climate-system response or 'fingerprint' of those factors responsible for the changes observed. At first sight, one may have expected reanalysis data to provide one of the keys to characterising many diagnostic aspects of the warming trend over the last 50 years. Bengtsson *et al.* (2004) show that this is by no means straightforward, the main stumbling block being changes through time in the observing systems. Results are also dependent on the assimilation system used. Reanalysis data have been developed largely for weather forecasting where these considerations are less important than they are in analysing long climate series. Although the analysis by Bengtsson *et al.* points to increases in lower-troposphere temperature and integrated atmospheric water-vapour content that are, once corrected, consistent with other measurements and with theory, their main conclusion is that more work is needed to make it possible to extract secure information on trends from reanalysis data.

Many attribution studies involve comparing a single, simulated spatial, vertical or temporal (diurnal or seasonal) pattern of change to the pattern recorded in observations. A wide range of studies is summarised by Mitchell *et al.* (2001), and the authors conclude from these that: 'All new single-pattern studies published since the second IPCC report in 1996 detect anthropogenic fingerprints in the global temperature observations, both at the surface and aloft'. Inclusion of both greenhouse gases and aerosols in models has led to simulations that correspond well with observed surface-temperature changes. The same applies to the 'free atmosphere' once the effects of ozone are also taken into account. Mitchell *et al.* (2001) also consider a range of more complex multiple-fingerprint studies, some using a combination of spatial and temporal patterns, others space-frequency combinations. These too reinforce the conclusions already stated above. Moreover, they help to refine estimates of the magnitude of anthropogenic forcing. Although the overall conclusions of Mitchell *et al.* echo all those proposed by the studies already referred to, they also stress the persistence of several uncertainties. These include those arising from discrepancies in model simulations of the difference between surface and free-air temperatures. Estimating internal climate variability for the pre-industrial period and calculating natural forcing for all but the last two decades also pose significant challenges. Doubts still surround the indirect effects of aerosols and most models have neglected the impact of processes such as biomass burning and land-cover change. Mitchell *et al.* also point out that the studies they summarise still leave open quite a wide range of possibilities for the sensitivity of climate to major increases in atmospheric CO_2 – a point that becomes of central importance in Section 13.1.3 where the issue of future warming is considered.

Braganza *et al.* (2004) present an analysis of attribution that links the concept of fingerprinting to the evaluation of recent trends. They do this by investigating the temporal evolution of several indices that they regard as spatial fingerprints. These include land–ocean temperature contrasts, the annual cycle of change in surface temperatures, the meridional temperature gradient in the northern hemisphere and the temperature contrast between hemispheres. For the period 1950 to 1999, the trends in all but the last index are inconsistent with natural forcing, but entirely consistent with forcing due to anthropogenic effects, including those of both greenhouse gases and sulphate aerosols. Their conclusion, once more, is that anthropogenic effects are responsible for almost all the changes in surface temperature observed since 1945, but that changes in the first half of the twentieth century were best explained by a combination of natural and anthropogenic forcing.

Further support for recent anthropogenic forcing comes from a remarkable study by Philipona *et al.* (2004). For the first time, they calculate and apportion cloud-free, long-wave, downward radiation to each contributing process for the period from 1997 to 2002 at eight sites in the Alps of central Europe. They conclude from their analysis of all the factors involved that an increase of $+1.8 \pm 0.8 \, \text{W m}^{-2}$ over the period can be confidently ascribed to enhanced greenhouse gases and associated water-vapour feedback. They regard their results as proof of greenhouse-gas warming by means of direct observations.

12.7   Concluding comments

This chapter has concentrated on recent climate change, its ecological consequences and its ascription to particular combinations of forcings and feedbacks, because these topics lie at the heart of the debate on global change. Not only do they form the basis for evaluating what is currently being experienced, they are a crucial platform from which to launch future projections. But whereas all projections will always be subject to unquantifiable uncertainty until experience tests them as the future unfolds, the evidence and data–model comparisons upon which much of the crucial science rests can be evaluated in more realistic ways. Wherever possible, in this chapter, the emphasis has been on evidence presented over the last three years. This is because the IPCC TAR, published in 2001, sets out a carefully qualified range of possible projections of future global warming and it is vital to understand to what extent all the evidence published in the wake of that reinforces or undermines the main conclusions. A wide range of individual studies has been briefly described before drawing together any conclusions.

On the basis of the evidence considered so far:

- There have been virtually no soundly based challenges to the reconstructions of past and current processes and changes upon which future global temperature projections are built.
- The most recent analyses of both surface and tropospheric temperature changes over the last 30 to 50 years include unambiguous evidence that they cannot be explained without recourse to the increasing influence of rising greenhouse-gas concentrations in the atmosphere. This conclusion holds good irrespective of the time-frame, models, or approaches to attribution used.
- Comparison of model simulations with instrumental, reanalysis and palaeo-records is providing essential constraints on the numbers used to quantify forcings and sensitivities in simulations of future climate change, though uncertainties still remain.
- There are strong indications that a significant oceanic response to this forcing is now under way, though its long-term consequences may be hard to predict.
- Although the natural forcings that probably dominated the record of global (and northern-hemispheric temperature change over most of the last millennium have continued to influence climate to the present day, their relative importance has declined over the last few decades, except in so far as they have probably served to counter, in part, the effects of greenhouse gases.
- The complex role of aerosols in ongoing climate change is still a matter of debate, though the main overall conclusion, that they have a net negative forcing effect, is not seriously in doubt.
- There are some indications that data–model comparisons spanning the last millennium are improved when the effects of land-cover changes are taken into account. More simulations need to be carried out with these effects included.
- Evidence spanning multi-decadal and longer timescales tends to support the view that changes in late-Holocene global mean sea-level have been related to changing climate, though more studies are needed before this can be confirmed unambiguously.
- The tide-gauge record of global mean sea-level increases of around 1.5–$2.0\,\mathrm{mm\,yr^{-1}}$ over the last century seems likely to be reliable, with the major factor being ice melt rather than steric effects. An acceleration of this trend to $c.\,3\,\mathrm{mm\,yr^{-1}}$ has probably occurred since 1990.

Virtually all of the attempts to detect a fingerprint of anthropogenic forcing in the most recent climate trends and climate-system characteristics point to the conclusion that free-air and surface warming, through increased atmospheric greenhouse-gas concentrations, is already significant.

Chapter 13
Future global mean temperatures and sea-level

Just as for the past, so for the future it is necessary to define the main timeframe of concern. This is, to some degree, predetermined by the approach used in the IPCC TAR (2001), the results of which provide points of reference and departure for much that follows in this and subsequent chapters. Most of the effort involved in that exercise has gone into establishing scenarios and projections through to 2100. A firm cut-off is, however, unrealistic, for what happens between now and then will have a huge influence on the subsequent course of events.

13.1 Future changes in global mean temperature

13.1.1 Dealing with uncertainties

Schneider (2002) points out that uncertainties, including both statistical and intrinsic uncertainties, cascade through multiplication into ever-widening ranges (Figure 13.1). Nevertheless, he strongly favours strenuous attempts to attach probabilities to alternative future scenarios. One of his key points counters the distinction made between biophysical and social systems by Grubler and Nakicenovic (2001). Schneider claims that in considerations of uncertainty there is no essential difference between 'natural' and 'social' systems in that they both include feedbacks and are to a greater or lesser extent path-dependent. He acknowledges, however, that social systems are harder to predict. Perhaps the best solution is to accept that there is a spectrum of predictability. At one extreme are systems characterised by such contingency-linked indeterminacy that probabilistic statements must remain entirely judgement-based and subjective. At the other extreme are systems with a sufficient degree of temporal and structural consistency as to allow the attachment of statistically determined probabilities. The cascade of uncertainties portrayed in Figure 13.1 includes and links both types. It begins with emission scenarios and these lie towards the less quantifiable end of the probability spectrum. The failure of the IPCC TAR (2001) to attach probabilities to their future-emission scenarios has been one of the triggers for the papers by Schneider already noted above.

229

Figure 13.1 The cascade of uncertainties involved in linking emission scenarios to the impacts of climate change at regional scale. Continuous bars relate mainly to Chapter 13, the dashed bar to Chapter 14 and the dot-dashed bar to Chapter 15. (Modified from Schneider, 2002.)

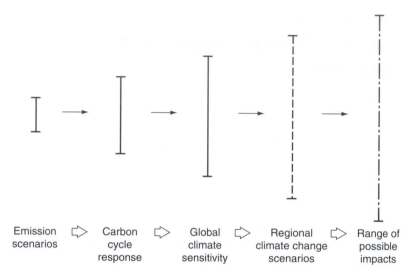

Emission scenarios ⇨ Carbon cycle response ⇨ Global climate sensitivity ⇨ Regional climate change scenarios ⇨ Range of possible impacts

13.1.2   Future-emission scenarios

The starting point for the critical cascade of uncertainties is the Special Report on Emission Scenarios (SRES) by Nakicenovic and Swart, (2001). The authors develop 35 scenarios, each claimed to be internally consistent and plausible. These fall into four families of scenarios, or storylines, out of which emerge six actual scenarios used to drive the succeeding steps in the cascade. The scenario families reflect a simple matrix expressing antitheses between high economic growth and environmental sustainability in one dimension, and between global equalisation and regional differentiation in the other. Scenario family A1 emphasises rapid economic growth, with population increasing to the middle of this century, then declining. Implicit in this group of scenarios is the theme of rapid technology transfer and upward convergence between economies, leading to generally higher per capita incomes over the world as a whole and less differentiation into 'haves' and 'have-nots'. It is the most optimistic family in terms of conventionally defined global economic growth and material prosperity. Within the family, three scenarios are proposed, one with heavy dependence on fossil fuel (A1FI), one with rapid expansion of non-fossil-fuel use (A1T) and one with a balance between the two (A1B). The single A2 storyline envisages a more fragmented world with little convergence and great regional differentiation. Fertility patterns remain divergent between regions and global population growth continues. The single B1 scenario is comparable to A1 in terms of demography, but social, economic and technological changes lead to a much greater emphasis on reducing dependence on materials, improving efficiency of resource use and promoting environmental sustainability. In B2, the final scenario, the themes that characterise B1 are retained, but with a much higher level of regional differentiation and continued population growth, though at

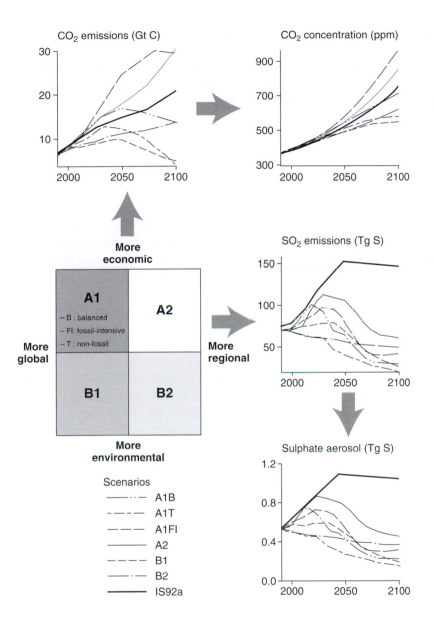

Figure 13.2 A summary of the IPCC TAR scenarios and their conversion to atmospheric concentrations in the case of CO_2 and sulphate aerosols. (From IPCC TAR, 2001.)

a slower rate than in A2. Figure 13.2 gives a schematic guide to the various scenarios referred in this and succeeding chapters.

Each socio-economic/demographic/technological scenario is translated into evolving emissions until the year 2100 for CO_2, CH_4, N_2O and SO_4. Figure 13.2 shows the results for all six scenarios, as well as the emission figures used in the 1992 equivalent of SRES for CO_2 and SO_4. The main change since 1992 has been the projections of a large but variable downturn in SO_2 from the middle of the century onwards. Some divergence between the SRES emission scenarios is apparent from the beginning of the projections, but becomes greatly emphasised from around 2030 onwards.

All the scientific evidence considered in this book has very little bearing on the question of which among these scenarios is more or less likely. Present-day demographic, socio-economic, geopolitical and technological trends suggest that, barring a major reversal of current economic and demographic forces, or some major discontinuities in the form of global war, famine or disease, the mid- to higher-range emission scenarios, with strong rather than reduced regional differentiation, are the more probable. Put another way, there is as yet no reason for projecting a future in which emission scenarios A1T, B1 or B2 will prevail, without huge efforts, coordinated on a global scale and effective for curbing energy-rich life styles and fossil-fuel use in the richest nations. The Kyoto agreement (Bolin, 1998; Grubb *et al.*, 1999) was designed to take a decisive first step in this direction, but it has had limited success. Notable is the refusal of the United States to subscribe, despite having signed the United Nations Framework Convention on Climate Change (UNFCCC), with its stated objective to stabilise greenhouse-gas concentrations at a level that prevents dangerous interference with the climate system. There is no justification for ignoring the higher-emission scenarios and their possible future consequences. At the same time, it may be unrealistic to opt for the A1F scenario without some qualifications, in view of (1) the widespread, albeit modest development of alternative-energy sources in a growing number of developed countries including the United Kingdom (Batchold, 2003), (2) the recent, but perhaps only temporary, emission reductions in at least one major developing country, China (Streets *et al.*, 2001) and (3) the five-fold or even greater acceleration in the rate of increase in atmospheric CO_2 concentrations that would be involved were the A1F scenario to unfold. If all the currently known fossil-fuel reserves were to be exhausted, with no additional attempts to capture and store emissions, CO_2 would eventually reach around 1000 ppmv, according to Lenton and Cannell (2002), a figure only slightly in excess of the A1F projection for 2100. One further point to bear in mind is that ongoing land-use change could significantly add to atmospheric CO_2 concentrations. In the model-based study of Gitz and Ciais (2003) replacement of forest by cultivated land tends to increase the atmospheric burden in both the short and longer term. The short-term loss arises from the release of carbon during deforestation; in the longer term there is a reduction in the turn-over time of carbon in biomass and soil, leading to shorter periods of sequestration, after deforestation. The likely size of this effect is impossible to predict in view of the degree to which it is strongly scenario and model dependent.

13.1.3   Deriving future global temperature projections

For each emission scenario, the figures for emission have to be translated into changing atmospheric concentrations through time (Figure 13.2). This involves

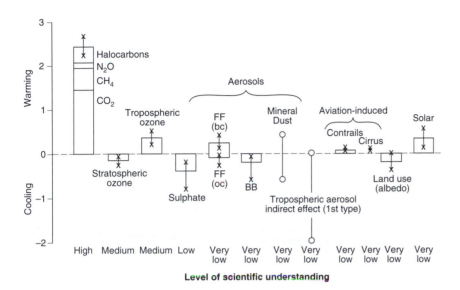

Figure 13.3 The global mean radiative forcing of the climate system in (W m^{-2}) for the year 2000 relative to 1750, with annotations to indicate the level of scientific understanding attached to each estimate. (From IPCC TAR, 2001.) Bars above (below) the line show the estimated positive (negative) effects associated with each of the main influences on radiative forcing over the 250 years. Each bar is therefore a function of the overall difference, between 1750 and 2000, in the strength of each type of forcing influence, and the estimated sensitivity of the radiation received at the Earth's surface to that influence. In the case of mineral dust and indirect tropospheric aerosol effects, the level of understanding is so low that no central values have been assigned.
FF – fossil fuel
bc – black carbon
oc – organic carbon
BB – biomass burning

taking full account of all that is known about the atmospheric residence times, sources, sinks and interactions of each of the chemical species involved. For each of these, as well as for other agents implicated in forcing and feedback processes within the Earth system, models designed to develop future trajectories must also incorporate soundly based estimates of the sensitivity of the climate to each of the processes and their interactions (Figure 13.3). The IPCC TAR first calculates and presents the net-radiative forcing estimated for each scenario, then aggregates all these effects into a series of global temperature trajectories based on model ensembles for each emission storyline (Figure 13.4). These projections are arrived at by means of a simple energy-balance climate model calibrated using the results from seven coupled atmosphere – ocean global climate models (AOGCMs) (see 2.3). Bars are added on the right hand side of the graph to encompass the full spread of temperatures simulated for 2100.

Translating emissions into atmospheric concentrations involves modelling for the future a range of processes about which there is significant uncertainty both in the past and at the present day. It is worth noting that despite these uncertainties, and setting aside, for the reasons outlined above, the three most optimistic storylines and the most fossil-fuel intensive, the range of concentrations of atmospheric CO_2 projected by the IPCC are virtually identical for the remaining scenarios right up to 2060, and methane concentrations only begin to diverge significantly after 2050. The projected increases in N_2O are more modest and less important from the standpoint of future radiative forcing. Values for sulphate aerosols have been revised downwards between 1992 and 2001, so that the later scenarios carry with them the implication of a reduced indirect aerosol effect.

Figure 13.4 The IPCC TAR (2001)-projected temperature changes to 2100, using all six emission scenarios (see Figure 13.2) and an emission scenario from the Second Assessment Report, IS92a (IPCC, 1996). Dark shading shows the full range of projections by the model ensemble; light shading spans individual model projections for all the scenarios. The bars to the right show the full range for each of six individual models. (From IPCC TAR 2001.)

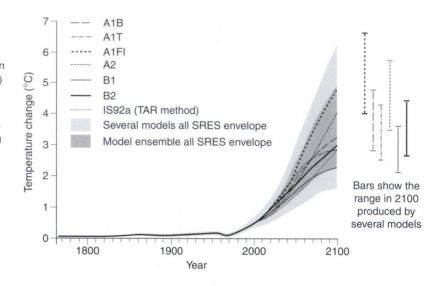

The direct radiative effects of each of the greenhouse gases in $W\,m^{-2}$ can be calculated theoretically from their physico-chemical properties, but there is still considerable doubt concerning their aggregate and indirect effects, especially in view of the problems involved in quantifying the role of future changes in water vapour and clouds. According to some authorities, water vapour should almost double the sensitivity of climate to increased greenhouse-gas concentrations. Inferred warming mainly occurs because warming increases the potential absolute humidity of the atmosphere and is also likely to force moist air to higher, colder altitudes from which less heat is radiated into space (Del Genio, 2002). The magnitude and even the sign of this effect have nevertheless generated a good deal of controversy.

From a comparison between model simulations and direct observations of the cooling and drying of the atmosphere that took place as a result of the Mount Pinatubo eruption in 1991, Soden *et al.* (2002) and Forster and Collins (2004) show that water-vapour feedback reinforced the cooling effect after the eruption. Positive feedback with warming is a logical corollary of this. They further suggest that global climate model (GCM) simulations are able to provide a realistic representation of this feedback. Because the forcing involved in the eruption was different from that involved in any future global warming, these results are not a perfect test in relation to future scenarios. They do, nevertheless, clearly reinforce the view that water vapour will have a positive feedback in a warming world and that models are capable of capturing this.

As Figure 13.4 shows, the final step in the global calculation is subject to the various uncertainties arising from, for example, the difficulty in quantifying the sensitivity of the climate system to changes in net radiative forcing, and from differences between models. According to the IPCC projections, the mean global temperature change is likely to be between 1 °C and 2 °C by the middle of this century and at least twice this by the end. The range of values for 2100, taking into account the full envelope of model projections, is between 1.4 °C and 5.8 °C. The upper value is significantly higher than the 3.5 °C estimated in the previous IPCC report (IPCC, 1996), partly because of the reduction in projected sulphate concentrations, partly because of the incorporation of feedbacks from the carbon cycle, improved treatment of greenhouse-gas forcing, including the effects of methane and tropospheric ozone, the direct use of AOGCM results, and different assumptions about the rate of change in thermohaline circulation (Wigley and Raper, 2001). The implied rate of warming is between two and ten times that recorded during the twentieth century (Pittock, 2002). Further projections follow, notably for sea-level, annual run-off, hence potential water availability, the likely future amplitude and frequency of extreme events, and the probable regional implications of the spatial patterns of change.

Of all the climate properties available for study, global mean temperature is the one most frequently quoted. It is also the one most extensively and rigorously quantified for the past, hence the one for which past values most effectively constrain future projections. For this reason alone, the recent IPCC TAR and subsequent global mean temperature projections call for detailed examination.

Each IPCC report sets a new and higher standard for estimating future global warming, but it is inevitable that many questions arise that merit further consideration. For example:

- Do the IPCC estimates encompass the full range of uncertainty?
- If we accept any given scenario as a starting point without an attached probability, how close can we get to attaching probability distributions to the temperature implications of that scenario?
- How robust are the climate sensitivities to greenhouse-gas forcing used by the IPCC?
- Is it possible to incorporate a wider range of forcings and feedbacks into the projections?
- Are the interactions and feedbacks between the climate system and biogeochemical changes fully considered and, if not, what difference might they make?

Some of these issues are considered briefly below.

One source of uncertainty arises from the failure of the IPCC TAR to include explicitly a full representation of the likely effects of land-cover change. The only feedback considered is the overall negative effect of albedo changes, estimated to have been between 0 and $-0.4\,\mathrm{W\,m^{-2}}$ since AD 1750. Pielke (2002) claims that the range of forcings that should have been considered includes reflectivity, evapotranspiration and hydrology. MacCracken (2002), however, argues that the increasing tendency for the most advanced models to include at least some of these processes as feedbacks may make Pielke's claim less valid. He further points out that the overall range of climate sensitivity to CO_2 even in earlier reports (1.5 to 4.5 °C for a doubling of atmospheric concentrations) is consistent with historical and palaeoclimatic evidence. This, he claims, suggests that the net forcing was not so far out even if all the contributory processes had not been explicitly included. The role of land-cover change as a driver of climate change past and present remains a controversial issue. Bergengren *et al*. (2001), by means of equilibrium vegetation models coupled interactively with a GCM, suggest that land-cover change would play an important role under conditions of doubled CO_2, with strong positive feedbacks especially where poleward migrating forests replace tundra. The resulting changes in albedo have consequences similar those described in Section 7.3 as the 'biome paradox'. Renssen *et al*. (2003) model the opposite effect by assuming global deforestation. In their 1750-year-long simulation the resulting increase in albedo leads to an expansion of sea-ice and eventually a strong non-linear ocean response, including weakening of thermohaline circulation and of northward heat transport in the Atlantic. Both studies suggest that extensive land-cover changes could have important impacts on the Earth system.

Historical data on global-temperature change play a key role in the IPCC process (see Chapter 12). In the IPCC TAR, models are included in the set used only if they adequately replicate the temporal pattern of changing temperatures during the twentieth century (see 12.6.1). The record of past temperature changes has also been used independently to calculate probability distributions of future global temperatures under given emission scenarios (Stott and Kettleborough, 2002). In their analysis they use the UK Hadley Centre model, HadCM3. The four SRES scenarios they use are A1F1, A2, B1 and B2. These together span pretty well the full range of emission sequences considered by the IPCC. Stott and Kettleborough carry out alternative ensembles of simulations, some with the full range of forcings included in the model and some with only well-mixed greenhouse gases. They also constrain the models by ensuring that they replicate the range of natural variability recorded over the last 140 years. One outcome of their analysis is the suggestion that up until 2030, global temperature is relatively insensitive to which of the SRES scenarios is chosen and the predicted increase lies between 0.9 and 1.9 °C. By the end of the century, the different scenarios lead to strongly divergent results, with, at the two extremes, A1F1 generating an increase of between 3.6 and 7.5 °C and B1, between 1.9 and 4.0 °C. They then ascribe probability density functions to the values for each projection (Figure 13.5). They make the general point that until *c*. 2040 uncertainties are dominated by uncertainties in climate response, not by differences in emission scenarios. This is partly the effect of increases in atmospheric CO_2 that have already taken place, but also partly because there is an element of self-cancelling between the differences in greenhouse-gas and aerosol emissions in each emission scenario. It is worth noting that the increase in global mean temperature of between 3.6 and 7.5 °C associated with the fossil-fuel-rich scenario was generated by one of the most sophisticated models available in terms of the range of forcings and feedbacks included.

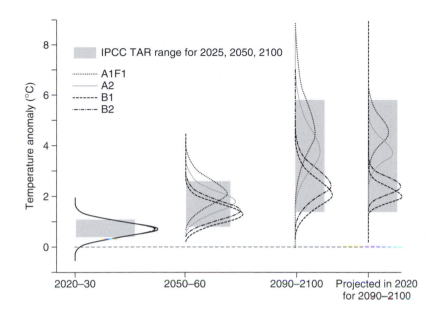

Figure 13.5 Probability density functions for three periods and four emission scenarios. Temperature anomalies are relative to 1990–2000 and calculated by constraining HadCM3 simulations to observed temperature changes 1900–99. No probabilities are attached to the superimposed IPCC TAR ranges. (Modified from Stott and Kettleborough, 2002.)

Wigley and Raper (2001) adopt a somewhat different approach to estimating probabilities. They begin by identifying the main sources of uncertainty, including the full range of SRES emission scenarios. They then represent each of these as probability density functions (PDFs) from which they select values to drive a climate model. Finally, the range of climate-model results is expressed as output PDFs. For this procedure to work realistically, it is necessary to select realistic values for the PDFs of many uncertain factors that have a bearing on future warming. The results of their analysis are shown in Figure 13.6. Like both the IPCC TAR (2001) and Stott and Kettleborough (2002), Wigley and Raper suggest that the calculations lead to only a very narrow band of uncertainty for global mean temperature by 2030. Their 90% probability range for the increase from 2000 to 2030 is between 0.5 and 1.2 °C, at the low end of the range indicated by previous authors. The spread of the 90% probability range for 2100 is between 1.7 and 4.9 °C. Both values lie within the IPCC TAR range, but their probability distribution suggests that both the highest and lowest IPCC TAR values are unlikely. Even so, their median warming rate from 1990 to 2100 is 0.28 °C per decade, around five times the rate over the past 100 years.

A third approach is that of Knutti *et al.* (2002, 2003). In both papers they use a model of reduced complexity (see 2.3) to provide independent constraints on radiative forcing and ultimately, future global mean temperatures. The first paper concentrates on estimating climate sensitivity to radiative forcing. The simulations from which the range of acceptable values are derived are constrained by the requirement that they match observed recent changes in surface warming and ocean heat uptake. Probabilities are then attached to the values obtained for total radiative forcing and indirect aerosol forcing.

Figure 13.6 Evolution of
uncertainties in global
mean warming shown as
probability density
functions for three periods
relative to 1990. No
probabilities are attached
to the superimposed IPCC
TAR range for 2100.
(Modified from Wigley and
Raper, 2001.)

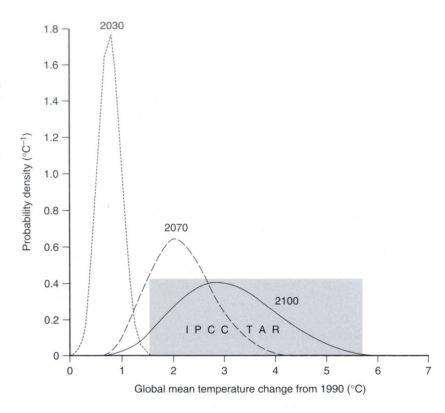

Using two contrasting scenarios from the SRES set (B1 and A2), they then derive probabilistic warming projections. The results suggest that there is a much stronger probability of temperatures exceeding the IPCC range (40 %) for the two scenarios modelled, than for them falling below the IPCC range (5 %). The issue of climate sensitivity to radiative forcing remains a key source of uncertainty, partly because it depends on the balance between positive 'greenhouse' forcing and likely negative aerosol forcing. Since they are mutually compensating, persisting uncertainties about the latter leave open the possibility of a wider range of values for the former. Uncertainties about sensitivity appear to have relatively little effect on projections for the first few decades, but become increasingly important towards the end of the century.

In the second paper, Knutti *et al.* (2003) use the same constraints on their model but greatly enhance the efficiency of their model ensembles by using neural networks. This approach allows for more comprehensive treatment of climate sensitivity, of each of the sources of uncertainty and of ocean and terrestrial-biosphere feedbacks through changes in the carbon cycle. It also significantly increases the rigour with which statistical analysis can be applied to the modelling results. First, the model is once more constrained by observational data on recent atmospheric and ocean warming. The calculations used to convert emissions to atmospheric concentrations include feedbacks from the carbon cycle in the form of reductions in ocean uptake as warmer conditions decrease CO_2 solubility, and reduced uptake by the terrestrial biosphere as a result of increased soil respiration and forest die-back. Albedo changes due to land-cover change and the effects of dust are not considered. Probability density functions are calculated for all ensembles that are consistent with recent observations and the ensemble size is increased until the PDF is stationary. As in the earlier paper, Knutti *et al.* (2003) use SRES scenarios A2 and B1 as the basis for translating their model output into future global mean temperature estimates to which PDFs are attached (Figure 13.7). As with the previously summarised projections, the differences generated by the two contrasting emission scenarios are negligible for the first 20–30 years, but begin to diverge significantly thereafter. Overall, the values suggested for the increase in temperature by 2100 are higher than those given in the IPCC TAR. The best-guess values are greater by 20 to 30 %, the upper limits of the uncertainty range, by 50 to 70 %. The higher values cannot be excluded, largely because higher climate sensitivities remain a possibility for as long as aerosol forcing is poorly constrained by the observational data.

Haung (2004) offers a quite different approach to constraining future anthropogenically driven global warming. First, he merges data from both a surface-observation series and those from borehole temperatures. He claims that this greatly improves the match between reconstructed temperatures and the history of radiative forcing by greenhouse gases, anthropogenic aerosols, volcanism and solar variability. From the merged data, validated in this way, he estimates that the transient response of global temperatures to forcing is 0.4–0.7 °C per $W m^{-2}$. From this it is possible to estimate the global temperature increase associated with the forcing implied by each of the IPCC scenarios. For an IPCC TAR mid-range increase in radiative forcing of 2.5 $W m^{-2}$ up to 2050, this suggests an increase in temperature of 1.0 to 1.8 °C by the middle of the century – a range broadly consistent with that proposed by the IPCC TAR.

For the present, the conclusions to be drawn from the IPCC TAR and subsequent studies rest on quite a strong measure of agreement regarding a likely increase of around 1 °C in global mean temperatures over the next 30 years or so, irrespective of the emission scenario used. Thereafter, different scenarios and different sensitivities give rise to significant divergences. Several post-IPCC TAR analyses tend to indicate that the IPCC estimates are more likely to be low than high for any given scenario. Although global mean temperature *per se* is not the main cause of concern in future climate change, it comes closest to being testable against past evidence. The projections made by models generating simulations that are consistent with this evidence therefore become, in most cases, the ones used as a basis for projecting other facets of future climate change and their likely effects.

Figure 13.7 Probability
density functions for
surface warming relative
to 1960–90 for scenarios
B1 (thin lines) and A2 (thick
lines). The dashed lines
show values constrained
by the sensitivities used in
IPCC TAR (1.5–4.5 °C);
solid lines show values
constrained
independently by
observations; shaded
areas show the extent to
which the range of values
constrained by
observations is
consistently higher for
each projected time slice
than that constrained by
the IPCC sensitivity values.
(Modified from Knutti
et al., 2003.)

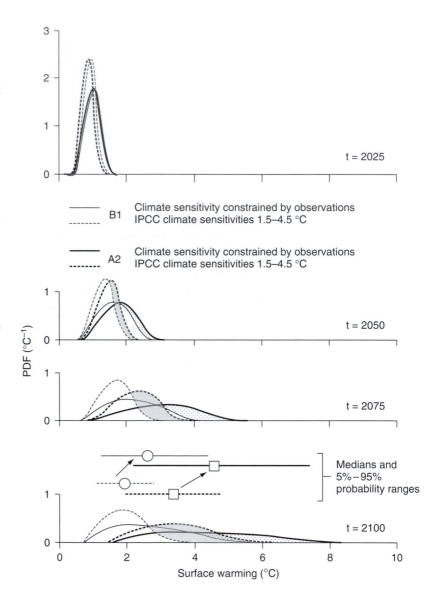

13.2   Future global mean sea-level

Although the precise links between global mean temperature and sea-level
during the late Holocene are still open to a good deal of uncertainty, none of
the evidence available seriously counters the view that warmer temperatures in
the future will be associated with higher sea-levels. All the factors considered in
the discussion of recent sea-level changes in Section 12.4 are considered in the
IPCC estimates in addition to possible changes in ground-water storage (Church
and Gregory, 2001). The projected rise by 2100 is estimated to be between 0.09

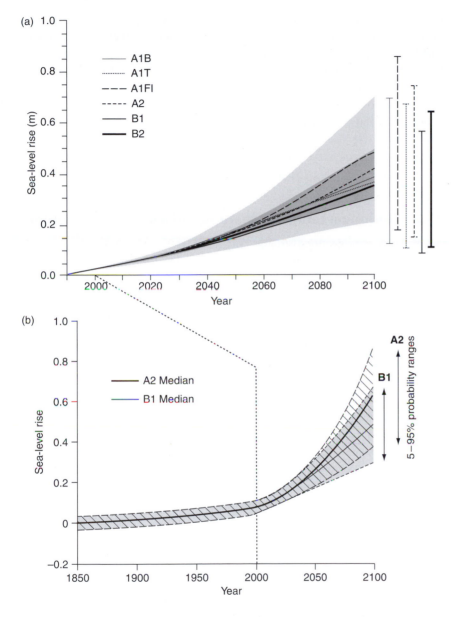

Figure 13.8 (a) Projected sea-level rise by 2100, using all six emission scenarios (see Figure 13.2). Dark shading shows the full range of projections by the model ensemble; light shading spans individual model projections for all the scenarios. The bars to the right show the full range for each of six individual models. (From IPCC TAR, 2001.) (b) Projected sea-level rise for emission scenarios B1 and A2, with 5–95% probability ranges The median values for these two scenarios are 12 to 18 cm higher than those in the IPCC TAR. (Modified from Knutti *et al.*, 2003.)

and 0.88 m for the full range of scenarios, with a mean value of 0.48 m. For each projected increase, about half is ascribed to the effect of thermal expansion, a process that will continue for several hundred years irrespective of the trend in future greenhouse-gas emissions, because of the slow mixing of heat in the oceans. As can be seen from Figure 13.8a the wide spread of possible values for each emission scenario greatly exceeds the difference between their mean values. This reflects the fact that there are large uncertainties additional to those arising from the temperature projections alone, notably uncertainties concerning the response of major glaciers and polar ice caps. Knutti *et al.* (2003) also calculated the projected sea-level rise by applying their simulations to SRES scenarios B1 and A2. Their median values of 0.48 m and 0.62 m respectively are significantly higher than those projected by the IPCC TAR (Figure 13.8b). Not all authorities agree. Mörner (2004), for example, takes a strongly opposed minority view. His analysis focuses mainly on observational records over the last 100 years and stresses the claim that between 1930 and 1950 mean sea-level fell. He also claims that the rate of sea-level rise in the late twentieth century failed to increase (but see 12.4.1). His projections for 2100 give values of $+10 \pm 10$ cm, or $+5 \pm 15$ cm.

13.3   The longer term – are there 'surprises' round the corner?

All the above projections strongly suggest that the trends in global mean temperature and sea-level from around the middle of the twenty-first century onwards will depend in part on the emission storyline actually followed. Looking further ahead, this becomes even more true, although few fully articulated projections have been presented. Even if an optimistic view is taken and CO_2 emissions are projected as declining at some point within the next 100 years, the projections presented in generic form in the IPCC TAR show that atmospheric concentrations will continue to rise and remain stable probably for centuries. All the models referred to above would translate this into a temperature increase continuing for several centuries before stabilising at a higher level. Sea-level would continue to rise for several millennia due both to ice melt and steric effects (Figure 13.9). In this formulation, the absolute values for sea-level and temperature depend largely on the level at which emissions peak. For any given long-term scenario, the uncertainties in temperature and sea-level projections widen even beyond the limits attached to the 2100 projections.

One of the frequently cited potential 'surprises' is a partial or even complete shutdown of the meridional overturning circulation in the Atlantic, as a

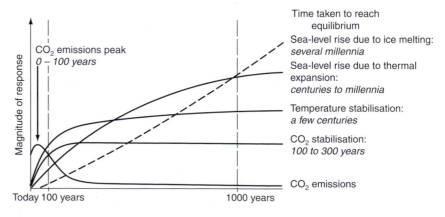

Figure 13.9 Projected relative atmospheric CO_2, temperature and sea-level trends beyond 2100, assuming that CO_2 emissions stabilise somewhere between 450 and 1000 ppm within the next 100 years. (From IPCC TAR, 2001.)

consequence of increased freshening of the North Atlantic and inhibition of North Atlantic deep water (NADW) formation (Manabe and Stouffer, 1995, 2000; Stocker and Schmittner, 1997; Rahmstorf and Ganapolski, 1999). Although the degree of freshening at the present day is still probably an order of magnitude less than that required for a complete shutdown, the continuation of current trends could eventually lead to the transgression of some critical, but as yet poorly quantified threshold. Vellinga and Wood (2002) use the HadCM3 model to estimate some of the likely consequences of such a shutdown. In their simulation using instantaneous freshening, overturning circulation collapses in ten years. This leads to a reduction in annual mean temperatures of 1–3 °C over Europe within a matter of 20–30 years. The simulated 3 °C cooling over Britain could bring mean annual temperatures closer to those characteristic of the Younger Dryas than to those prevailing during the coldest decade of the Little Ice Age.

The probability of a cessation in NADW formation and consequent complete shutdown in thermohaline circulation during this century is now deemed low by most authorities (see summary in Knutti et al., 2003), but there are wide variations in the projected percentage reduction in Atlantic meridional over-turning circulation (MOC) for any time in the future. Variations arise from the difficulty in simulating many of the factors upon which its stability depends. These include the response of the hydrological cycle to increased global temperatures, especially in the high-latitude regions surrounding the North Atlantic and draining into the Arctic ocean. Ocean-mixing processes have also proved difficult to parameterise, as has the initial strength of the MOC. All these generate uncertainties additional to those intrinsic to the global-temperature projections themselves. There is also the possibility that the MOC might be stabilised by increased wind-driven evaporation leading to increased surface salinity (Delworth and Dixon, 2000), as well as by changes in El Niño southern oscillation (ENSO) (Latif et al., 2000; Schmittner et al., 2000). Knutti et al. (2003) present an analysis based on their neural-network study applied to emission scenario B1, the more optimistic of the two they consider. In their analysis the projected strength of Atlantic overturning relative to the pre-industrial state shows a wide range of possible values, with the 5 to 95 % probability range for 2100 lying between 15 and 85 % of its current strength. At the very least, their results provide no grounds for complacency. Nor do those of Schaeffer et al. (2002). They use a model in which a rather advanced ocean component is linked to an 'intermediate-complexity' atmosphere. The IPCC SRES scenarios A1B, B1 and a modification of the latter were quantified in terms of atmospheric CO_2 concentrations using the IMAGE model (IMAGE-team, 2001). The resulting CO_2-equivalent concentrations ranged from 580 to 1050 ppmv for 2100. For the A1B scenario, all the simulations generate a sharp transition, at some time between 2040 and 2080, to conditions in which a collapse of convection occurs in the ocean around Spitzbergen. The resulting feedbacks are quite complex. The export of NADW is reduced, but only by some 12 %. Reductions in surface-air temperature also occur, especially in Greenland and Scandinavia. In all the simulations generated by the scenario with the lowest increase in CO_2-equivalent concentrations, (which depends on urgent and successful mitigation), no transition occurs, whilst in the intermediate case (B1), only a third of the simulations carried out generated a transition before 2100 and the changes are more gradual. The timing of transitions within each ensemble of simulations is strongly dependent on small perturbations of the kind resulting from natural variability. There are several important conclusions. Even with a transition to a much more spatially confined collapse of ocean convection, rather than a complete shutdown in NADW formation, there are likely to be significant consequences for surface temperatures at a regional scale. The likelihood of such a transition is strongly scenario dependent, with only the most mitigation-oriented scenario avoiding all indications of collapse before 2100. The strong sensitivity to small perturbations as the transition threshold is approached makes it all the more important to understand the likely future interactions between warming trends and natural variability. It also seriously limits predictability. One response to this problem is to identify diagnostic properties, changes in which can be detected readily during routine monitoring. Vellinga and Wood (2004) propose up to 15 observations that, on the basis of their simulations using the HadCM3 model, are likely, when used together, to lead to improved detection of changes in MOC. If simulations using other AOGCMs reinforce these conclusions, the probability of early detection may be significantly improved.

A second major 'surprise' anticipated by some authorities is the rapid collapse of the west Greenland ice sheet. Recent evidence for marginal instability over the last decade has been noted in Section 12.3.4. Oppenheimer's (1998) analysis of the likelihood of a collapse leading to rapid sea-level rise tends to downplays the probability of such an event within the present century. Rather, he favours a scenario in which collapse takes place gradually over 500 to 700 years, with basal melting and thinning of ice streams contributing 60 to 120 cm to sea-level rise towards the end of the time. More catastrophic scenarios leading to rapid collapse on shorter timescales are mapped out as possibilities and not entirely precluded. Should the maximum volume of ice loss envisaged by Oppenheimer eventually occur, the resulting sea-level rise would be of the order of 4–6 m. In a more recent analysis, Oppenheimer and Alley (2004) point out that one of the main processes that could lead to rapid melt – marginal ice-stream dynamics – has never been accurately modelled. If the recent, rapid disintegration of the Larsen ice sheet proved to be the precursor of widespread disintegration, there would be serious consequences for global sea-level. The authors estimate that a doubling in atmospheric CO_2 concentrations might be sufficient to produce the rise in temperature required to increase the risk of widespread melting and potential instability beyond acceptable levels. In addition, warming of circum-polar water could lead to thinning and ice-shelf loss with a smaller increase in air temperatures. There are indications that this is already occurring and leading to loss of grounded ice. Their conclusions include a note of warning as well as a call to quantify better the various lines of inference used in their evaluation.

Fear of catastrophic rise in sea-level has focused attention on Antarctica, but Schiermeier (2004c) and Gregory *et al.* (2004) summarise evidence suggesting that melting of the Greenland ice cap is also likely to contribute to rising sea-level in the future by up to 7 m over the next 1000 years. Such a scenario would depend mainly on the degree of warming in summer, since this largely determines the extent to which enhanced snowfall is exceeded by melting. In the projections by Gregory *et al.*, using a range of AOGCMs, melting exceeds snowfall once annual average warming exceeds 2.7 °C. In almost 70% of the combinations of AOGCM simulations and emission scenarios this figure is exceeded. Although the most extreme outcomes may be unlikely, even the simulations based on rather conservative projections imply a slow but highly significant long-term rise in sea-level in the absence of any substantial reduction in emissions.

13.4 Concluding comments and questions arising

From the present chapter we can draw some tentative conclusions about the likely future behaviour of global mean temperature, sea-level, large-scale ocean circulation and the least stable parts of the Antarctic and Greenland ice sheets. Pending a review of the most sceptical views (16.2), we may conclude that:

- Although the IPCC TAR and subsequent studies all make projections tied to the SRES emissions scenarios, the way in which these scenarios are treated in post-IPCC modelling exercises introduces a strong element of independent evaluation.
- Over the next 30 years or so global mean temperature is likely to rise by around 1 °C. This appears to be a robust conclusion irrespective of the kind of climate/Earth-system model used and also to be rather insensitive to the trajectory actually taken by emissions of greenhouse gases and sulphate aerosols.
- In the analyses reviewed here, there is no strong disagreement with the IPCC TAR projections that by the end of the century, global mean temperature will probably have risen by between about 1.4 and 5.8 °C.
- Such divergences with the IPCC TAR as have been revealed by the subsequent studies reported here mostly favour the view that for any given emission scenario, the IPCC TAR projections are more likely to be low than high, though generally the most extreme values are given a rather low probability.
- The rate of global mean temperature rise during the present century is therefore likely to be between two and ten times that which occurred during the twentieth century and probably at least five times.

- Although there are many areas of uncertainty, narrowing the range of temperature projections beyond *c*. 2030 will depend above all on narrowing the range of emission scenarios and refining estimates of the sensitivity of the climate system.
- The course of emissions during this century will have a major effect on the evolution of global mean temperatures in the centuries beyond 2100.
- Although it is acknowledged that future sea-level rise is already 'built in' by virtue of the slow rate of thermal mixing in the oceans, estimates of the rate of rise are still subject to major uncertainties additional to those linked to temperature projections.
- At least one post-IPCC TAR projection of sea-level rise suggests that the IPCC TAR range (0.09 and 0.88 m for the full range of scenarios, with a median value of 0.48 m) may be on the low side.
- The likelihood of either a shutdown of the thermohaline circulation or a collapse of the west Antarctic ice sheet during the present century is probably low, but neither possibility can be precluded, especially during subsequent centuries, if there is no early curb to the increase in greenhouse-gas emissions.
- Opinions differ as to the wisdom of attaching probabilities to emission scenarios, though there is a growing tendency to attach probability density functions to the projections of global mean temperature and sea-level that each scenario generates.

These are important conclusions regarding the future functioning of the Earth system, but what do they tell us about the likely consequences for human populations? How will these projected changes interact with the myriad other effects human activities are having in almost every corner of the planet? To explore these questions, we have to look beyond global mean temperature and sea-level. The former is of much less significance for human populations than are major changes in hydrological regimes at regional scales. The latter, for any given stretch of coast, needs to be translated into local impacts that reflect many factors other than global mean levels. Having established some tentative conclusions that lie at the heart of much of the global-change debate, we are still far from answering many of the key questions from the standpoint of ecosystem functioning, environmental sustainability and human vulnerability.

Chapter 14
From the global to the specific

14.1 Introduction

Global mean temperatures, whether for times in the past or the future, have become a touchstone for 'global change'. As the previous chapter has shown, the fact that they have been reconstructed for the past and modelled in a reasonably well-constrained, probabilistic way for the future gives them special value. As with temperature, so with sea-level; the global picture makes a realistic point of departure for further analyses. Once we move beyond global temperature and sea-level and seek either spatially to disaggregate global values or consider aspects of climate other than temperature, everything becomes much more complicated and even more uncertain.

This chapter looks at the ways in which future projections of climate change may be resolved zonally and regionally, paying particular attention to the likely implications of future climate change for the hydrological cycle. Ideally, down-scaling future climate scenarios to the regional level should involve a logical sequence of inferences that include the testing and calibration of regional models against the instrumental record from the recent past at the very least. Model projections might then be developed for any chosen emission scenario, with attached probabilities, for target times and places. These steps would pave the way for a closer look at the future interactions between climate and the full range of human activities. They are, however, far from simple and they continue to provide formidable challenges in an area where, despite impressive progress, there is still a very long way to go.

Approaches to developing regional projections of future climate have been quite diverse (Gyalistras *et al.*, 1998) The purely empirical/historical approach builds on the possibility that the record from the past, both instrumental and proxy-based, may contain analogues for future conditions. Bauer *et al.* (2003) examine this possibility in relation to the future climate and vegetation of the Sahara where the early-Holocene period of greater moisture availability, extensive lakes and abundant vegetation (see 7.4) has been regarded by some as a possible analogue for conditions under future global warming. Their Earth-system models of intermediate complexity (EMIC) simulations for the early–mid Holocene

(Figure 2.3) and for the future, under a range of atmospheric CO_2 concentrations show that the past is likely to be a poor analogue because the global pattern of climate change is quite different, as are the combinations of forcings and feedbacks involved at regional level. Spatial analogues have also been used on the assumption that the climate of another region may be used as a template for the future climate in the region of interest. These types of analogue approach have been used with varying degrees of sophistication and, in some cases, combined with modelling in an effort to improve the process of downscaling to regional level for example. Increasingly, the modelling approach, constrained by recent climate observations, has come to dominate the field. Gyalistras *et al.* (1998) compare the projections generated for the region of the Alps using a wide range of approaches, but with no firm conclusions as to the relative merits of each. In the present account, the focus is on modelling approaches and mainly on the most recent examples, since the field is developing so rapidly, as computing power increases, global climate models become more comprehensive and detailed, and the need to develop regional scenarios becomes ever more urgent.

We begin by looking at the results of some of the most recent attempts to test simulations of rainfall and evaporation and to develop future scenarios for major climate regimes like the southeast Asian monsoon and for regional climates elsewhere.

14.2   Future precipitation, evaporation and runoff

The IPCC TAR envisages a modest increase in global mean precipitation and evaporation during the present century, but with strong regional differences. The projected overall increase is in line with the sequence of processes outlined in Figure 14.1. This does not imply that precipitation can be modelled with any high degree of reliability. Climate models still fail to reproduce the present-day diurnal distribution of rainfall. Allen and Ingram (2002) show that even at the level of mean annual precipitation over land on a global basis, without any differentiation into regions or seasons, models provide only a rough approximation of the observed variability over the last 55 years (Figure 14.2). They point out that the changes in precipitation over this period do not follow the temperature trend but are more closely linked to natural variability than to greenhouse-gas forcing. They infer that, Figure 14.1 notwithstanding, the anthropogenic influence on precipitation is still relatively weak (or at least difficult to detect at a global level). One of their most important conclusions is that uncertainty ranges derived from inter-model comparisons as applied to future precipitation projections are likely seriously to underestimate true uncertainty. This cautious view is reinforced by Lambert *et al.* (2004) who conclude from their detection and attribution study that 'Predictions of twenty-first-century hydrology based on sensitivity to twentieth-century temperature changes are likely to be inaccurate'.

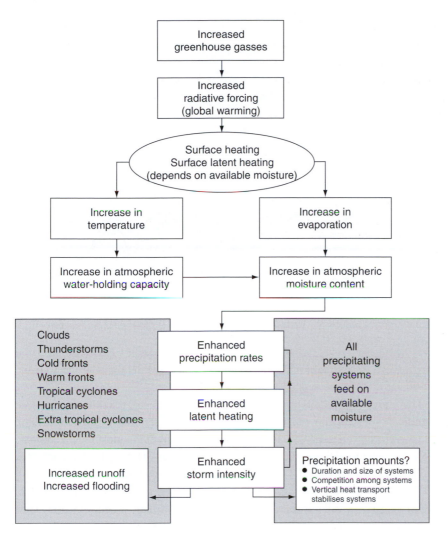

Figure 14.1 A schematic outline of the likely effects of warming on atmospheric-moisture content (specific, not relative humidity), evaporation and precipitation rates. (Modified from Trenberth, 2000.)

Despite differences between the results generated by the various models used in projections of future precipitation, there are significant areas of agreement, especially on a simple zonal basis, with areas of currently high precipitation expected to experience increases and areas of high evaporation (for example continental interiors) decreases. Precipitation is expected to increase in northern mid- to high-latitudes, as well as in Antarctica, during winter months. The pattern in lower latitudes is spatially somewhat less coherent. By using climate-model projections as input to a hydrological model, the IPCC TAR presents a partial picture of the balance between precipitation and evaporation as expressed through runoff. Among the more robust results of this exercise, based on two versions of the UK Hadley Centre atmosphere–ocean global climate model (AOGCM), are projected increases in runoff in high latitudes

Figure 14.2 Changes in observed global mean temperature (a) and precipitation (b) since 1945, compared with simulations using the HadCM3 model. The bold continuous lines show the observed values; the bold pecked lines show the ensemble mean values. The shaded area represents the range of values covered by the four simulations used. (Modified from Allen and Ingram, 2002.)

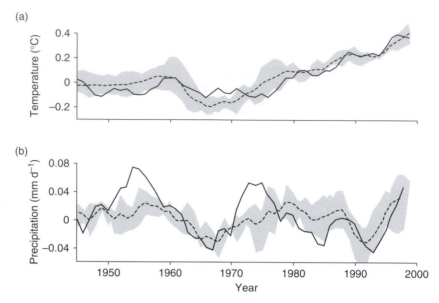

and southeast Asia, alongside decreases in central Asia, around the Mediterranean, southern Africa and Australia. The results for other parts of the world are much more scenario- and model-dependent.

The above indications are, in so far as they may be reliable, much more significant for human populations than are estimates of global mean temperature. We therefore need to explore their reliability in light of the most recent comparisons between model output and climate observations. Harvey (2003) compares the output from eight AOGCMs, with observed precipitation for the period 1979–99, compiled from rain-gauge measurements, satellite data and NCAR-NCEP reanalysis. All but two of the models have flux adjustments and they vary by a factor of four in overall spatial resolution. The models vary greatly in the skill with which they reproduce the observation-based precipitation climatology, especially in the tropical and southern Pacific. Three out of eight fail to simulate the precipitation maximum associated with the ICTZ (inter-tropical convergence zone) and four represent poorly the major band of precipitation associated with convergence in the southern Pacific region. Only three models do well in simulating both these areas of heavy precipitation. The non-flux-adjusted models generate two zones of ICTZ-linked precipitation maxima rather than one. In other respects, the models simulate precipitation correctly if often crudely. The pattern correlations (R^2) between each model and the observed values range from 0.56 to 0.85 for annual, 0.56 to 0.82 for December/January/February and 0.49 to 0.75 for June/July/August precipitation. Figure 14.3a compares the variations of mean daily precipitation with latitude for each model against the maximum and minimum observed mean daily values for the period.

(a)

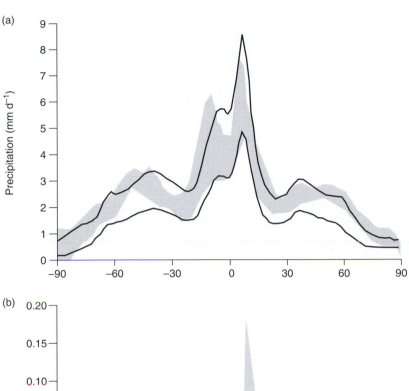

(b)

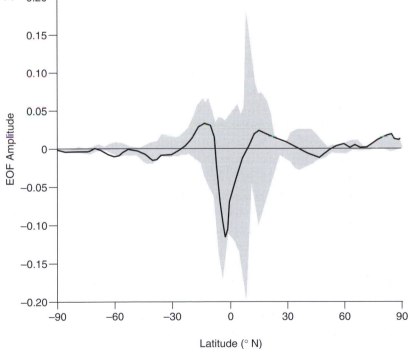

Latitude (° N)

Figure 14.3 (a) A comparison between observation-based and simulated zonal mean annual-precipitation values for the period 1979–99 using eight AOGCMs. The shaded area represents the total span of values for all models. The two bold lines represent the maximum and minimum observed values. (b) Values for the first EOF of precipitation vs. latitude for the period 1979–99. The shaded area shows the range of values for all eight AOGCMs The bold line relates to the observed values. (Modified from Harvey, 2003.)

A more sophisticated basis for comparison uses empirical orthogonal functions (EOFs) (a modified form of principle-component analysis applicable to spatio-temporal variability) as a basis for evaluating the differences between models and the observations. Taking both the annual and seasonal results as a whole, correlations (R^2) between models and observations for EOF_1 (linked to El Niño southern Oscillation, ENSO) range from 0.012 to 0.61. Figure 14.3b expresses this scatter of model simulations relative to observations as zonally averaged values. Overall, the results are mixed and the author concludes that much needs to be done to improve model climatology. It is important to measure the relative success of the model against the kind of key issue that future projections of precipitation need to be able to project realistically. The future of the Amazon rainforest is of vital importance from many perspectives, including biodiversity, feedbacks into carbon cycle and regional if not global climatology. Harvey points out that whereas in the simulations using the flux-adjusted HadCM2 model, the rainforest survives to 2100, albeit with reduced productivity, with the non-flux-adjusted HadCM3 model it is replaced by semi-desert.

14.3   Modelling monsoons and ENSO

One of the key issues for the future is the likely behaviour of the monsoon systems that are responsible for delivering water to a large proportion of the human population. Walliser *et al.* (2003) present the results of a monsoon global climate model (GCM) intercomparison project involving ten groups, each of which carried out a set of ten ensemble simulations for the two-year period from 1 September 1996, using identical prescribed sea-surface temperatures (SSTs) but different initial conditions. The study focuses on within-season rainfall variability and the skill with which the models are able to simulate spatial and temporal changes during a monsoon season. A major goal of the study was to assess the degree to which the spatial propagation of the monsoon system through the season, and associated precipitation patterns, may be predictable using state of the art models. The authors conclude that 'of the ten models, only a few exhibit a somewhat realistic north-eastward propagating character' and that 'none of the models demonstrate a great amount of fidelity in simulating all or most of the details of the spatio-temporal pattern well'. Teleconnections beyond the regions most affected by the monsoon are poorly represented, as indeed is mean annual rainfall in the equatorial Indian Ocean. Clearly, current models fail to capture adequately the complex processes driving major systems like the southeast Asian monsoon even over a short period within the span of instrumental observations. When we consider the greater range of variability such systems have exhibited on century to millennial timescales, the problem becomes compounded. Nevertheless, there may be some rather robust, albeit rather generalised and qualitative conclusions to be drawn from future projections of south Asian monsoon behaviour. In the ensemble simulations with two and four times the current CO_2 atmospheric concentrations carried out by Meehl and Arblaster (2003), using a coupled model run with a range of SST values, mean monsoon precipitation is increased. This is due largely to warmer Indian Ocean SSTs providing an enhanced moisture source. Inter-annual variability in precipitation also increases at higher CO_2 levels, mainly as a result of warmer Pacific Ocean SSTs giving rise to

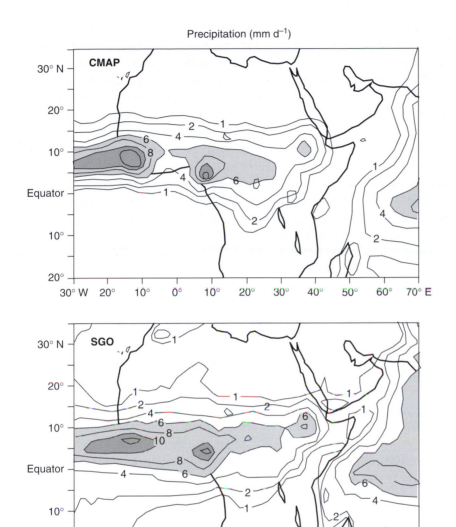

Figure 14.4 Comparison between the observed and simulated geographical distribution of precipitation in July, August and September in equatorial Africa and adjacent regions. The CMAP map refers to a data base 1979–1996. The SGO map refers to a simulation by a climate model (the Arpege-climat SGO) for the period 1980–2010. (Modified from Maynard *et al.*, 2002.)

greater variability in evaporation. In conclusion, although such broad scale projections may be possible with reasonable confidence, current models are still a long way from being able to translate these into credible projections for particular regions in terms of timing, quantities and inter-annual variability.

Maynard *et al*. (2002) in their study of the west African summer monsoon use a coupled ocean/sea-ice/land-surface/atmosphere model, which is generally quite well constrained by comparison with reanalysis data over a 30-year control period (Figure 14.4). Their future scenario is for 2070–2100 and is based on

SRES B2 with a storyline intermediate in terms of emissions, and regionally diverse in terms of demographic and technological change. Several of their future projections echo those proposed in earlier studies – greater warming over land than over the sea, an enhanced hydrological cycle, higher monsoon precipitation over west Africa and a poleward shift in the monsoon regime. As the authors point out, however, the results are 'not sufficient to assess the regional impacts of global warming'. As in the previous cases, the simulations fail to resolve changes at regional level, partly because the spatial resolution of the model used is too coarse to allow inclusion of geographical variations in land-surface feedbacks. Moreover, although the authors state that the increase in atmospheric-water content exceeds inter-annual variability in rainy-season precipitation, their timeframe for such a claim, as in all the cases considered in this chapter, may be far too short to encompass the full range of variability experienced during the late Holocene.

Timmerman *et al.* (1999), using a GCM that simulates ENSO-like variability reasonably well, claim that the warmer climate simulated for future decades may lead to more persistent El Niño-like conditions, with increased inter-annual variability and the likelihood of increased drought incidence. This view is reinforced by Boer *et al.* (2004), on the basis of both model simulations and comparisons between ENSO in recent times and at the LGM. The analysis by Boer *et al.*, however, neglects all the evidence for El Niño events during the first half of the Holocene (Moy *et al.*, 2002; Gagan *et al.*, 2004; McGregor and Gagan, 2004), as well as the indications of volcanic forcing noted by Adams *et al.* (2003). These considerations, along with the inferred non-linearity of the atmospheric responses to SST variability during the Holocene (McGregor and Gagan, 2004) add to the challenge of linking probabilities to projections for the vast area in which climate is partly controlled by ENSO variability.

14.4   Circum-Arctic climates

One of the most likely outcomes of any future climate change is strong warming over the northern polar regions. Models differ, however, in the projected strength of any future 'polar amplification' and its regional expression. As we have seen in Sections 12.3.1 and 12.5, there is already strong evidence for widespread changes in sea-ice and in arctic ecosystems, supporting the view that responses to warming are already under way. Less sure is the extent to which these changes reflect the effects of anthropogenic greenhouse gases or changes in the phase of the closely linked North Atlantic and Arctic oscillations (NAO and AO, respectively) described in Section 7.7. Dorn *et al.* (2003) explore this question using a downscaled regional climate model linked to an AOGCM for both control and future simulations. The control simulations are broadly comparable to the observed record for both the NAO and AO, so some confidence may be attached to the regionalised scenarios they generate.

For future projections Dorn *et al.* use the HadCM2 and ECHAM4 models and they present the simulated future climates up to 2050, then select, for downscaling to regional level, time intervals in which the effects of elevated CO_2 coincide with extreme opposite phases of the modes of variability. They simulate temperature, precipitation and sea-level pressure at regional scale for what are hypothetical future climates in which either high or low NAO/AO indices coincide with the effects of elevated CO_2. Among the several points they make based on the results of their simulations is the suggestion that for some circum-Arctic regions the amplitude of NAO/AO variability during the first half of the present century may be at least comparable to, and perhaps exceed, the effects of any 'greenhouse warming'. They find no evidence for an increasing trend in the NAO index, but their simulations show that the results are highly model dependent as well as very sensitive to the initial conditions used in the simulation. This latter point suggests that the modes of variability may respond non-linearly to small variations in the mean climate. In some regions the simulated changes exceed any possible linear response to global atmospheric composition. None of these suggestions constitutes a prediction about the future climate of any circum-polar region. They are credible simulations for particular combinations of atmospheric composition and variability index. Nevertheless, they provide interesting indications of the possible future interaction between global mean temperature trends and one of the dominant modes of climate variability.

14.5   High-resolution simulations

Moving from global to regional values for future climate projections calls for higher spatial (horizontal) resolution, either through decreasing the grid-cell size of the global model or downscaling the results for a nested set of grid cells. Duffy *et al.* (2003) use an AGCM, the Canadian Climate Center Model CCM3, at a range of spatial resolutions to explore the extent to which higher spatial resolution leads to improved simulations at regional level. Increasing resolution carries with it the need to adjust parameters for a wide range of sub-grid-scale processes, such as convection, cloudiness or precipitation. This and other limitations mean that higher resolution does not necessarily lead to enhanced performance. Their simulations, using forcings and boundary conditions designed to allow comparison with the annual climate of the year 1998, clearly show that increasing spatial resolution coupled with retuning improves the degree to which simulations capture almost all the features of observed regional climate, including precipitation. On the scale of the USA, the spatial pattern and the intensity of extreme events are more realistic in the high-resolution simulations. Simulations of tropical variability are much less successful at all scales of resolution. The final conclusion is that the results obtained may be strongly model dependent, implying that the effects of changing resolution may vary from model to model.

In a follow-up paper, Govindaswamy *et al.* (2003) proceed to apply CCM3 model simulations at different resolutions to the future climate at 2100. Only a single model is used and SSTs are prescribed. The exercise is designed to explore differences between simulations at different resolutions rather than to provide future projections, the specific characteristics of which have been tested for their likely reliability. The conclusions are thus limited and include the demonstration that higher-resolution simulations make little difference to the representation of large-scale patterns and climate sensitivity, but capture regional, orographic effects more effectively. One of the most intransigent problems in regional dowscaling arises from the fact that any biases in the global model are inherited through the nesting process by which the regional model is developed.

Raisanen *et al.* (2003) come closer to providing probabilistic, regional projections in that their study uses two contrasted emission scenarios (A2 and B2) and a regional climate model (RCM, with a grid scale of 49 km) developed by the Rossby Centre in Sweden and driven by two global models, HadAM3H and ECHAM4/OPYC3. The control runs span 30 years (1961–1990), as do the projected scenario runs (2071–2100). Their research strategy therefore generates four scenarios of possible future regional climate. These lie within the mid-range of the total range projected by the IPCC TAR (Cubasch *et al.*, 2001), with the two global driving models used giving a temperature increase between the control and the future-scenario time windows of 3.2–3.4 °C for A2, and 2.3–2.6 °C for B2.

In the control runs, both models simulate mean surface air temperature rather well for 1961–1990, but with a warm bias averaging *c*. 1 °C that is strongest in the summer simulations for southeastern Europe, and for winter simulations in northern Scandinavia. Cloudiness and precipitation are less well simulated, with an underestimate of summer precipitation in southern Europe. There is also a tendency for the models to generate unrealistically high levels of cloudiness in northern Europe and the converse in the south. The diurnal temperature range in northern Europe is underestimated and extreme rainfall events and wind speeds are not well simulated, perhaps because the grid scale is till too coarse at 49 km^2. Despite some limitations of the control simulations, they provide one of the best bases available so far for exploring the effects of different emission scenarios, and alternative models. They constitute an important, albeit still tentative step towards developing future climate scenarios for Europe at a regional scale. In their simulations, precipitation generally increases in northern Europe, especially in winter, but decreases in central and southern Europe in summer, though the quantities and spatial patterns involved are strongly model dependent, partly as a result of model-generated differences in atmospheric circulation. Projected extreme-temperature values are in proportion to the mean-temperature increases, except in the case of increased summer maxima in the south and higher winter minima over much of the rest of continent. In both cases, the extremes show greater divergence from the control values than do the mean annual temperatures. Extreme daily precipitation increases over almost the whole area, which implies that in the south reduced precipitation is linked to fewer rainy days rather than reduced rainfall intensity. The results include some that are robust across both scenarios and both models. In northern Europe, the strongest warming is in winter or late autumn, but in central and southern Europe, peak warming occurs in summer, when local increases can range between 6 °C and 10 °C, depending on the chosen combination of scenario and model. Recent temperature trends around the Mediterranean may be a first expression of such a change (Saaroni *et al.*, 2003), as may the extreme summer temperatures in much of Europe in 2003, if the model-based analysis by Schär *et al.* (2004) of their statistical probability under different conditions of forcing is reliable. This view is further reinforced by Pal *et al.* (2004), who find that there is a high degree of consistency between the model-based projections of future summer atmospheric-circulation and precipitation regimes in Europe, and the observed changes over the last 25 years, culminating in the summers of 2002 and 2003. On the basis of their analysis, they suggest that the central Mediterranean and central/western Europe are likely to be especially vulnerable to both summer droughts and floods in future.

The UKCIP (2002) climate-change scenarios generated for the UK are broadly comparable to the Raisanen *et al.* (2003) study in that they use a downscaling procedure and a regional climate model to generate interpretable projections at 50 km resolution. They use output from only one global model, (HadCM3) to drive a high-resolution model (HadAM3H) which in turn provides the boundary conditions for a regional climate model. The model drives the simulations with four emission scenarios and is even capable of providing descriptions of projected future climate at 5 km resolution. This particular approach to future climate scenarios is considered further in Section 16.6.

Mearns *et al.* (2003) present a rather less developed set of scenarios for the southeastern USA, using a model nested in only a single GCM and two times the current CO_2 concentrations. Having demonstrated in their study that the nested regional model matches observations for the region more effectively than does the driving GCM, they outline a series of changes under two times the CO_2 concentrations that are common to both scales of resolution. These include large decreases in summer and increases in spring rainfall. Other changes are more spatially and seasonally differentiated in the scenarios generated by the regional model as compared with those using the GCM.

Benestad (2003) presents a careful evaluation of model simulations of past and projected future regional climate change in southwest Scandinavia. He contrasts the poor performance of single-model derived scenarios with the much more reliable performance of multi-model ensembles when compared with temperature data for Oslo from 1890 to 1999. The scenarios are derived from downscaling seven GCMs. On the basis of the good performance by the multi-model ensemble, he concludes that the results 'suggest that multi-model ensembles are capable of predicting most of the local climate changes related to an enhanced greenhouse effect'.

14.6   Floods in the future

By now, the need to develop projections of the changing magnitude and frequency of climate extremes, in addition to mean values, has been widely acknowledged. In their consideration of extreme-precipitation events, Allen and Ingram (2002) suggest that one of the most significant changes is likely to be an increase in the frequency of the heaviest rainfall events as absolute humidity increases along with temperatures. Even if such increases were to remain quite modest, they could nevertheless substantially shorten the return period of extreme events.

Walsh *et al.* (2003) have attempted an analysis of this effect for cyclones and tropical storms over eastern Australia. Using a model with 30 km horizontal resolution, they first simulate cyclone and storm occurrence for a 30-year period, beginning in 1967. Although their model tends to overestimate occurrence during the control period, it provides a reasonable basis for attempting to estimate the direction and scale of possible future changes. Using a scenario with three times the current CO_2 concentrations the authors find little indication of a significant change in cyclone occurrence, but a potentially large increase in the incidence of tropical storms. The projections are very sensitive to the thresholds, criteria and spatial resolution used and no probabilities are attached to the results. Rather similar conclusions with regard to future trends in hurricane and storm activity under global warming are reached for the coastal United States by Bengtsson (2001).

Palmer and Raisanen (2002) and Milly *et al.* (2002) provide two pioneering studies that include estimates of statistical probability. Palmer and Raisanen use, as their starting point, 80-year integrations from a multi-model ensemble of 19 coupled models. Their control-ensemble run is based on constant atmospheric CO_2 concentrations of 330 ppmv; the 'greenhouse' projection is driven by a transient increase of 1 % per year. This gives rise to a doubling of atmospheric CO_2 concentrations in 70 years, which is in the mid-range of the IPCC TAR projections. Probabilities of events of a given magnitude and frequency were established for the control run and for the years 61–80 of the 'greenhouse' run. The resulting calculations point to a five-fold increase in the probability of exceptionally wet winters (two standard deviations above the mean) in the UK over the next 50–100 years. Higher probabilities of similarly exceptional summer-monsoon precipitation in the catchment basins of the Brahmaputra, Ganges and Meghna rivers are also envisaged, pointing to a likely increase in the risk of flooding in Bangladesh.

Milly *et al.* (2002) couple their climate simulations to a hydrological model applied to river basins greater than 200 000 km^2 in extent, a scale appropriate to the resolution of the climate model used. As a first step, they assess the probability that the observed increase in the frequency of 100-year floods (i.e., those with a probability of 0.01 in any given year) over the last few decades has been entirely random. For the world as a whole, they show that the distribution of events through time has only a 1.3 % probability of being random. For extra-tropical regions, the probability is 3.5 %. This they regard as strong evidence for the increasing incidence of major floods over the last few decades. The authors then develop future projections based on a 1% increase in atmospheric CO_2 concentrations for 140 years, followed by stable concentrations at a level four times the control values. In all but one of the major river basins considered, the projections indicate an increased frequency for the '100-year' flood. In half of the basins, the increase is eight-fold or more, implying a decrease in return period from 100 years to less than 12.5. The conclusions as regards both the recent trend and future projections are tentative, especially since the modelled climate lacks any representation of many of the forcings and feedbacks that are likely to affect climate. In Clarke's (2003) view, most future projections of extreme events using extrapolation from current hydrological regimes become extremely hazardous when the statistical stationarity of magnitude/frequency relations is called into question, and once changes in both climate and land use are involved.

The likelihood of a significant increase in flood magnitude or frequency in the lower valleys of the Brahmaputra, Ganges and Meghna rivers is signalled in both the above papers. Mirza *et al.* (2003) give this possibility further consideration. By linking empirical river discharge and hydrodynamic models to four GCMs, they seek to characterise likely changes in the peak discharge of the three rivers, in order to assess the extent to which the projections are model dependent, and to translate the projected changes into impacts on flood regimes in Bangladesh. They consider the cases for global mean temperature increases of 2, 4 and 6 °C using each of the four models. The authors translate these results into changes in inundated area in several categories based on topography and current inundation characteristics. They conclude by reinforcing the view that the higher peak discharges projected for the future may indeed lead to more serious flooding in Bangladesh, with consequent increased vulnerability for human populations. The possibility that the situation may be further exacerbated by rising sea-level is not considered.

Taking the above studies as a whole, one thing that emerges is the difference between developing probabilistic models proposing likely changes relative to a modelled control period, and generating simulations that may provide absolute estimates of consequences in terms of flood hazard and human risk. So far, attaching probabilities to the latter seems to be proving elusive.

14.7   Future droughts

The various sections of text above bring us closer to a consideration of future climate change in relation to human welfare. This is the main focus of the next chapter. This section, by briefly considering water scarcity, begins to build a bridge towards a more rounded view of the future impacts of all aspects of global change. It is important to bear in mind distinctions between meteorological (deficit of precipitation), agricultural (largely soil-moisture deficit) and hydro-logical (low lake-levels and river flow) definitions of drought (Trenberth *et al.*, 2004), as well as to recall that a warming atmosphere (by increasing both evaporation and absolute humidity) can increase the chances of both drought and floods in the same, or neighbouring regions.

The record of past human migrations (Dillehay, 2002), welfare and societal collapse is rich in examples of economic and cultural decline coinciding with prolonged drought (Haberle and Lusty, 2000). While always acknowledging that the effects of environmental stresses on human populations are mediated by a whole complex of interacting social factors, and that it is therefore unacceptable to invoke climate alone as the cause of civlisation collapse (Rosen and Rosen, 2001; Dahlin, B. H., 2002; Shennan, 2003; Dillehay *et al.*, 2004), the coincides are still impressive, confirming that in many situations drought has been one of the factors contributing to major declines in civilisations as diverse as the Maya (Hodell, 1995; Haug *et al.*, 2003), Anasazi (Larson *et al.*, 1996), Hohokam (Nials *et al.*, 1989), Tiwanaku (Chepstow-Lusty *et al.*, 1997) and prehistoric cultures in the Atacama and Andean Altiplano (Nunez *et al.*, 2002) in the New World; likewise the Akkadian (Weiss *et al.*, 1997; Weiss and Bradley, 2001) and Harrapan empires (Singh *et al.*, 1990; Staubwasser *et al.*, 2003), and groups in the east Mediterranean (Rosen, 1995), the Sahara (Hoelzmann *et al.*, 2001; Nicoll, 2004), South Africa (Tyson *et al.*, 2002) and China (Huang *et al.*, 2003) in the Old. Continuous records of recurrent drought linked to periodic cultural decline (and vice versa) come from east Africa (Verschuren, 2000). At the present day, much of the world's population suffers from serious water stress in some form or other and over the next two decades, rising-water demands linked to population growth and economic development will play a much greater role than climate change in defining the status of water resources (Vörösmarty *et al.*, 2000). All these considerations prompt a closer look at the projected future availability of water in relation to human needs. Few authorities doubt the size and urgency of the problem (see, e.g., Arnell, 1999; Parry *et al.*, 2001; Anon, 2003a), though signs that it is beginning to generate effective global action, as distinct from further study, are all too sparse.

In view of the difficulties involved in modelling the spatial distribution of precipitation for the present day, itself only one part of the challenge, it is hardly surprising that the task of developing hydrological models that can be used for projecting the potential availability of water resources in the future is fraught with a multitude of uncertainties. Arnell's (1999) analysis illustrates well the problems involved when models generate widely different climate-change and water-demand scenarios. Some progress has been made by applying improved climate models to well characterised past drought events. Giannini *et al.*

(2003) show that the major twentieth-century droughts in the North American 'dust bowl' and in the Sahel region can be simulated in atmospheric models by specifying SSTs. In their analysis, ocean forcing is more important than changes in land cover, though the latter may generate feedbacks that prolong the drought.

Doll (2002) attempts to evaluate the impact of climate change and variability on irrigation. She uses a global model to compute irrigation requirements under a range of conditions. The research strategy is designed to make possible comparisons between the effects of 'climate change' (in the future), and those of 'climate variability' (in the recent past). The former is defined in terms of differences between the 'baseline' climate (1961–1990) and decadal simulations for the 2020s and 2070s. The latter is described in terms both of inter-annual variability over the timespan 1901 to 1995, and long-term multi-decadal variability, defined by comparing values for 1901–30 and 1931–60. The two GCMs used, HadCM3 and ECHAM4 generate rather different future projections, with the former indicating more extensive areas of reduced annual precipitation in north Africa, around the Mediterranean and in the Middle East. Despite the careful approach taken, uncertainties abound in addition to those already cited for simulations of present-day precipitation. They span the whole range from neglected processes in the climate models to uncertainties about future crops and their growing seasons, and about the effects of increased atmospheric CO_2 on crop physiology. Moreover, only for the largest river basins is the horizontal resolution of the models used appropriate. Tentative overall conclusions include an indication that around two thirds of the area currently equipped for irrigation may suffer from increased water requirements and that for up to half the total area the negative impacts of 'climate change' are more significant than those of climate' variability', as both are defined above. The latter observation is somewhat conditional on how variability is defined. This is one of several papers comparing 'change' with 'variability' (see, e.g., Hulme et al., 1999). The very exercise highlights two limitations in much of the current research on future climate change. First, variability, whether termed 'natural' or not, is defined in terms of part or parts of the all too short instrumental record, much of which lies within the period of growing anthropogenic impacts. As we have seen in Section 7.8, proxy records for the late Holocene show that for many parts of the world, especially when we consider rainfall, hydrology and drought incidence, this short period fails to capture the full range of variability that has occurred, both on inter-annual and, more especially, multi-decadal timescales. Furthermore, change and variability should not be seen as competing concepts but rather as complex interacting processes that may sometimes act in mutually compensating ways, but may also be mutually reinforcing to the point of driving future changes over critical thresholds and generating strongly non-linear responses.

Chapter 15
Impacts and vulnerability

15.1 Introduction

This chapter briefly considers some of the methods involved in and outcomes arising from analyses of future environmental impacts and human vulnerability.

As we have seen in the two previous chapters, all projections of future climate and climate impacts, whether at global or regional scale, are, in large measure, contingent upon emission scenarios that are a function of human actions. So far, we have largely side-stepped the issue of estimating the likely consequences for human populations of any given scenario and the projections arising from it, but it is at this level of evaluation, with all its additional uncertainties, that the scientific research and the concerns of most policy makers meet at all spatial and organisational scales from planetary to local. At one extreme, we have, for example, concerns about the type and extent of intervention in global energy fluxes or biogeochemistry that may be necessary and justified in mitigating the effects of a long-term build up in atmospheric greenhouse gases. At the other extreme, every human population, irrespective of its size or location, is faced with the following questions. What kind of climate are we likely to experience in the future? How different will it be from the recent past? What implications will this have for our life support and welfare? Can we do anything to prepare for and adapt to the likely changes if and when they can be confidently predicted? Those who bear responsibility for designing policies to ensure the long-term conservation of particular sites or habitats must face similar questions. How should the combination of future climate change and ongoing human impacts mould conservation policy? What might be realistic conservation goals bearing in mind past history and future projections? Which potential goals are realistic in both ecological and economic terms? What needs to be done to achieve them? Despite the huge uncertainties inherent in all attempts to answer such questions, the urge to provide answers has led to the emergence of *impact assessment* as a major field of research, building on the kinds of projection we have already considered, but often integrating them into much more complex models that include the socio-economic and technological process with which they will interact.

In Schneider's (2002) scheme (Figure 13.1) the task of developing impact assessments at any spatial or organisational scale involves a cascade (some would say an explosion) of uncertainties. At the outset, it is useful to consider more closely the nature of this cascade, for, as Van Asselt and Rotmans (2002) point out, types and sources of uncertainty are quite diverse. In light of their own analysis, they develop a modelling framework (TARGETS – Tool to Assess Regional and Global Environmental and Health Targets) that recognises that some aspects of uncertainty are more realistically explored through acknowledging differing human perspectives, than through statistical analysis. They develop a series of impact scenarios linked to alternative perspectives. This approach acknowledges at the outset that the inputs to models that include many of the types of uncertainty noted above cannot be value free. They therefore include qualitative and subjective interpretations that are explicitly characterised in the alternative perspectives they use.

The text that follows presents two types of impact assessment. In the first, particular sectors of human concern – forestry or agriculture for example – are considered separately. In the second, assessments that attempt to deal with a much wider range of impacts in an integrated way are considered on spatial/organisational scales ranging from global to regional. Following this, an approach that begins by addressing issues of vulnerability for a given population or region is briefly considered.

15.2   Sectoral impacts and impacts on ecosystems

The account below mainly follows the subheadings used in the IPCC TAR; it summarises some of the conclusions of the report, and amplifies some of the issues dealt with in subsequent papers.

15.2.1   Water resources

One of the most confidently stated projections envisages a continued retreat of glaciers and a shift from snow to rainfall in many regions where winter climate is marginal for snowfall. These changes would lead to a shift in stream flow from spring to winter. This in turn may require changes in water-storage strategies where populations depend on glaciers and snow pack for seasonal storage. Although many other projected changes in rainfall and runoff are strongly model dependent, the relative consistency with which models project future reductions in runoff in southern Europe and many currently water-stressed semi-arid areas is a major cause for concern. Even where mean annual rainfall is projected to increase, the likely concomitant increase in flood magnitude and frequency will probably lead to increased vulnerability in some densely populated regions. The IPCC TAR projects a three-fold increase by 2025 (from 1.7 to 5 billion people) in the number of people living in countries that are water stressed. Doll's (2002) attempt to estimate future irrigation demand and the likely impact of climate change on irrigation provision (see 14.7) illustrates the difficulties involved in making such estimates on the basis of conflicting models with inadequate spatial resolution, but it also reinforces the view that this aspect of water demand is likely to exceed supply for an increasing number of people as populations grow and runoff probably declines.

The study by Vörösmarty *et al.* (2000) embraces all aspects of future water demand (domestic, industrial and agricultural) and places the projections in the context of future climate simulations for the year 2025. The stresses arising from increased human demand, driven by demography and economic development, are much greater than those arising from projected climate change. Their estimates of the number of people living under conditions of moderate or severe water stress rise from 2.2 to 4.0 billion, an increase broadly in line with that presented in the IPCC TAR. The projected increases are most severe in Africa, Asia and South America. These kinds of projections are, in part, already 'built in' by current demographic patterns and development trajectories. Beyond the first few decades of the century, population growth is less certain and possible climate changes are more likely to generate events, including droughts, falling beyond the range recently experienced. Climate change may therefore become an increasingly dominant controlling factor in determining the degrees and distribution of water stress around the world.

For the west of the United States, changing climate is likely to play an important role in future water supplies even during the first half of this century, the period for which projected global mean warming is quite modest and not so strongly scenario- or model-dependent. Barnett *et al.* (2004) use a coupled ocean–atmosphere model to generate three projections of future global climate change which are then downscaled to provide projections of impacts on climate-sensitive environmental systems in the west of the USA. Their main results include indications that:

- The Colorado River will not be able to meet all the demand placed on it.
- The Central Valley of California will experience serious depletion of freshwater resources.
- The Sacramento Delta is likely to become more saline.
- Changes in seasonal flows linked to reduced snow-pack will impose difficult choices on managers of resources in the Columbia River system.
- In smaller catchments in the northwest, even more extreme changes in seasonal flow will reduce the availability of summer-irrigation water.
- Trends towards increased summer temperatures and reduced humidity are likely.
- In much of the region future changes are likely to extend the fire season and increase fire danger.

In the above and all such studies, the likely effects remain conjectural until better simulations of recent rainfall patterns and variability have been achieved, and models are available that can build on these simulations at much higher levels of horizontal resolution than are currently possible.

Finally, it is worth reiterating that in many regions currently experiencing severe water stress, the late-Holocene record of droughts includes some that are much more severe and of longer duration than those captured in the instrumental record and used in 'validation' exercises for climate models. Even without 'global change' as currently conceived, and in the absence of what seem to be inevitable major increases in population, in many water-stressed areas there would be cause for concern.

15.2.2   Terrestrial and freshwater ecosystems

Chapters 8 to 11 have outlined many of the already growing threats to ecosystems as a result of human activities, and Section 12.5 includes several examples of the changes already taking place in boreal ecosystems especially as a result of recent warming. The combined impacts of future human activities and climate change on terrestrial ecosystems are likely to be unprecedented. Two aspects are of special concern: the implications for future changes in the carbon cycle and the likely consequences for all aspects of ecosystem function and biodiversity.

Overpeck *et al.* (2003) consider changes in terrestrial ecosystems in response to past and future climate changes. One way of looking at the potential effects of future climate change is to carry out equilibrium simulations of future biome and species distributions once the models used have been tested against recorded distributions under present- and past-climate conditions. Retrospective testing brings out the importance of biosphere feedbacks. Without these feedbacks, biome models forced by climate alone underestimate the response to climate change. Even models that do not incorporate biosphere feedbacks project major shifts in the future distribution of biomes, ecotones (the transitional regions between biomes) and species. Overpeck *et al.* (2003) show examples of simulations made by an equilibrium model, but these ignore the question of how and how quickly, in this case, forest trees may achieve a distribution reflecting the new climate space. Even if we accept only the lowest projected global mean temperature increase by 2100, the rate of global climate change envisaged for the twentieth century is remarkably rapid when compared with the mean rate of global warming during the whole of the transition from full glacial to 'peak' interglacial conditions. During this period of around 9000 years, global temperature rose by *c.* 5 °C – a mean rate of change between 20 and 30 times slower. There were, however, periods (the opening of the Bølling Allerød and the Holocene – see 6.1 and 7.1) when, at least for parts of the northern hemisphere, including western Europe, the rate of temperature increase was more comparable to that expected in the coming decades. Close study of the responses of plants and animals to climate change during these periods may be as near as we can get to gaining some impression of the way in which species and ecosystems respond. Hughen *et al.* (2004), using biomarkers for tropical vegetation alongside climate proxies in the high-resolution record from the Cariaco Basin, show that the response of tropical ecosystems to climate change during the last deglaciation lagged climate by only a few decades, and certainly by less than a century. The evidence summarised by Amman (2000) suggests that the recovery of birch woodland at the end of the Younger Dryas period in the Swiss Alps did not lag behind the climate shift by more than thirty years. On the other hand, full re-establishment of boreal forest in northern Sweden, in the region of Abisko, appears to have taken around 2000 years (Barnekow, 2000). In the case of the Swiss Alps, local survival of birch in favoured habitats was probably conducive to rapid recolonisation. Moreover, re-establishment of birch woodland probably did not signal the development of an ecosystem in equilibrium with climate. The contrast between the two cases therefore probably reflects differences in topography and the availability nearby of parent plants producing widely dispersed seeds, as well as different criteria for assessing the 'completeness' of the response of to climate change.

Rates of response depend on more than the rate of temperature change. Since species tend to respond in individualistic ways to external forcing, climate change is likely to lead to a re-sorting of assemblages and the development of newly constituted ecosystems. Some sense of this can be gained from the analysis by Pearson *et al.* (2002) aimed at assessing the impacts of climate change on species distributions using a model (SPECIES) in which an artificial neural network is coupled to a climate-hydrological process model. Their model simulates remarkably well the contemporary British distributions of most of the 32 species they consider. Their conclusions regarding the high level of individuality in species responses to future climate changes are therefore rather credible. Species can migrate at widely differing rates, with, at one extreme, mobile vertebrates and rapidly invasive ruderal plants, at the other, many forest trees.

In addition to rates of change and of migration, several other factors come into play. For many species, successful migration into new areas will depend on the creation of suitable habitats through disturbances such as fire, drought or storm damage. Models designed to simulate future changes in distribution therefore need to incorporate changing disturbance regimes. Even where there may be partial analogues for rapid warming from the opening of the Holocene, for example, the base state of the climate is now vastly different, as are soils and existing biome distributions. In addition, anthropogenic impacts have transformed the landscapes over which changes in distributions and interactions are likely to take place. Many ecosystems with high levels of biodiversity or endemicity, including several of the biodiversity hotspot identified by Myers *et al.* (2000) are restricted in extent, separated by the effects of human activities from areas that might be vital for the future survival of key constituent species, and already subject to major environmental stresses from time to time. Midgley *et al.* (2002) consider one such biodiversity hotspot, the fynbos biome in the Cape region of South Africa. This is a unique vegetation type dominated by Proteaceae and Ericaceae many of which are endemic to the region. Using the HadCM2 and CSM climate models to generate scenarios to 2050, they assess the extent to which, for any given climate projection (and resulting reduction in fynbos area), the future range of individual species would lie outside and beyond their current range. Depending on the scenario used, the loss of the fynbos area is projected to be between 51 and 65 %. Around 10 % of the Proteaceae studied have ranges lying entirely within the 'lost' area. These the authors identify as the most vulnerable species and the ones most likely to give early indications of climate-change impacts.

Extreme fragmentation through human activities can exacerbate the problem of maintaining or enhancing current levels of biodiversity. A further complicating factor is the possible effect of increasing atmospheric CO_2 concentrations on the competitive relations between species. Evidence for changes in overall productivity arising from the possible fertilising effect of higher levels of CO_2 is rather inconclusive. It is mostly based on controlled FACE (free air carbon enrichment) experiments of limited extent and duration, with the results showing a high level of dependence on the type of ecosystem, soil-nutrient status and a range of other factors (Mooney *et al.*, 1999). One rather robust conclusion, however, is that elevated CO_2 concentrations change the competitive relations between species (Norby, 2004).

Just as future projections of species and biome distributions ultimately need to simulate trajectories in addition to hypothetical equilibrium states, they also need to include an awareness of past dynamics. Models designed to develop projections of the effects of future climate change on ecosystems, whether or not the models used take into account ongoing human impacts, often take the present day as their baseline. But, since the state of an ecosystem is contingent on past events and processes and all baselines are *dynamic*, this may seriously limit future projections if no attempt is made to establish the antecedents of the present-day ecosystems, and to identify the past events and processes that have generated what we can observe and measure today. This is well illustrated by several papers dealing with landscape and wildlife dynamics in the northeast of the United States, where the status of present-day forest ecosystems is related both to climatic gradients and to the history of recent human activities in the region (Foster *et al.*, 2002a; 2002b; Hall *et al.*, 2002). The human history since European settlement involved a period of much more extensive agriculture, followed by farm abandonment and forest regrowth. Irrespective of the possible impacts of changing climate, conservation policy for the future must build on the legacy from the past in order to accommodate the dynamic nature of the ecosystems to be conserved or promoted.

A simple but crucial example of the kind of errors that can occur if past history and dynamics are ignored comes from estimates of the present and future carbon balance of forest ecosystems in the east of the United States. Upscaled, site-based measurements made during the 1990s led to the assertion that these forests served as major areas of net carbon sequestration, thereby holding out some promise of serving as long-term sinks in any calculations of the overall carbon budget of the country. This view had to be seriously qualified once it was realised that the sites used were regrowth forests in which the carbon budget reflected the early stage of development of the stands.

Many of the ecosystems of greatest concern are in mountain regions. Potential impacts include those on water resources, isolated ecosystems, biological diversity, natural hazards, health issues and tourism (Beniston, 2003). Warming, by raising the altitudinal boundaries of each vegetation belt, could seriously reduce or even eliminate suitable habitats. In many temperate mountain regions the warmer temperatures of the mid Holocene also led to increased stress for species and biotic communities dependent on high light levels, hence absence of forest cover. Tree lines reached higher altitudes and light-demanding taxa had to survive in greatly restricted 'refugia' habitats. The message from the survivors of these earlier periods of warmer climate must be to identify and protect these kinds of 'refugia' habitats for mountain plants and animals in the future.

All the points made above should be set alongside the discussion of biodiversity in Chapter 11. Clearly, the combined stresses of climate change, habitat destruction and invasions by exotic species pose literally incalculable threats to biodiversity. Loss of diversity at the species level is likely to be compounded by reductions in diversity at the population level within species (Harte *et al.* 2004). Both processes tend to reduce the degree of redundancy in many ecosystems. As we have seen, this is believed to play a key role in ecosystem resilience. In addition, one of the effects of future climate change will almost certainly be to favour invasive taxa and opportunistic species typical of early-successional stages in ecosystem development, at the expense of late-successional species. Although it is already possible to outline the threats to ecosystems and biodiversity at a rather general and qualitative level, translating this into concrete, quantitative, probabilistic projections remains contentious. Thomas *et al.* (2004) predict that on the basis of mid-range climate warming scenarios for 2050, between 15 and 37% of the terrestrial species in the regions they studied will be 'committed to extinction'. They claim that anthropogenic climate change may prove to be just as great a threat to biodiversity as the trends in land use and habitat destruction considered in Chapters 9 and 11, while acknowledging that the two drivers of change will inevitably interact. Their analysis suggests that reducing the rate of global warming would bring significant benefits in terms of extinction rates. Their projections rely rather heavily on a putative link between extinction risk and geographical-range size for any given species, and this is one of several aspects of the study challenged by subsequent communications (Thuiller *et al.*, 2004; Buckley and Roughgarden, 2004).

The potential impact of future climate change on freshwater ecosystems has received much less attention. Warming is likely to exacerbate the effects of eutrophication in many lakes. Any changes in effective moisture will have effects on lake-levels, water residence times and water chemistry, especially in areas of high evaporation. Changes in the thermal regime of lakes will impact aquatic communities and species distributions, but some of the impacts may be indirect and quite subtle. Temperature changes throughout the year largely control the stratification of lakes. Quite small changes have the potential to change the thermal structure of the water column at key times of the year and replace stratification with the mixing of the whole water column and vice versa – a simple and significant example of non-linear change once a critical threshold has been transgressed. This in turn affects oxygen and nutrient supply, hence trophic patterns within the aquatic ecosystem. Finally, it is important to remember that any significant changes that occur in the fluxes of nutrients, soils and sediments within the terrestrial ecosystems forming the catchments of a lake will in turn impact the lake ecosystem. Terrestrial and aquatic ecosystems are closely coupled and the latter cannot be considered as functionally independent from the former.

15.2.3   Food production

In this vital area of activity, future climate change will interact with a vast range of other factors, some geochemical, like the possible fertilising effect of higher levels of atmospheric CO_2 and available nitrogen, some linked to changing management regimes and technology, for example, improved water-use efficiency, selective breeding and genetic engineering. The scope of the challenge can be readily assessed by considering world-population growth. During the second half of the twentieth century, the period of accelerating global change in every sense, the world's population grew from 2.5 to 6 billion. By 2025, it is expected to reach 8 billion. During this period, the urban population in developing countries is expected to double. This implies a massive increase in the demand for food. Since 1950, cereal production has kept pace with the rising population, though persistent under-nourishment and large-scale famines have not been avoided. Over the next two decades, cereal production will need to increase by at least a further 35 to 70 %, partly for direct human consumption, partly for livestock (Dyson, 1996; Delgrado, 1999; Gregory and Ingram, 2000a).

Models designed to generate future projections, even at the smallest scale of the individual plot or field, need to encompass much more than climatic effects (Gregory and Ingram, 2000b). Soil properties such as nutrient- and water-availability must be included, as well as biotic interactions involving the changing effects of pests, diseases and weeds. Scaling up projections to regional or national levels brings in a whole range of additional factors, cultural, economic and political. Nevertheless, some assessment of all these interacting influences in the future will be necessary if the worst impacts on food security are to be avoided and the potential for enhanced production in potentially favoured areas is to be realised. Gregory *et al.* (2002) claim that there is now little scope for meeting the ever-increasing global demand for food by extending food production to new areas, unless vast swathes of tropical forest are cleared and farmed, with dire consequences for biodiversity, greenhouse-gas emissions, regional climate and hydrology. The only alternative is greater intensification, mainly by increasing yields and, to a lesser extent, by increasing the number of crops grown during the year. Intensification carries with it environmental consequences some of which seriously threaten sustainability. These include soil erosion, the depletion of soil nutrients and organic matter, reduced water-retention capacity and the increased 'export' of nutrients and pesticides in quantities that can damage aquatic ecosystems downstream (see 10.2.1). Much depends on the type of intensification adopted. Pilot projects summarised by Sanchez (2000) and by Pretty *et al.* (2002), involving, for example, agroforestry, improved cropping practices, pesticide-free pest control and greater water-use efficiency, suggest that some of the necessary increases in yield can be achieved with minimal external inputs and lead to reduced environmental degradation and enhanced sustainability; however, Gregory *et al.* (2002) claim that intensification with low external inputs of energy and materials is unlikely to deliver the increased yields that will be required on a global scale. Future management policy, they claim, must include high levels of external input, but focus on increasingly efficient and environmentally benign use of technologies designed to maximise yields.

Projections are no more secure in the area of future food production in pastures and rangelands (Campbell *et al.*, 2000). Higher concentrations of atmospheric CO_2 are expected to lead to modest increases in overall grassland productivity, especially in moisture-limited regions. Associated changes in species composition and the possibility of shrub invasion in drier areas are among the many factors that make it difficult to translate estimates of increased biological productivity into projected changes in animal production. Campbell *et al.* emphasise, above all, the many uncertainties still impeding any credible projections in these aspects of food production, especially in regions that are potentially subject to increased stress from higher temperatures, greater variability in precipitation and likely water shortages.

The global challenges involved in increasing food production to meet projected needs and outlined by Gregory *et al.* (2002) are immense even without taking into account projected climate changes beyond the first two decades of the present century. As for the likely effects of future climate change, it is difficult to do more than identify, in broad terms, regions potentially advantaged by the combination and others most probably not. Most of the former lie in developed, cool temperate regions of the northern hemisphere. Many of the latter lie in the tropics and sub-tropics and include mainly less developed countries (Shah, 2002). The IPCC TAR consider that in much of Africa, many countries in

Asia and much of Latin America, climate change is more likely to diminish than to enhance the potential for food production. In all these cases, they also consider the adaptive capacity of human systems to be low; in some countries, China (Chameides *et al.*, 1999) and Pakistan (Maggs *et al.*, 1995), for example, there are strong indications that regional haze and tropospheric ozone are already giving rise to significant declines in crop yields. Model-based projections indicate that the current gap in cereal production of around 10 million tonnes for the 400 million people undernourished at the present day could increase by a factor of 13 to 15 by 2080 (Steffen *et al.*, 2004). No matter which emission scenarios, climate models or projections of food production are used, the contrast between the populations most at risk and those responsible for the highest levels of CO_2 emissions over the last 50 years remains stark and troubling in a world in which the concept of equity is gaining ground more in rhetoric than in reality.

15.2.4   Forestry

The challenge of projecting future climate-change impacts on forestry is at an early stage of development. On a global scale, problems arise from the heterogeneity of forests and forest-site conditions and the consequent diversity of resources and uses to which they give rise. Effective impact assessment thus needs to include a wide variety of inputs, ranging from biophysical disciplines to conservation, management, wood-product development and resource economics. The long timespans over which forests develop, mature and undergo changes in distribution, together with the unknown incidence and effects of future disturbance regimes such as fire, storms and other weather extremes, pests and diseases, each of which can have rapid impacts, pose special difficulties for modelling forest behaviour under changed environmental conditions. On a regional scale, studies aimed at testing forest-succession models against the past record of climate variability and observed forest response include encouraging case studies (e.g., Bugmann and Pfister, 2000; Cowling *et al.*, 2001) that may eventually increase confidence in more widely applicable future projections. Further progress in this area is urgently needed, for managed forests not only play diverse but often significant roles in regional and national economies, and they may play an increasingly important role in modulating the global carbon budget.

15.2.5   Coastal areas

Defining the coastal zone as land lying within 100 km of the coast and less than 100 m above mean sea-level, Nicholls and Small (2002) estimate that 23 % of the world's population live there and point to the relatively high population densities over this part of the land surface. Almost 300 million people live within the limits of the 5 m contour and some 400 million below 10 m. Despite the high densities, only a minority of coastal dwellers live in large, urban environments.

So far, most of the impacts of global change on the coastal zone have been the result of human activities of the 'cumulative' kind – changes in sediment supply as a result either of accelerated erosion or impoundments upstream, changes in nutrient flux, residential and industrial developments, other types of construction, transformation of coastal ecosystems for exploitation. There is little sign that these kinds of impact will be reduced in future, though the balance between them may change with changing demography and shifting economic priorities. The key question concerns the likely future impact of rising sea-level. If we accept for the purpose of this discussion the majority view as summarised in the IPCC TAR and Section 13.2, a modest sea-level rise during the present century seems highly likely, with the strong prospect of a further rise beyond 2100. The way any rise in global mean sea-level affects a given stretch of coast depends on a vast range of additional factors some of which will be peculiar to the locality. One of the simplest ways of approaching the problem is in terms of the link between mean values and the magnitude and frequency of extreme events. Figure 15.1, which can be applied to many other aspects of changing environmental regimes in addition to sea-level (see, e.g., flood frequencies in 14.6), presents an idealised scheme in which extreme events have a normal Gaussian distribution irrespective of the mean value. It shows how the distribution shifts and the frequency of extreme events increases as the mean value increases. Lowe *et al.* (2001) carry out simulations of future changes in the occurrence of storm surges in the UK. They include meteorological forcing, mean sea-level rise (+50 cm) and changes in storm propagation to derive the results shown in Figure 15.2. The end result is a significant reduction in return period at most coastal sites as sea-level rises.

The implication of these graphs is that modest rises in mean sea-level can have important consequences for the incidence of extreme events, increasing both their likely frequency and the probability that the most severe ones may exceed the engineering specifications of sea defences. At any given location on the coast, many other factors will influence this probability – movement of the land itself (e.g., Shennan, 1989), coastal lithology and configuration, sediment supply and transport (Zhang *et al.*, 2003: Stive, 2003), tidal regime, wind and wave conditions, atmospheric pressure and the phase of modes of climate variability such as El Niño southern oscillation (ENSO). Depending on whether El Niño or La Niña conditions prevail, sea-levels on either side of the equatorial Pacific can vary by tens of centimetres, that is to say, over roughly the same range as the projected mean sea-level rise (Goodwin, 2003). Although rising sea-level clearly increases exposure to potential hazards for coastal populations, there is no simple, generally applicable rule that can assess likely impact from region to region. Specific knowledge, including detailed histories of past coastal change, will always be required (Pilkey and Cooper, 2004).

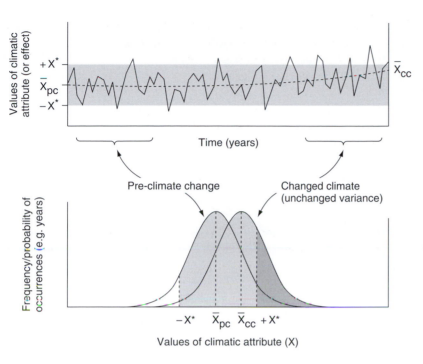

Figure 15.1 The relationship between changing mean values and extremes in system with unchanged variance. A notional 'coping range' is added. (From IPCC TAR, 2001.)

------ Trend in mean value of X (20 year running mean)

$\bar{X}_{pc}$ Mean value of climatic attribute (X) at start of time series (pre-climate change)

$\bar{X}_{cc}$ Mean value of climatic attribute (X) at end of time series (climatic change)

$+X^*$ Upper critical value of X for system of interest: values $> +X^*$ are problematic and considered 'extreme' or beyond 'damage threshold'

$-X^*$ Lower critical value of X for system of interest: values $< -X^*$ are problematic and considered 'extreme' or beyond 'damage threshold'

▢   Coping range or zone of minimal hazard potential for system of interest

▢   Probability of 'extreme' events (i.e. climatic attribute values $> +X^*$)

Much concern has been expressed about the future fate of low-lying island states. Nurse and Sem (2001) paint a rather gloomy picture for small, low-lying island states where the combined threats of sea-level rise, a higher incidence of extreme weather events and a range of associated adverse impacts are seen as potentially deleterious. Barnett and Adger (2003) focus their attention on atoll states and present a sea-level-rise vulnerability index that takes into account coastline length, demography and GDP as a simple proxy for adaptive capacity. Although their analysis is highly simplified and fails to take into account some of the key geomorphological factors that will affect coastal vulnerability, they identify at least five states, including the Maldives, as highly vulnerable. By contrast, and considering sea-level only, Mörner *et al.* (2004) present evidence to suggest that the people of the Maldives have survived sea-levels around 50–60 cm

Figure 15.2 The effects of a 50 cm rise in mean sea-level (MSL) and changed climate-related forcing on the recurrence interval of storm surges, illustrated by the simulated consequences for Immingham on the east coast of England. The continuous line shows the climate-related changes without any rise in sea-level. The upper curves also take into account the rise in sea-level. Note the effect the full combination of changes has on the recurrence intervals of 1.5 m and 2.0 m surge heights. In both cases, the recurrence interval declines by over an order of magnitude. (Modified from Lowe *et al.*, 2001.)

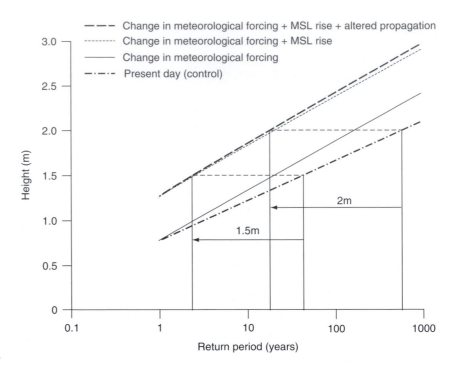

higher than present. Moreover, the data they present indicate that the islands are not currently experiencing a rise in sea-level. Reconciling these contrasted perspectives will require a stronger inferential framework for projecting future sea-level on a regional basis, as well as a recognition that, as Barnett and Adger (2003) suggest, sea-level, indeed the purely physical aspects of the Earth system alone, may not be the only, or even the main threat to sustainability in such situations.

15.2.6   Marine ecosystem: fisheries and coral reefs

Marine fisheries have been an increasingly important source of protein, especially for people in developing countries. Although human populations are set to rise, fish catches have levelled off in recent years and there is a strong probability that marine fisheries will fail to meet future requirements even without the complicating effects of climate change (Steffen *et al.*, 2004). Several of the impacts on the coastal zone have an adverse effect on marine fisheries since they involve loss of habitats and food supplies for exploited species. The various threats to coral reefs already noted (see 10.4) also imply a likely depletion of fish stocks in the areas affected, as coral reefs are also important habitats for fish species. Bryant *et al.* (1998) estimate that coral reefs provide food for around a billion people in Asia alone. The combined effects of regional bleaching and over-exploitation are already severe. The future is likely to bring changes in ocean chemistry due to higher concentrations of CO_2, as well as more frequent severe tropical storms. Whereas coral reefs unstressed by human pressures tend to recover from the latter, in recent times human impacts have tended to impair their capacity to recover, leading, in extreme cases, to a shift in state from dominance by corals to dominance by seaweed (Bellwood *et al.*, 2004). Hughes *et al.* (2003) conclude their review of the status and future of coral reefs by suggesting that, at the global scale, reefs will undergo major changes in structure and species composition, rather than total disappearance. Preserving reefs under threat from the combination of climatic and anthropogenic stresses at the regional scale will pose severe challenges to research and management, the more so since what Bellwood *et al.* term 'cryptic loss of resilience' has often occurred without its implications being either diagnosed or countered effectively. The challenge is greatest in regions like the Caribbean where functional diversity (see 11.3) is least (Bellwood *et al.*, 2004).

In many parts of the world it is becoming apparent that climate and sea-surface temperature variability play an important role in modulating the size and location of fish populations in many environments. Until the future behaviour of this and other modes of climate variability is captured more reliably in climate models, it will remain impossible to say to what extent future climate change will moderate or exacerbate what are already sources of concern and contention at national and international level.

Increases in CO_2 concentrations in the world's oceans may well have consequences well beyond those outlined above. Using the projections given by Sabine *et al.* (2004), a doubling of atmospheric CO_2 would imply a 30% decrease in carbonate-ion concentrations in the world's oceans and a 60% increase in hydrogen-ion concentration. Such acidification would have major impacts on marine ecosystems. Moreover, these and associated chemical changes in the oceans would almost certainly have overall positive feedbacks on atmospheric concentrations of CO_2.

15.3   Impacts, risk and vulnerability assessment

As the previous section illustrates, most of the projected sectoral impacts that are closely linked to human welfare are strongly differentiated on a regional basis. The assessment of risk, environmental impacts and vulnerability in an integrated way, spanning cultural and biophysical variables and their interactions, has quickly burgeoned into a huge, multi-faceted field of study, with almost as many different regionally, nationally or globally oriented projects and approaches as there are research groups.

On of the most comprehensive and ambitious schemes on a global scale is the IMAGE 2 (Integrated Model to Assess the Global Environment) modelling framework (Leemans, 2003). The model IMAGES 2 provides a framework for integrating   socio-economic   and   energy-use   scenarios   with   aggregated atmospheric and ocean-circulation model simulations and calculations of changes in land use and terrestrial-carbon retention. Simulations are calibrated against data for the period 1970 to 1995. The integrated model includes a range of realistic feedbacks such as CO_2 fertilisation and land-use change resulting from climate change. A modelling framework of this kind, within which 'what if' questions can be explored in a realistically complex and interactive context, allows some assessment of the likely effects of alternative mitigation and adaptation strategies. This approach goes beyond the IPCC TAR by including feedback loops from processes such as changed land-use patterns, deforestation rates and terrestrial-carbon fluxes.

Most integrated impact schemes seek to address questions on a smaller, national or regional scale. Jones (2001) outlines an approach to climate-driven risk assessment using climate scenarios of the kind developed by the IPCC as a starting point. In his formulation, a major aim is to reduce risk by limiting the probability of transgressing damaging thresholds. Risk involves two factors in combination – the *probability* of an adverse event and its likely *consequences*. The assessment process outlined acknowledges the cascade of uncertainties already referred to in Section 13.1.1. While accepting that many of these reflect inherent indeterminacy, the approach outlined seeks to develop response strategies notwithstanding. To this end, potential impacts are expressed as conditional probabilities; conditional because they depend ultimately on emission scenarios. Conditional probabilities are attached both to the climate scenarios, and to the likelihood of defined impact thresholds being exceeded. The latter are defined using a suitable impact model. Finally, and in light of the evaluation, strategies for mitigation or adaptation can be assessed and options defined. One of the key questions in this type of approach hinges on the way in which human responses to risk have tended to change the distribution curve of probabilities. As Fagan (2004) states: 'Like many civilisations before us, we've simply traded up in scale, accepting vulnerability to the big, rare disaster in exchange for better ability to handle the smaller, more common stresses such as short-term droughts and exceptionally rainy years'. There is a case for applying more widely the concept put forward by Messerli *et al.* (2000) of 'trajectories of vulnerability'.

Harremoes and Turner (2001) identify two essential components of what has become known as integrated assessment. The assessment must be integrated over a range of relevant disciplines spanning the environmental sciences and policy problems. It must also provide suitable information for policy making. Among the core objectives that they list are the need to reduce risk, to mitigate the effects of uncertainty, and to improve social inclusion and equity. Central to the strategy they outline is the involvement of stakeholders (individuals and groups with a vested interest in the outcomes) alongside experts throughout the whole process. They then envisage a sequence of processes, proceeding from identifying the issues involved, through structuring the issues, analysing the relationships involved, data collection and monitoring, assessment of results and feedback.

Much of the rapidly growing published literature on integrated assessment is methodological and all the substantive rather than purely illustrative projections are necessarily subject to the cascading uncertainties already characterised in previous sections. The most comprehensive attempt to summarise the projected regional impacts of future climate change within the context of credible political, technological, cultural and socio-economic scenarios is the second volume of the IPCC TAR (Watson *et al.*, 2001) and it is not within the scope of this book to summarise this account. Although it is debatable to what extent the value of these assessments is reduced by the limited number of climate-model ensembles used and the lack of any full appraisal of possible feedbacks between climate change and societal responses at regional level, they provide a broadly based consensus view that highlights the high levels of future vulnerability in many of the economically poorer regions of the world where most aspects of global change, howsoever defined, are probably set to compound the threats to sustainability that already exist.

One approach to the problem is to target a particular group and a specified range of activities, then try to assess possible outcomes depending on both climatic and socio-economic variables. This is essentially what Steffen *et al.* (2004) recommend when they refer to a 'vulnerability oriented approach'. They claim that this may provide more realistic estimates of future impacts. Rather than focus on a single environmental stress, as is often the case in impact assessment, vulnerability assessment (Turner *et al.*, 2003) focuses on a target group or unit of concern, whether it be a defined human population, economic group, ecosystem or biome. It then seeks to identify those stresses most likely to give rise to negative impacts and those factors likely to reduce the adaptive capacity of the unit of concern. Key concepts in vulnerability assessment include the following:

Exposure – the extent to which the unit of concern comes into contact with a defined stress.
Sensitivity – the extent to which a given set of stresses impacts the unit.
Resilience – the capacity of the unit to withstand or recover from the adverse effects imposed by the combination of stresses envisaged. This is quite a complex concept involving the degree to which a unit may be prepared for changing patterns of stress. It may also reflect the existence of reserves that the unit may have; for example, capital or human resources, or in the case of ecosystems, functional redundancy (see 11.3). Linked to this is the concept of adaptive capacity, which is also seen as making an essential contribution to resilience.

Several approaches are possible within the overall framework of vulnerability assessment, including the use of indicators (e.g., Langeweg and Gutierrez-Espeleta, 2001; Riley, 2001a; 2001b), the concept of 'tolerable windows' (Bruckner *et al.*, 1999) and analyses in terms of 'syndromes' (Petschel-Held, 2001). Clark and Dickson (2003) and Jäger, (2004) herald the emergence of what they describe as 'sustainability science'. The former identify the challenge of reconciling 'society's developmental goals with the planet's environmental limits over the long term'. They see this as calling for a closer partnership between scholars and practitioners. Translating the goals of sustainability science into achieving what they term a 'sustainability transition', Kates and Parris (2003) outline the current trends in a wide range of indicators, biophysical, cultural and technological. From this, they conclude that achieving such a transition by the middle of the present millennium is unlikely. In a subsequent paper (Parris and Kates, 2003) they identify a series of goals and quantitative targets on the way to achieving such a transition.

One aspect of vulnerability often considered separately is the issue of future global change and human health (see, e.g., Martens, 1998). Clearly, research in this field must seek to link epidemiology (itself a complex field involving multiple interactions between human and biophysical processes, each subject to unknown future changes) to projections of changing climate at regional level. As Jaenisch and Patz (2002) note, in summarising the outcome of a comparative review: 'a paucity of long-term data and complexity of relationships among factors affecting climate variability and change and human health rendered projections of current predictive models inadequate as the basis for policy decisions.' Until regional climate projections gain greater credibility, the field remains speculative. Even where there is good agreement between climate-model projections, the health impacts may still be contentious (see, e.g., Wilson, 2000). The situation is made even more complex by demonstrations that links between current epidemiological trends and climate change are far from easy to interpret (see, e.g., Hay et al., 2002).

The themes outlined in this chapter bring us to the point where we must try to address the questions posed at the beginning of the book in the light of what has been learned from evidence for past environmental changes, from present-day observations and from future projections.

Chapter 16
Sceptics, responses and partial answers

The yearning to believe in oracles has always been with us, but so too has scepticism

Holland (2003)

16.1 Introduction

Before attempting to bring all the research considered so far to bear on the questions posed at the beginning of the book, we must first deal with the views and assertions of those who are sceptical about the future projections outlined in the last three chapters. Over the last decade, 'global warming' sceptics have presented a variety of arguments challenging the claim that future climate change is a major concern for human populations. Dismissing the claim often carries with it the corollary of discharging the present generation from any responsibility to take action designed to limit economic growth based on the energy provided by fossil-fuel combustion.

There is no single sceptical viewpoint and sceptical responses to IPCC-based policy proposals are quite varied. A crucial distinction is between those who challenge the science (hence also all that flows from its projections) and those who accept at least the more moderate research-based projections, but disagree with pro-mitigation scientists and policy makers about the degree to which future changes should be a cause for concern and urgent action. Von Storch and Stehr (2000) express this view neatly: 'We are convinced that greenhouse gases are accumulating in the air and strongly believe that near-surface temperatures are rising in response. But we are not convinced that present and future climate change will have a significant impact on society and global systems.' At the more moderate end of the range are cautious suggestions that more and harder evidence on key aspects of the climate system may be needed before actions are taken that risk limiting the short-term rewards of economic growth. At the other extreme there are assertions that the IPCC-TAR (2001) rests on 'junk-science', that policies based on it should be rejected and that its use by leading governments should be suppressed. The text that follows can be no more than a personal, inevitably value-loaded attempt to identify and deal with some of the main points at issue. At the very least, those who are interested should be aware of viewpoints as contrasted as those outlined in Lomborg (2001) and in Steffen *et al.* (2004).

Several points need to be made at the outset:

- The fact that the views summarised in the IPCC TAR (2001) and most subsequent papers on the subject, and endorsed by the American Geophysical Union (AGU Council, 2003) are majority views does not guarantee that they will turn out to be correct. At least two concepts that are now central to Earth-system science were minority views less than 50 years ago. Until it was demonstrated that the concept of plate tectonics accounted for many previously puzzling aspects of structural geology, the notion of 'continental drift' was regarded by most authorities with great scepticism. Equally, until the strong coherence between climate change and changes in the Earth's orbit was conclusively demonstrated, the ideas proposed by Milankovitch (1941), were championed by only a minority of Quaternary scientists.
- Since future projections are not subject to proof, or even to rejection through rigorous testing, they will always be vulnerable to criticism, until future experience establishes the extent to which they were justified.
- Most of the scientists closely involved in climate or Earth-system modelling are well aware of, and deal explicitly with, the limitations of their models and available data. Piecemeal criticism of model simulations therefore does not necessarily provide insights additional, or contrary to those expressed in the criticised publications.
- Conversely, presenting the results of simulations second hand as though they were *not* subject to significant uncertainties simply provokes misunderstanding, as well as criticism.
- Also, as Broecker (2004) points out, exaggerated, or alarmist projections that are weakly underpinned (see, e.g., Schwartz and Randall, 2004) play into the hands of those sceptics with a personal or political agenda, who use every opportunity to discredit even cautious and well-qualified projections that reinforce the need for mitigation of any kind.

16.2   Some sceptical science-based arguments

Some of the issues raised by sceptics are considered below:

16.2.1   Allegedly incorrect representation and interpretation of climate change over the last millennium

The representation by Mann *et al.* (1999) of past climate change in particular has come in for a succession of challenges. It has become caricatured as the 'hockey stick' graph of global temperature change and attracted much criticism from studies such as those of Soon and Baliunas (2003) and Soon *et al.* (2003), who conclude from their analyses that the existing data are insufficient to permit any global reconstructions, that many regional reconstructions run counter to the alleged global trends, and that twentieth-century temperatures were not exceptional relative to the peak warmth of the Medieval Warm Period (see 7.9). Their approach and assertions are robustly refuted in a multi-authored article by Mann *et al.* (2003).

McIntyre and McKitrick (2003) claim to have corrected the Mann *et al.* (1999) paper and thereby demonstrated that the Medieval Warm Period included multi-decadal periods warmer than the twentieth century. They state that the original paper contains 'collation errors, unjustifiable truncation or extrapolation of source data, obsolete data, geographical-location errors, incorrect calculation of principal components and other quality control defects'. Once more, these serious allegations and 'corrections' are in the process of being strongly refuted (see, e.g., www.realclimate.org). Several errors in the list of data used by Mann *et al.* (1998) have been acknowledged, but have no effect on the published results (Mann *et al.* 2004). McIntyre and McKitrick's failure to attach realistic error bars to their reconstruction, to set this alongside the measured temperature trend over the last 30 years, and to address the period post 1970, rather than generalise about the twentieth century as a whole, reduces the credibility of their assertions and especially those of their followers who use statements like: 'Hockey stick slapped' (Murray, 2003a). As Esper *et al.* (2004) show, there is some doubt about the shape of the low-frequency changes in global mean temperature over the last 1000 years, but these doubts do not add up to the kind of criticism made by sceptics of late-twentieth-century warming. Indeed Huang's (2004) integration of the published time series from both proxy-based surface reconstructions and borehole records leads to a reaffirmation of anthropogenic forcing over the last three decades, as well as to estimates of future global warming consistent with those of the IPCC TAR.

16.2.2   Limitations in model-based representations of the climates of the present day and recent past

One of the essential planks of successive IPCC assessments and the many projections of future climate that have been published over the last few years has been the relative skill with which global, or hemispheric temperature trends during the late nineteenth and twentieth centuries can be reproduced by climate models. The openly avowed limitations of these models have been used to discredit them as tools for developing future projections. This has mainly been done by focusing on aspects of climate for which the model performance is still questionable (precipitation, for example), choosing areas and time intervals for which selected global climate models (GCMs) make a relatively poor job of calculating past climate, and claiming that where *post hoc* scenarios appear to match observationally based reconstructions, mutually cancelling errors are involved.

16.2.3   Casting doubt on the effects of feedbacks and the role of greenhouse gases

This has been done by basing judgements on studies that may be interpreted as questioning the role and magnitude of greenhouse warming (see, e.g., Lewis, 2003). This strategy includes:

- Claiming that the effects of urban heat islands and land-cover change have been underestimated and provide an alternative explanation for recent warming (Kalnay and Cai, 2003). This view has been challenged by Trenberth (2004) and by Vose *et al.* (2004). Any suggestion that the effects Kalnay and Cai claim are responsible for recent warming trends is difficult to sustain in light of all the arguments outlined in Chapters 12 and 13, though some of the points raised in the analysis by de Laat and Maurellis (2004) of the regional expression of recent temperature changes may require further evaluation.

- Citing the recorded *decrease* in evaporation from standard-evaporation pans at many sites during the second half of the twentieth century. At first sight one would expect evaporation to increase if significant warming were occurring. This 'pan evaporation paradox' is dealt with by Roderick *et al.* (2001) and Roderick and Farquhar (2004) and shown to be a function of changes in sunlight, not temperature.

- Proposing that recent ocean-circulation changes are 'natural' and are acting as *drivers* of recent climate change, rather than as responses or potential feedbacks (cf. Bratcher and Giese, 2002), counter to most of the studies cited in section 12.3.5. Given the complex interactive nature of the Earth system and the importance of internal feedbacks it will always be possible to use simplistic cause–effect arguments as a basis for proposing alternative patterns of interaction between driving processes and responses. At least some changes, for example high-latitude freshening in the North Atlantic, cannot easily be characterised simply as ocean-generated drivers of climate change.

- Recalculating future climate projections using different parameterisations for the effects of aerosols, black carbon and water vapour and claiming a *negative* effect for the latter via low-latitude cloud feedbacks. As noted in section 13.1.3, the analysis by Soden *et al.* (2002) of the water-vapour feedback in the wake of the Mount Pinatubo eruption, as well as the model–data comparisons of post-Pinatubo cooling by Forster and Collins (2004), tend to counter this last point.

- Challenging the range of sensitivities used by the IPCC. Lindzen *et al.* (2001), for example, on the basis of observations, infer a possible adaptive mechanism whereby higher sea-surface temperatures (SSTs) reduce high-level cloudiness, thereby allowing infrared cooling. This allows them to estimate sensitivity to CO_2 doubling at between 0.64 and 1.6 °C, rather than between 1.5 and 4.5 °C. The independent, post-IPCC TAR studies summarised in section 13.1.3 all tend either to support the IPCC sensitivity range or lean towards the higher values.

- Pointing to the apparent discrepancy between the degree of recent warming in the troposphere relative to that at the Earth's surface. As a result of recent re-evaluation of remotely sensed tropospheric and stratospheric temperature changes, this anomaly may prove to be much less significant, though still worthy of much further research (see 12.6.1).

- Highlighting what are seen as unrealistic assumptions, both in economic and biophysical terms, underlying the more extreme of the Special Report on Emission Scenarios (SRES) 'storylines', In so far as these latter involve steep increases in fossil-fuel use and atmospheric concentrations, relative to current trends, these criticisms have some credibility. At the same time, the continuing emphasis on growth in all the world's economies, the projected increases in population, and the strong posibility that the carbon sink in the terrestrial biosphere may saturate during the present century must all be borne in mind.

- Interpreting the changes in atmospheric greenhouse-gas concentrations during glacial – interglacial transitions as entirely the consequence of other forcings and feedbacks, and not as having any positive role to play in climate change. Although doubts remain as to the precise phasing of greenhouse-gas increases at some transitions from glacial to interglacial conditions (see 5.2, 6.4 and 6.5), this view runs counter to every analysis by palaeo-scientists familiar with the full range of data from ice-core, marine and terrestrial archives.

16.2.4   Stressing the supposed triviality of low levels of global warming

In part, this is done by pointing out that the increase in temperature over the last century is less than the inter-annual variability (Murray, 2003b). In this argument, no link is made between changing mean values and associated changes in the magnitude and frequency of extreme events (see Figure 15.1). Nor does it address the fact that even without greater extremes in terms of peak magnitude, changes in the persistence or frequency of droughts, for example, can have severe effects. Comparisons, such as those by Wigley and Raper (2001), between the mean rate of change over the last century and that projected for the next 30 to 50 years are ignored, as are all the many lines of evidence for the significant impacts on ecosystems that are already occurring.

16.2.5   Ascribing most of the recent changes in stratospheric ozone to natural variability rather than human activities

This clearly simplifies and distorts the combination of factors currently responsible for changing stratospheric-ozone concentrations (see 8.4). Using the argument is tempting for sceptics, for it appears to carry with it the corollary that current and likely future changes will also be dominated by natural rather than anthropogenic influences (Lieberman, 2003).

16.2.6   Oversimplifying the causes of global sea-level variation

This is done by focusing on ice caps and glaciers as the sole factor and claiming that because some few selected glaciers are expanding, there is no global mean trend. Murray (2003c) thereby implies that estimates of sea-level rise are exaggerated and also concludes that 'from the point of view of the world's need for more potable water, it might be regarded as a good thing that less freshwater be locked up in glaciers'.

16.2.7   Ascribing the dominant role in all past and future climate changes to solar variability

This argument hinges on the many examples of inferred links between past climate changes and solar variability (see especially 7.11.1), the recently proposed mechanisms for amplifying solar effects on the Earth's climate, and the possibility that uncertainties in estimates of the sensitivity of climate to other processes may have led to an underestimate of the role of solar variability. Solanki and Krivova's (2002) analysis is one of several papers effectively disposing of the proposition that solar variability is responsible for global warming since 1970. They scale solar effects to the maximum value statistically permitted by the record of temperature changes for the period 1856 to 1970 and show that irrespective of the basis for solar–climate linkages, solar variability alone cannot possibly account for more than 30% of the post-1970 rise in temperature. Broccoli *et al.* (2003), using a somewhat different approach, estimate a net solar forcing of $+0.23\,°C$ during the twentieth century and only $+0.15\,°C$ for the period 1940–1997. The comparable figures for greenhouse gases are $+0.94\,°C$ and $+1.37\,°C$, respectively. Only during the period 1900–1940 were the two forcings of comparable inferred magnitude.

16.2.8   The normality of change and the resilience of the Earth system

This view is exemplified by the writings of Philip Stott (see, e.g., www.probio-tech.fsnet.co.uk/) whose brand of scepticism has been summarised in numerous widely scattered publications. It stresses the perpetual nature of climate change with or without anthropogenic impacts, and the importance of maintaining ecological function rather than the integrity of human constructs such as 'tropical rainforest'. It emphasises the dangers of interfering with a system as complex and non-linear as the Earth system, accepts the need for higher energy inputs to developing economies and sees enhancing adaptive capacity as the appropriate response to any likely future changes.

16.3   Some non-scientific assertions

To the scientific and somewhat less scientific arguments illustrated above are added a range of entirely non-scientific arguments. Many of these focus as much on the IPCC *process* as on the outcomes, and seek to identify flaws at each stage. Lomborg (2001) draws most of the arguments together and presents them alongside a summary of many of the sceptical views outlined in the previous section. Other sceptics have resorted to discrediting key scientists involved in the IPCC climate reconstruction and climate modelling. The Competitive Enterprise Institute, CEI (www.cei.org) in particular has published articles in which key protagonists of global warming and their arguments have been caricatured, their arguments over-simplified, or misrepresented, and their motives called into question either directly, or by subtle implication (Georgia, 2002a, b; Anon., 2003b; Murray, 2003a). These attacks follow earlier ones on the lead authors of the second IPCC assessment (IPCC, 1996).

Other lines of argument hinge on perceptions of energy use, growth economics, notions of freedom and particular world views. Many are expressed in the publications of the CEI and the Capital Research Center, www.capitalresearch. org (see especially Osorio, 2003). They include an emphasis on the need for fossil-fuel-based economic development world-wide; the inappropriateness of any attempt to curtail energy-based growth nationally or internationally, by treaty, legislation, trade credits or tax incentives; the dismissal of all equity-based concepts save those that can be 'cured' by energy-linked economic growth; the promotion of large automobiles on the basis of greater road safety; the characterisation of third-world poverty as owing more to corrupt regimes and overpopulation than to any other factors, environmental , fiscal or political, and a visceral antipathy to all the United Nations-linked non-governmental organisations (NGOs) with any concern for environmental protection. These points of view are generally held alongside the belief that, 'as in the past', technological solutions can be found for any problems that the environment throws up (as well as any that the technological solutions themselves may generate). All these arguments are presented with varying degrees of coherence and balance. Lomborg (2001) presents a comprehensive case, underpinned by a wealth of statistical information, in support of the view that, given continued access to widely available energy, the world will only get better overall. He sees the 'environmentalist' arguments for caution, reduced energy use and mitigation as fundamentally flawed.

16.4   Key questions and some tentative responses

16.4.1   What will be the amplitude and rate of global climate changes and the likely response of sea-level over the next century?

Such consensus as there is among all but the most extreme sceptics would accept projections of global mean temperatures some 1 to 1.5 °C above current levels by the middle of this century. The range of increase by the end of the century is most unlikely to be less than 1.5 °C, unless virtually all the essential components of current state of the art atmosphere–ocean global climate models are seriously wrong, or currently unverified processes contrive to generate massive negative feedbacks. The possibility of an increase significantly in excess of 5 °C cannot be precluded, but on the basis of current understanding one or more of the following would probably be required: a serious underestimate of the sensitivity of climate to greenhouse-gas forcing; conversion of current terrestrial and marine sinks of carbon into major sources; an acceleration in the annual rate of increase of atmospheric greenhouse-gas concentrations; virtual exhaustion of all known fossil-fuel reserves by the end of the century.

Projections of global precipitation and precipitation/evaporation are much less secure. In so far as global generalisations are possible, higher total precipitation, earlier snow-melt, greater extremes and a higher level of variability seem likely. To the extent to which there is any convergence between model projections, they would suggest moister conditions in higher northern latitudes, a greater likelihood of droughts in sub-tropical semi-arid regions, but more chance of higher levels of precipitation and potential flooding in some regions affected by summer-monsoon rainfall. The coexistence of higher risks of both drought and floods may be partly explained by the kind of mechanism outlined by Neelin *et al.* (2003). Using a climate model of intermediate complexity they show that as surface warming increases, regional reductions in precipitation occur around the margins of convection zones. This is because the higher temperatures in the troposphere increase the amount of moisture needed in the surface-boundary layer for convection to occur. Where enough moisture is present, precipitation is maintained, but one consequence of this is less moisture in surrounding regions of air subsidence. Regions affected by this subsiding air suffer reduced rainfall. The authors claim that this is a major factor in the simulation of future low-latitude droughts in global-warming scenarios, as well as one of the dominant causes of drought during El Niño events.

In the present state of knowledge, it is unrealistic to venture far beyond such spatially non-specific generalisations without a better sense of the ways in which current modes of climate variability are likely to interact with anthropogenic forcings. More confidence is needed in the ability of climate models to be able to generate simulations that can, at the very least, effectively replicate the key aspects of both present- and past-climate *variability* at sub-global level.

The impact of future climate change on global sea-level has proved extremely difficult to quantify and the range given in the IPCC TAR (9–88 cm by 2100, relative to 1990) accommodates most of the estimates given by all but the most conservative projections. Several post-IPCC TAR studies suggest that the IPCC projections for any given emission scenario are more likely to be under- than over-estimates. Virtually all longer-term projections beyond 2100 have sea-level continuing to rise. Those based on higher-emission scenarios project sea-level rises that would be catastrophic in human terms. Even the modest rises projected for the next few decades require careful evaluation for any given coastline before they can be translated into the full range of implications for human populations and coastal ecosystems.

16.4.2 How will the changes in global mean climate be expressed in terms of extremes (droughts and floods for example) at continental, national and regional level?

All these questions depend on secure climate projections at sub-global level and these, for the most part, are simply not available with either the degree of confidence, or the attachment of sufficiently narrow probability ranges, to be of practical use as guides to the direction and degree of impact on ecosystems, human welfare and human activities. This is not simply a personal view. Schiermeier (2004a), in summarising the outcomes of a meeting of regional climate modellers, quotes one of the senior participants as saying: 'We're not yet at the promised level where regional climate models can really influence policy making.' Although this may come as no surprise to readers of Chapter 14, it would be too pessimistic to conclude that no guidance can be offered towards identifying which measures for protection or adaptation should be considered in the light of a plausible range of scenarios. In this connection it is important to remember lessons from the past about greater extremes even under natural forcing, to take careful note of many of the more consensus-based generic statements about likely future extremes that do not depend on regional models, and to be aware of the progress towards developing credible sub-global models that has already been made over the last decade.

16.4.3 What are the future implications of other ongoing processes resulting from human activities as they proceed alongside climate change, e.g., loss of biodiversity, changes in ecosystem functioning, soil degradation, deforestation?

In the preceding chapters, and in much of the literature on future environmental change, climate has been dealt with largely in isolation from the many cumulative impacts of human activities, many of which feed back into climate change. This is unrealistic. It reflects the intellectual and pedagogic need to separate and explain, not the reality of interwoven processes, effects and multiple feedbacks. The most urgent need is to define those interactions and synergies that carry maximum risk, always remembering that risk reflects both probability and potential consequences. Schellnhuber (2002) seeks to do this by defining 'switch' or 'choke' points that represent key locations and functions modification of which would have major significance for the Earth system as a whole. One of the key points he defines is Amazonia, since model simulations suggest that complete loss of forest cover could have massive consequences for climate over a wide area. He also identifies biophysical processes that are linked to human activities mainly through the impacts of the latter on climate. This ignores those situations where the complex of interacting biophysical and human processes pose increasing threats to ecosystems, irrespective of future changes in climate. For the most part, these must be considered on a case-by-case basis. The limitations of regional climate projections, alongside the indeterminacy inherent in human affairs, make all such studies subject to huge uncertainties. To attempt a comprehensive survey of these would be to repeat too much of what is presented in Watson *et al.* (2001) and illustrated by a wide range of examples in Steffen *et al.* (2004).

16.4.4 How will the interacting complex of changes encompassed in all the above affect key issues of vulnerability and sustainability for human populations and the resources upon which they depend?

The ever-growing industry of impact assessment deals with these kinds of question, but the claim made here is that the most such activities can lead to is a better sense of the range of possible futures to which adaptive capacity may be oriented. Building adaptive capacity through characterising the range of future possibilities, and strengthening those aspects of societies that can serve to enhance adaptive capacity at all levels of social and political organisation, seems wise and necessary. Pretending, in our present state of knowledge, that by combining regional climate models with, for example, models of ecological or agronomic responses, and others for human decision making, it is possible to produce a concrete basis for choosing a narrowly defined direction of adaptation is trusting to an illusion. It seems more likely to limit adaptive capacity than to enhance it. Should any nexus develop within which unrealistic expectations on the part of policy makers are 'satisfied' by unjustified or spurious predictions on the part of researchers, the results would be an affront to science and a betrayal of the public at large. Even if it were possible to downscale global climate models more effectively, several key stumbling blocks would remain, notably those arising from our inability either to predict the behaviour of complex, interactive, non-linear systems or to overcome the inherent indeterminacy in human affairs. The most realistic approach is to develop conditional scenarios and use these to explore system resilience and vulnerability in linked model studies that incorporate feedbacks between all aspects of the biophysical–human complex. In this context, systems may be ecosystems, schemes for water management or flood protection, or indeed any coherent attempt to define management options for an uncertain future.

16.4.5   What is the likelihood that future changes will include major perturbations of the Earth system such as a reorganisation of ocean circulation patterns or a collapse of the west Antarctic ice sheet?

Section 4 of Chapter 13 suggests that, in all probability, the issue of major perturbations of the kind implied may be not so much a question of 'if', but rather of 'how soon, how sudden and how dramatic?' To answer these questions calls for much more effective modelling that incorporates well-constrained representations of many processes currently only crudely parameterised. Meanwhile, especially for anyone inclined to take seriously the possible consequences of present and near-future actions on the societies of the twenty-second century and beyond, there is no cause for complacency.

16.5   A personal perspective on mitigation and adaptation

From the necessarily personal responses outlined above, and the summary of the science that underpins them in previous chapters, comes a conviction that the outcomes of all but the most conservative future emission scenarios are likely to include serious implications for many human populations on decade-to-century timescales. This is consistent with the December 2003 'position statement' made by the Council of the American Geophysical Union (AGU, 2003), which includes the following: 'The global climate is changing and human activities are contributing to that change'. Even those who minimise the implications of global change over the next few decades would be well advised to consider the longer term. They might also contemplate the disturbing paradox that those who are most dismissive are those who are recommending a pathway that may turn out to maximise the likelihood of worst-case outcomes.

It follows that proposals to rule out all mitigation strategies are in conflict with the scientific evidence and its likely future implications. This book is not about proposing, outlining or evaluating mitigation strategies, but several points seem self-evident:

- The degree of global coordination, shared will and commonality of purpose required to promulgate and sustain a policy designed to *minimise* future greenhouse-gas emissions world-wide is neither attainable, nor, given the degree of central direction (and hubris) it would require, or the damaging human consequences it could entail, politically or ethically desirable. This is not to deny the importance of efforts to negotiate consensus-based protocols for emission reduction.
- The conflict between emission reduction and the demand for energy-based growth is most unlikely to be resolved in favour of the former on a global scale in the foreseeable future. The most one may hope for is a growing recognition of the need to respect the precautionary principle and to develop 'guardrails' against the most damaging future outcomes (WGBU, 1996). This implies a willingness to act decisively as future projections come closer to identifying the levels and rates of increase in emissions most likely to threaten existing human populations, infrastructure and life-support systems. In this connection, identifying indicators that can be used to diagnose the onset of damaging changes in Earth-system function (see, e.g., Vellinga and Wood, 2004) may prove to be especially important.
- Pro-energy sceptics often equate future growth in energy use with future growth in welfare, without any differentiation on the basis of the huge contrasts in levels of economic development and human welfare that actually exist across the world. This leads to arguments that are both disingenuous and unrealistic. There is an urgent need to reconcile welfare imperatives (and their implications in terms of higher-energy demands) for currently impoverished human populations, with emission controls, constraints on wasteful energy use, especially in the developed world, and acceptable mitigation strategies.
- Only under the most extreme and as yet highly unlikely circumstances would acceptable mitigation strategies include the types of strongly interventionist geo-engineering schemes outlined by Keith (2001) and Schneider (2001b). There is clearly scope for major improvements in sequestration technology (Lackner, 2003). Additional initiatives that go much beyond improved emission controls, and increased carbon sequestration in the terrestrial biomass through tree planting, as a temporary palliative, would probably be either ineffective, as in the case of iron fertilisation of areas of the ocean, or incalculably dangerous, as in the case of various schemes to modify the energy balance of the atmosphere. Even temporary sequestration through afforestation is far from straightforward. As Körner (2003) shows, net sequestration involves a delicate balance between slow gains through growth, and rapid losses through fire and other types of disturbance. Where afforestation increases vulnerability to the latter, the result may be a net loss of carbon.

All the above points should be viewed in the light of two further observations:

1. What is in question is not the survival of the Earth system, nor indeed the human species, but rather the future welfare of a significant proportion of the human population and the resource systems upon which it depends.

2. The combined effects of the processes linked to human-population growth, aspirations and demands on the one hand, and those arising from likely climate change on the other, make the problems to be faced in the future quite unique.

The former point disposes of the view that the demonstrated past resilience of the Earth system provides a basis for an essentially optimistic forward look; the second point counters the view that human survival of past changes lends an air of normality to the challenges posed by any changes that lie ahead.

Given the above rather pessimistic evaluation of the prospects for early and effective mitigation, there is a clear need to examine the issue of adaptation very seriously, yet the thrust of much of Chapters 14 and 15 is to doubt the wisdom of prescribing specific actions to meet projected future impacts, no matter how sophisticated or comprehensive the scheme of reasoning may be. The logic of this is that the emphasis should not be on pre-emptive adaptation to a particular set of scenarios, but on raising adaptive capacity in the face of future uncertainty. Articulating and constraining that uncertainty through the development of a realistic range of plausible scenarios seems valid and necessary. Alongside this, there is also the need to apply models that explore the sensitivity of existing systems to the range of scenarios generated. To go beyond this and adopt narrowly defined responses linked to specific scenarios underpinned by projections that rest on the present state of scientific knowledge still seems unwise at best. One of the most intransigent aspects of impact/vulnerability assessments over the world as a whole is the degree to which many of the societies most at risk from future global change are ones with only a limited repertoire of adaptive options (Watson *et al.*, 2001).

16.6   A final key issue

Schiermeier's (2004a) quotation, casting doubt on the ability of existing regional climate models to support policy making, raises challenging questions when one considers the extent to which policy making in fields such as countryside management, conservation, regional land-use planning, coastal and flood defences and water-resource management require future climate scenarios (UKCIP, 2002), and the fact that all impact assessments must include a future climate element applicable at regional or national level. The contention here is that several of the key questions posed at the beginning of the book can only be given tentative and highly qualified answers, usually with many attached uncertainties, some more or less quantifiable, some essentially indeterminate. Out of the response to the key issues outlined in Chapter 1 emerges a further question of outstanding importance: how may we best respond to the dilemma that arises from this troubling gap between what can currently be underpinned by well-validated science and what is urgently needed by policy makers?

The UKCIP (2002) briefing report contains state of the art examples of regional climate scenarios for time slices within the present century. It is deliberately couched in clear and jargon-free language, but makes no secret of the serious limitations in the projections provided. It lists the main uncertainties associated with both the emission scenarios and the science involved at each stage in the process; these include those associated with using only a single global model, with additional uncertainties arising from downscaling to regional level and from the use of a 'pattern scaling' procedure to generate scenarios for time intervals between 1961–90 (the control period) and 2071–100, the future time slice for which the regional model was run. In every section of the text, the report cautions against using the present projections as predictions. In the chapter on 'Further Work and Developments in Research' many pathways for improving projections are outlined, including the provision of more detail, higher resolution, a wider range of information and probabilistic predictions. The main approach to validation seems to be through developing alternative models and testing them against observations for a single, recent time slice, then using the output from models that have been acceptably validated to generate probability density functions. Most of the other future developments seem to lean towards stake-holder requirements, rather than validation. Is this the best way forwards?

One of the key demonstrations in the projection of future global climate is the skill with which models that combine both natural and anthropogenic forcing can replicate the sequence of changes in global temperature over the past 150 years. Despite uncertainties arising from the mutually compensating effects of forcing by greenhouse gases and aerosols, this is one of the firmer planks upon which future projections are built. It signals a competence in realistically simulating, to a good first approximation, *changes through time*. Despite great progress in recent years, nothing in the business of developing future scenarios at regional scale matches this at present, yet the observational data are available, along with, for the last 60 years or so, the reanalysis data too. To someone looking in from outside the regional-modelling community, there seems to be both a need and an opportunity to try to get closer to more rigorous testing and temporal 'calibration'. Even to reach the point where a simulation of regional climate for a period early in the instrumental series could be compared with one close to the present day would surely be a useful advance, though there is also scope for testing against spatially well-resolved proxy-based reconstructions too (see, e.g., Luterbacher *et al.*, 2004). It might test the skill with which regional models can measure change, rather than replicate observations for a single time slice. It might also give a better idea of how well or otherwise downscaling can cope with changes in natural forcing (solar and volcanic), which are currently discounted in the procedures used. In the tension between underpinning the science as securely as possible and responding to stake-holder requirements, where, within a finite budget, does the best balance lie? As Covey (2000) states, in relation to climate scenarios: 'Such an exercise cannot prove models correct but can prove them incorrect if they fail to reproduce the observations while keeping adjustable parameters within acceptable bounds'. We are obliged to accept that more rigorous testing will not guarantee reliability, but it should at least help to detect the basis for, quantify, or even reduce uncertainty.

Applied to the wider realms of environmental change, involving ecological processes and human decisions, Covey's statement, with strong Popperian echoes, hits against a distinction made neatly by Francis and Hare (1994). They contrast two approaches to science, an 'experimental-predictive' and a 'historical-descriptive' approach. For us, the crucial characteristic of the former is 'that all times can be treated alike'. By contrast, in the latter mode of study, what we record is 'unique, i.e., dependent, or contingent upon everything that came before'. As we have seen in section 1.2.2, contingency is one of the key characteristics of many aspects of the Earth system, especially as human interventions become increasingly dominant. Somehow then, we have to find ways of reconciling this with the need to create and evaluate future scenarios, despite the fact that in many critical respects, all times cannot be treated alike. Harte (2002), who makes a similar distinction to that made by Francis and Hare, nevertheless believes that the impediments to generalisation imposed by contingency and particularity can often be overcome without venturing beyond familiar concepts of science. He cites scaling laws in ecology and nutrient mass-balance models as examples. An additional, or alternative approach could involve the use of cellular-automaton type models, as briefly outlined in section 2.4.3, since they are able to incorporate ongoing feedback and hence replicate contingency, through the introduction of simple rules.

Further questions arise concerning the extent to which bridging the gap between science and policy requirements constitutes a fundamental shift in the nature of science. Haug and Kaupenjohann (2001) and Saloranta (2001), following Funtowicz and Ravetz (1992, 1993), express the view that the only type of research that is sufficiently responsive to the combination of urgency and uncertainty provoked by fears of the consequence of global change is what they term post-normal science. They highlight what they see as a discontinuity between the traditional scientist role and the role in which scientists responsive to the demands of future policy makers are cast. They envisage that science may split into an 'academic branch' and a 'managerial, public-policy branch'; and that modelling for science per se and modelling for decision making may diverge. Their analysis thus disconnects future-oriented, hence policy-oriented science from traditional science by stressing aspects that lie outside the traditional scientific realm.

The thrust of this book is against such a proposition. Moreover, the approach exemplified by the UKCIP (2002) report is also counter to it, for the need for scientific underpinning is never sidelined or neglected. My only questions are about the nature of validation and the allocation of resources between competing priorities. A future-oriented science is necessarily different in terms of research strategies, methodologies, verifiability and application, but that only decouples it from the more traditional approaches to science if groups of practitioners within in it dissociate themselves from the fundamental concepts of testing for consistency and refining inferential schemes in the light of new results. In the context of designing future scenarios at regional scale, one of the barriers to retaining the union between science 'proper' and the provision of guidance to policy makers is the likelihood that testing against what we know of past processes and interactions is more likely to constrain scenario development and impair its fluency, than to facilitate it. As Diaz (2001) has noted, 'Considerable optimism and a certain disdain for all things past are distinctive characteristics of quickly developing fields of science'. Indeed, Don DeLillo the novelist goes well beyond this with his bold view of the future: 'Computer power eliminates doubt. All doubt rises from past experience. But past experience is disappearing. We used to know the past, but not the future. This is changing.' (Burn, 2003). Both of the latter two views find many echoes in the burgeoning scientific and not so scientific literature about the future.

Where the only criteria used to assess value are apparent credibility and acceptance by users, testing against independently derived empirical data runs the risk of becoming redundant. Whenever this happens, all ultimately lose out – scientists for their lack of rigour and intellectual honesty; policy makers for the flimsy basis underpinning their planning. Scenarios generated for policy makers should be conditional, constrained and, in so far as possible, probabilistic; conditional in the sense of depending, for example, on given emission scenarios; constrained by tests against past observations using a strategy that focuses on tests of the fidelity with which past *change* can be replicated; probabilistic at least in relation to the statistical properties of the range of simulations generated. Anything short of this falls into the category of coherent myth – useful perhaps, but to a greater or lesser degree mythical nonetheless.

Anyone who has worked in the field of past environmental change for half a century is familiar with the nature of coherent myths, having in all probability both shared and contributed to a succession of them. What redeems this activity is the dogged skill with which the best practitioners test, refine, revise and transcend the limitations of each successive interpretative scheme. My best hope is that by developing an equal partnership between the empirical and the model-based, between past changes and future scenarios and between biophysical and human processes, the same may be said for attempts to project the climates and the linked human–environmental interactions of the future. Those who develop projections of the future nature of the Earth system are in a line of business that has much in common with macro-economic forecasting, but the lead times are longer and the potential for non-linear responses and threshold-linked behaviour no less. The longer lead times reduce the scope for observed behaviour to feed back into revised projections. Thus, if, as is generally supposed, the likely changes over the next 30 years or so have largely been 'built-in' and will, in any case, be relatively modest, that in itself constitutes a potential problem. It could easily encourage the view that there's not much point in taking any action and that anyhow, things are not changing so much. Yet the next 30 years will be critical for the longer term.

   Much emphasis has been placed on the need to make science responsive to future policy making. One of the corollaries of the foregoing analysis is that there is an equal need for policy makers and the general public to whom they are responsible to have a better grasp of what, in any given state of knowledge and understanding, can actually be supported by sound research and what remains largely unsubstantiated. I have already outlined the need to bridge three methodological gaps, between contemporary and paleo-research, between empirical and modelling approaches and between biophysical and socio-economic perspectives. As in the other three cases, bridging the fourth and final gap considered here – that between well-validated research and policy making – requires understanding and respect from both sides.

References

Achard, F., Eva, H. D., Stibig, H. -J., *et al*. (2002). Determination of deforestation rates of the world's humid tropical forests. *Science* **297**, 999–1002.

Achard, F., Eva, H. D., Mayaux, P., Stibig, H.-J. and Delward, A. (2004). Improved estimates of net carbon emissions from land cover change in the tropics from the 1990s. *Global Biogeochemical Cycles* **18**, GB2008.

Ackert, R. P. (2003). An ice sheet remembers. *Science* **299**, 57–8.

Adams, J. B., Mann, M. E. and Amman, C. M. (2003). Proxy evidence for an El Niño-like response to volcanic forcing. *Nature* **426**, 274–8.

Adams, J. M. and Piovesan, G. (2002). Uncertainties in the role of land vegetation in the carbon cycle. *Chemosphere* **49**, 805–19.

Adelson, J. M. and Helz, G. R. (2001). Reconstructing the rise of recent coastal anoxia: molybdenum in Chesapeake Bay sediments. *Geochimica & Cosmochimica acta* **65**, 237–52.

AGU Council (2003). American Geophysical Union position statement on human impact on climate. *EOS* **84** (51), 574.

Aharon, P. (2003). Meltwater flooding events in the Gulf of Mexico revisited: implications for rapid climate changes during the last glaciation. *Paleoceanography* **18**, 3–1–3–14.

Ahn, J., Wahlen, M., Deck, B. L. *et al*. (2004). A record of atmospheric CO_2 during the last 40 000 years from the Siple Dome, Antarctica ice core. *Journal of Geophysical Research* **109**, D13305.

Allen, M. R. and Ingram, W. J. (2002). Constraints on future changes in climate and hydrologic cycle. *Nature* **419**, 224–32.

Alley, R. B., (2000). The Younger Dryas cold interval as viewed from central Greenland. *Quaternary Science Reviews* **19**, 213–26.

Alley, R. B., Mayewski, P. A., Sowers, T., *et al*. (1997). Holocene climate instability: a prominent widespread event 8200 years ago. *Geology* **25**, 483–6.

Alley, R. B., Meese, D. A., Shuman, C. A. *et al*. (1993). Abrupt increase in Greenland snow accumulation at the end of the Younger Dryas event. *Nature* **362**, 527–9.

Alverson, K. and Oldfield, F. (2000). Past global changes and their significance for the future: an introduction. *Quaternary Science Reviews* **19**, 3–7.

Alverson, K., Bradley, R. S. and Pedersen, T. F. (2001). Environmental Variability and Climate Change. IGBP Science 3. Stockholm, IGBP.

(eds.) (2003). Paleoclimate, Global Change and the Future. Berlin, Springer Verlag.

Amman, B. (ed.) (2000). Biotic responses to rapid climatic changes around the Younger Dryas. *Palaeogeography, Palaeoclimatology, Palaeoecology* **159**.

Amman, B. and Oldfield, F. (2000). Preface: Rapid warming project. In Amman, B. (ed.). Biotic responses to rapid climatic changes around the Younger Dryas. *Palaeogeography, Palaeoclimatology, Palaeoecology* **15**, pp. v–vii.

Amman, C. M., Meehl, G. A. and Washington, W. M. (2003). A monthly and latitudinally varying volcanic forcing dataset in simulations of twentieth century climate. *Geophysical Research Letters* **30** (12), 1657.

An, Z. (2000). The history and variability of the east Asian paleomonsoon climate. *Quaternary Science Reviews* **19**, 171–87.

An, Z. and Porter, S. C. (1997). Millennial-scale oscillations during the last interglaciation in central China. *Geology* **25**, 603–6.

Anderson, C., Koc, N., Jennings, A. and Andrews, J. T. (2004). Non-uniform response of the major surface currents in the Nordic seas to insolation forcing: implications for the Holocene climate variability. *Paleoceanography* **19**, PA2003, 1–16.

Anderson, L., Abbott, M. B. and Finney, B. P. (2001). Holocene climate inferred from oxygen isotope ratios in lake sediments, central Brooks Range, Alaska. *Quaternary Research*, **55**, 313–21.

Anderson, P. M., Bartlein, P. J., Brubaker, L. B., Gajewski, K. and Ritchie, J. C. (1991). Vegetation–climate–pollen relationships for the arcto-boreal regions of North America and Greenland. *Journal of Biogeography* **18**, 565–82.

Anderson, T. L., Charlson, R. J., Schwartz, S. E. *et al.* (2003). Climate forcing by aerosols – a hazy picture. *Science* **300**, 1103–4.

Andreae, M. O., Rosenfield, D., Artaxo, P. *et al.* (2004). Smoking rain clouds over the Amazon. *Science* **303**, 1337–42.

Andren, E., Andren, T. and Kunzendorf, H. (2000). Holocene history of the Baltic Sea as a background for assessing records of human impact in the sediments of the Gotland basin. *The Holocene* **10**, 687–702.

Andren, E., Shimmield, G. and Brand, T. (1999). Environmental changes of the last three centuries indicated by siliceous microfossil records from the Baltic Sea. *The Holocene* **9**, 25–38.

Andrews, J. T., Hardadottir, J., Stoner, J., Mann, M. E., Krisjansdottir, B. and Koc, N. (2003). Decadal to millennial-scale periodicities in north Iceland shelf sediments over the last 12 000 cal. yr: long-term North Atlantic variability and solar forcing. *Earth and Planetary Science Letters* **210**, 453–65.

Angert, A., Biraud, S., Bonfils, C., Buermann, W. and Fung, I. (2004). CO_2 seasonality indicates origins of post-Pinatubo sink. *Geophysical Research Letters* **31**, L11103.

Anon. (2003a). How to slake a planet's thirst. *Nature* **422**, 243.

Anon. (2003b). Two wrongs not right . CO_2 and climate: world climate alerts. At www.co2andclimate.org/wca/2003/wca.

Appleby, P. G. and Oldfield, F. (1992). Applications of Pb-210 to sedimentation studies. In Ivanovich, M. and Harmon, R. (eds.). Uranium Series Disequilibrium: Applications to Earth, Marine and Environmental Studies. Oxford, Clarendon Press, pp. 731–78.

Appleby, P. G., Richardson, N., Nolan, P. J. and Oldfield, F. (1990). Radiometric dating of the United Kingdom SWAP sites. *Philosophical Transactions of the Royal Society B.*, **327**, 7–12.

Appleby, P. G., Oldfield, F., Thompson, R., Huttunen, P. and Tolonen, K. (1979). Pb-210 dating of annually laminated lake sediment from Finland. *Nature*, **280**, 53–5.

Arendt, A. A., Echelmeyer, K. A., Harrison, W. D., Lingle, C. S. and Valentine, V. B. (2002). Rapid wastage of Alaskan glaciers and their contribution to rising sea level. *Science* **297**, 382–6.

Ariztegui, D., Chondrogianni, C., Wolff, G. *et al.* (1996). Palaeotemperature and palaeosalinity history of the Meso Adriatic Depression (MAD) during the late Quaternary: a stable isotope and alkenones study. In Guilizzoni, P. and Oldfield, F. (eds.). Palaeoenvironmental Analysis of Italian Crater Lake and Adriatic sediments (PALICLAS). Memorie dell'Istituto Italiano di Idrobiologia, 55. Verbania Pallanza, Italy, Istituto Italiano di Idrobiogia, pp. 219–30.

Arnell, N. W. (1999). Climate change and global water resources. *Global Environmental Change* **9**, S31–S49.

Artaxo, P. (2003). Land-use change, aerosol production and climate: size matters!! LBA Science Highlight. At www//igbp.Kva.se/.

Asioli, A. (1996). High resolution foraminifera biostratigraphy in the central Adriatic basin during the last deglaciation: a contribution to the PALICLAS project. In Guilizzoni, P. and Oldfield, F. (eds.). Palaeoenvironmental Analysis of Italian Crater Lake and Adriatic Sediments (PALICLAS). Memorie dell 'Istituto Italiano di Idrobiologia, 55. Verbania Pallanza, Italy, Istituto Italiano di Idrobiologia, pp. 197–217.

Atkinson, T. C., Briffa, K. R. and Coope, G. R. (1987). Seasonal temperatures in Britain during the last 22 000 years, reconstructed using beetle remains. *Nature* **325**, 587–92.

Bahn, P. and Flenley, J. R. (1992). Easter Island, Earth Island: A Message from Our Past for the Future of Our Planet. London, Thames and Hudson.

Baillie, M. G. L. and Brown, D. M. (2003). Dendrochronology and the reconstruction of fine-resolution environmental change in the Holocene. In Mackay, A. W., Battarbee, R. W., Birks, H. J. B. and Oldfield, F. (eds.). Global Change in the Holocene. London, Arnold, pp. 75–91.

Banks, H. and Wood, R. (2002). Where to look for anthropogenic climate change in the ocean. *Journal of Climate Change* **15**, 879–91.

Barber, D. C., Dyke, A., Hillaire-Marcel, C. *et al.* (1999). Forcing of the cold event of 8200 years ago by catastrophic drainage of Laurentide lakes. *Nature* **400**, 344–48.

Barber, K. E. and Charman, D. (2003). Holocene Paleoclimate records from peatlands. In Mackay, A. W., Battarbee, R. W., Birks, H. J. B. and Oldfield, F. (eds.). Global Change in the Holocene. London, Arnold, pp. 210–26.

Bard, E., Arnold, M., Hamelin, B., Tisnerat-Laborde, N. and Cabioch, G. (1998). Radiocarbon calibration by means of mass spectrometric ^{230}Th/^{234}U and ^{14}C ages of corals: an updated database including samples from Barbados, Muraroa and Tahiti. *Radiocarbon* **40**, 1085–92.

Bard, E., Fairbanks, R. G., Arnold, M. *et al.* (1989). Sea-level estimates during the last deglaciation based on δ^{18}O and accelerator mass spectrometry ^{14}C ages measured in *Globigerina bulloides. Quaternary Research* **31**, 381–91.

Bard, E., Hamelin, B., Arnold, M. *et al.* (1996). Deglacial sea-level record from Tahiti corals and the timing of global melt-water discharge. *Nature* **382**, 405–10.

Bard, E., Rostek, F. and Menot-Combes, G. (2004). Radiocarbon calibration beyond 20 000 ^{14}C yr BP by means of planktonic foraminifera of the Iberian margin. *Quaternary Research* **61**, 204–14.

Bar-Matthews, M., Ayalon, A. and Kaufman, A. (1997). Late Quaternary paleoclimate in the eastern Mediterranean region from stable isotope analysis of speleothems at Soreq Cave, Israel. *Quaternary Research* **47**, 155–68.

Barnekow, L. (2000). Holocene regional and local vegetation history and lake-level changes in the Tornetrask area, northern Sweden. *Journal of Paleolimnology* **23**, 399–420.

Barnett, J. and Adger, W. N. (2003). Climate dangers and atoll countries. *Climatic Change* **61**, 321–37.

Barnett, T. P., Pierce, D. W. and Schnurr, R. (2001). Detection of anthropogenic climate change in the world's oceans. *Science* **292**, 270–4.

Barnett, T., Malone, R., Pennell, W. *et al*. (2004). The effects of climate change on water resources in the West: introduction and overview. *Climatic Change* **62**, 1–11.

Barnola, J.-M., Pimienta, P., Raynaud, D. and Korotkevich, Y. S. (1991). CO_2-climate relationship as deduced from the Vostok ice core: a re-examination based on new measurements and on a re-evaluation of the air dating. *Tellus* **43**, 83–90.

Bartlein, P. J., Webb, J. T., III and Fleri, E. (1984). Holocene climatic change in the northern midwest: pollen derived estimates. *Quaternary Research* **22**, 361–74.

Batchold, D. (2003). Britain to cut CO_2 without relying on nuclear power. *Science* **299**, 1291.

Bates, N. R., Pequignet, A. C., Jonson, R. J. and Gruber, N. (2002). A short-term sink for atmospheric CO_2 in subtropical mode water of the North Atlantic Ocean. *Nature* **420**, 489–93.

Battarbee, R. W. (1978). Observations on the recent history of Lough Neagh and its drainage basin. *Philosophical Transactions of the Royal Society B* **281**, 303–45.

Battarbee, R. W. (1990). The causes of lake acidification, with special reference to the role of acid deposition. *Philosophical Transactions of the Royal Society of London Series B* **327**, 339–47.

Battarbee, R. W. (1998). Lake management: the role of palaeolimnology. In Harper, D. M. (ed.). The Ecological Basis for Lake and Reservoir Management. London, Wiley.

Battarbee, R. W., Gasse, F. and Stickley, C. E. (eds.) (2005). Past Climate Variability through Europe and Africa. Dordrecht, Kluwer.

Battarbee, R. W., Mason, J., Renberg, I. and Talling, J. F. (eds.) (1990). Palaeolimnology and lake acidification. Philosophical Transactions of the Royal Society of London B **327**, 223–445.

Bauer, E., Claussen, M. and Brovkin, V. (2003). Assessing climate forcings of the Earth system for the past millennium. *Geophysical Research Letters* **2** (6), 9–1–9–4.

Beer, J, Mende, W. and Stellmacher, R. (2000). The role of the sun in climate forcing. *Quaternary Science Reviews* **19**, 403–15.

Behl, R. J. and Kennett, J. P. (1996). Evidence for brief interstadial events in the Santa Barbara basin, NE Pacific during the past 60 kyr. *Nature* **379**, 243–6.

Bellwood, D. R., Hughes, T. P., Folke, C. and Nystrom, M. (2004). Confronting the coral reef crisis. *Nature* **429**, 827–33.

Beltrami, W., Wang, J. F. and Bras, R. L. (2000). Energy balance at the Earth's surface: heat flux history in eastern Canada. *Geophysical Research Letters* **27**, 3385–8.

Bender, M. L. (2002). Orbital tuning chronology for the Vostok climate record supported by trapped gas composition. *Earth and Planetary Science Letters* **204**, 275–89.

Bender, M. L. (2003). Climate–biosphere interactions on glacial–interglacial timescales. *Global Biogeochemical Cycles* **17**, 8–1–8–10.

Benestad, R. E. (2003). What can climate models tell us about climate change? *Climatic Change* **59**, 311–31.

Bengtsson, L. (2001). Hurricane threats. *Science* **293**, 440–441.

Bengtsson, L., Hagemann, S. and Hodges, K. I. (2004). Can climate trends be calculated from reanalysis data? *Journal of Geophysical Research* **109**. D11111.

Beniston, M. (2003). Climatic change in mountain regions: a review of possible impacts. *Climatic Change* **59**, 5–31.

Bennett, K. D. (1997). Evolution and Ecology. The Pace of Life. Cambridge, Cambridge University Press.

Benson, L., Kashgarian, M., Rye, R. *et al.* (2002). Holocene multidecadal and multicentennial droughts affecting northern California and Nevada. *Quaternary Science Reviews* **21**, 659–82.

Bergengren, J. C., Thompson, S. L., Pollard, D. and DeConto, R. M. (2001). Modelling global climate–vegetation interactions in a doubled CO_2 world. *Climatic Change* **50**, 31–75.

Berger, A. (1979). Insolation signatures of Quaternary climatic changes. *Il nuovo cimento* **2**, 63–87.

Berger, A. and Loutre, M. F. (1991). Insolation values for the climate of the last 10 million years. *Quaternary Science Reviews* **10**, 297–318.

Bianchi, G. G. and McCave, I. N. (1999). Holocene periodicity in North Atlantic climate and deep ocean flow south of Iceland. *Nature* **397**, 515–17.

Bigler, C., Larocque, I. Peglar, S. M., Birks, H. J. B. and Hall, R. I. (2002). Quantitative multiproxy assessment of long-term patterns of Holocene environmental change from a small lake near Åbisko, northern Sweden. *The Holocene* **12**, 481–96.

Billen, G. and Garnier, J. (1999). Nitrogen transfers through the Seine drainage network: a budget based on the application of the 'Riverstrahler' model. *Hydrobiologia* **410**, 139–50.

Bindschalder, R., Diner, D. J. and Rignot, E. (2002). West Antarctic ice sheet releases new icebergs. *EOS* **83** (9), 85–93.

Biondi, F., Isaacs, C., Hughes, M. K., Cayan, D. R. and Berger, W. H. (2000). The near-1600 dry/wet knockout: linking terrestrial and near-shore ecosystems. In Proceedings of the Twenty-Fourth Annual Climate Diagnostics and Prediction Workshop. Washington DC, US Department of Commerce, NOAA.

Biondi, F., Lange, C. B., Hughes, M. K. and Berger, W. H. (1997). Interdecadal signals during the last millennium (AD 1117–1992) in the varve record of Santa Barbara Basin, California. *Geophysical Research Letters* **24**, 193–6.

Birks, H. J. B. (2003). Quantitative palaeoenvironmental reconstructions from Holocene biological data. In Mackay, A. W., Battarbee, R. W., Birks, H. J. B. and Oldfield, F. (eds.). Global Change in the Holocene. London, Arnold, pp. 107–23.

Birks, H. H., Battarbee, R. W. and Briks, H. J. B. (2000) The development of the aquatic ecosystem at Kråkenes Lake, western Norway, during the late glacial and early Holocene – a synthesis. *Journal of Paleolimnology* **23**, 91–114.

Bishop, J. K. R., Wood, T. J., Davis, R. E. and Sherman, J. T. (2004). Robotic observations of enhanced carbon biomass and export at 55 °S during SOFeX. *Science* **304**, 417–18.

Björck, S., Koc, N. and Skog, G. (2003). Consistently large marine reservoir ages in the Norwegian Sea during the last deglaciation. *Quaternary Science Reviews* **22**, 429–35.

Blaauw, M. (2003). An investigation of Holocene sun-climate relationships using numerical C-14 wiggle-match dating of peat deposits. Ph.D. thesis, University of Amsterdam.

Blaauw, M., van Geel, B. and van der Plicht, J. (2004). Solar forcing of climate change during the mid-Holocene: indications from raised bogs in the Netherlands. *The Holocene* **14**, 35–44.

Blaauw, M., Heuvelink, G. B. M., Mauquoy, D., van der Plicht, J. and van Geel, B. (2003). A numerical approach to ^{14}C wiggle-match dating of organic deposits: best fits and confidence intervals. *Quaternary Science Reviews* **22**, 1485–1500.

Blunier, T. and Brook, E. J. (2001). Timing of millennial-scale climate change in Antarctica and Greenland during the last glacial period. *Science* **291**, 109–12.

Blunier, T., Chappellaz, J., Schwander, J. *et al.* (1997). Asynchrony of Antarctic and Greenland climate change during the last glacial period. *Nature* **394**, 739–43.

Boer, G. J., Yu, S -J., Kim, B. and Flato, G. M. (2004). Is there observational support for an El Niño-like pattern of future global warming? *Geophysical Research Letters* **31**, L06201.

Boersma, A. (1978). Foraminifera. In Haq, B. U. and Boersma, A. (eds.). Introduction to Marine Micropaleontology, New York, Elsevier, pp. 19–77.

Bolger, T. (2001). The functional value of species biodiversity – a review. *Proceedings of the Royal Irish Academy* **1018**, 119–142.

Bolin, B. (1998). The Kyoto negotiation on climate change: a scientific perspective. *Science* **279**, 330–1.

Bond, G., Kramer, B., Beer, J. *et al.* (2001). Persistent solar influence on North Atlantic climate during the Holocene. *Science* **294**, 2130–6.

Bond, G. and Lotti, R. (1995) Iceberg discharges into the North Atlantic on millennial timescales during the last glaciation. *Science* **267**, 1005–10.

Bond, G., Showers, W., Cheseby, M. *et al.* (1997). A pervasive millennial-scale cycle in North Atlantic Holocene and glacial climates. *Science* **278**, 1257–66.

Bond, G., Showers, W., Elliot, M. *et al.* (1999). The North Atlantic's 1–2 kyr climate rhythm: relation to Heinrich events, Dansgaard/Oeschger cycles and the Little Ice Age. In Clark, P. U., Webb, R. S. and Keigwin, L. D. (eds.). Mechanisms of Global Climate Change at Millennial Time Scales. Washington DC, AGU, pp. 35–58.

Bond, N. A., Overland, J. E., Spillane, M. and Stabeno, P. (2003). Recent shifts in the state of the North Pacific. *Geophysical Research Letters* **30** (23), 1–1–1–4.

Bopp, L., Kohfeld, K. E. Le Quéré, C. and Aumont, O. (2003) Dust impact on marine biota and atmospheric CO_2 during glacial periods. *Paleoceanography*, **18** (2), 24–1–24–9.

Bopp, L., Le Quéré, Heimann, M. Manning, A. C. and Monfray, P. (2002). Climate induced ocean-oxygen fluxes: implications for the contemporary carbon budget. *Global Geochemical Cycles* **16**, 6–1–6–14.

Boyd, P. W., Law, C. S., Wong, C. S. *et al.* (2004). The decline and fate of an iron-induced subarctic phytoplankton bloom. *Nature* **428**, 549–53.

Boyd, P. W., Watson, A., Law, C. *et al.* (2000). A mesoscale phytoplankton bloom in the polar Southern Ocean stimulated by iron fertilization. *Nature* **407**, 695–702.

Braconnot, P., Joussaume, S., Marti, O. and de Noblet, P. (1999). Synergistic feedbacks from ocean and vegetation on the African monsoon response to mid-Holocene insolation. *Geophysical Research Letters* **26**, 2481–4.

Bradley, R. S. (1999). Paleoclimatology: Reconstructing Climate of the Quaternary, 2nd. edn. San Diego, Harcourt Academic Press, ch. 11.

Bradley, R. S. (2003) Climate forcing during the Holocene. In Mackay, A., Battarbee, R. W., Birks, H. J. B. and Oldfield, F. (eds.). Global Change in the Holocene. London, Arnold, pp. 10–19.

Bradley. R. S., Briffa, K. R., Cole, J., Hughes, M. K. and Osborn, T. J. (2003). The climate of the last millennium. In Alverson, K., Bradley, R. S. and Pedersen, T. F. (eds.). Paleoclimate, Global Change and the Future. Berlin, Springer Verlag, pp. 105–41.

Bradley, R. S. and Jones, P. D. (1992). When was the 'Little Ice Age'? In Mikami, T. (ed.). Proceedings of the International Symposium on the Little Ice Age Climate. Tokyo, Dept. of Geography, Tokyo Metropolitan University, pp. 1–4.

Braganza, K., Karoly, D. J., Hirst, A. C. (2004). Simple indices of global climate variability and change part II: attribution of climate change during the twentieth century. *Climate Dynamics*, **22**, 823–38.

Brasseur, G. (2003). An integrated view of the Causes and Impacts of atmospheric changes. In Brasseur, G. P., Prinn, R. G. and Pszenny, A. P. (eds.). Atmospheric Chemistry in a Changing World. Berlin , Springer Verlag, pp. 207–29.

Bratcher, A. J. and Giese, B. S. (2002). Tropical Pacific decadal variability and global warming. *Geophysical Research Letters* **29**, No.19.

Bray, R. C., Price, B. B., Clow, G. D. and Glow, A. J. (2001). Climate logging with new rapid optical technique at Siple Dome. *Geophysical Research Letters* **28**, 4635–8.

Brazdil, R., Glaser, R., Pfister, C. and Stangl, H. (2002). Floods in Europe – a look into the past. *PAGES Newsletter* **10** (3), 21–3.

Briffa, K. R. (2000). Annual climate variability in the Holocene: interpreting the message from ancient trees. *Quaternary Science Reviews* **19**, 87–105.

Briffa, K. R., Jones, P. D., Schweingruber, F. H. and Osborn, T. J. (1998). Influence of volcanic eruptions on northern hemisphere summer temperature over the past 600 years. *Nature* **393**, 450–5.

Briffa, K. R., Osborn, T. J., Schweingruber, F. H. *et al.* (2001). Low frequency temperature variations from a northern tree ring density network. *Journal of Geophysical Research* **106**, 2929–41.

Briffa, K. R., Osborn, T. J., Schweingruber, F. H. *et al.* (2002a). Tree-ring width and density data around the northern hemisphere: part 1, local and regional climate signals. *The Holocene* **12**, 737–57.

Briffa, K. R., Osborn, T. J., Schweingruber, F. H. *et al.* (2002b). Tree-ring width and density data around the Northern Hemisphere: part 2, spatio-temporal variability and associated climate patterns. *The Holocene* **12**, 759– 89.

Briffa, K. R., Schweingruber, F. H., Jones, P. D. *et al.*(1998). Reduced sensitivity of tree-rings to temperatures at high northern latitudes. *Nature* **391**, 678–82.

Briffa, K. R., Schweingruber, F. H., Jones, P. D. *et al.* (1999). Trees tell of past climates, but are they speaking less clearly today? *Philosophical Transactions of the Royal Society of London, B* **353**, 65–73.

Broccoli, A. J., Dixon, K. W., Delworth, T. L., Knutson, T. R. and Stouffer, R. J. (2003). Twentieth-century temperature and precipitation trends in ensemble climate simulations including natural and anthropogenic forcing. *Journal of Geophysical Research* **108**, ACL 16–1–16–13.

Broecker, W. S. (1989). The salinity contrast between the Atlantic and Pacific oceans during glacial time. *Paleoceanography* **4**, 207–12.

Broecker, W. S. (1998). Paleocean circulation during the last deglaciation: a bipolar seesaw? *Paleoceanography* **13**, 119–21.

Broecker, W. S. (2000). Abrupt climate change: causal constraints provided by the paleoclimatic record. *Earth Science Reviews* **51**, 137–54.

Broecker, W. S. (2003). Does the trigger for abrupt climate change reside in the ocean or in the atmosphere? *Science* **300**, 1519–22.

Broecker, W. S. (2004). Future global warming scenarios. *Science* **304**, 388.

Broecker, W. S., Bond, G., Klas, M., Bonani, G. and Wolfi, W. (1990). A salt oscillator in the glacial Atlantic? 1. The concept. *Paleoceanography* **3**, 659–69.

Broström, A., Coe, M. T., Harrison, S. P *et al*. (1998a). Land surface feedbacks and palaeomonsoons in northern Africa. *Geophysical Research Letters* **25**, 3615–18.

Broström, A., Gaillard, M. -J., Ihse, M. and Odgaard, B. (1998b). Pollen–landscape relationships in modern analogues of ancient cultural landscapes in southern Sweden – a first step towards quantifying vegetation openness in the past. *Vegetation History and Archaeobotany* **7**, 189–201.

Bruckner, T., Petschel-Held, G., Toth, F. L., Fuessel, H. -M., Helm, C. and Leimbach, M. (1999). Climate change decision-support and the tolerable windows approach. *Environmental Modelling and Assessment* **4**, 217–34.

Bryant, D., Burke, L., McManus, J. and Spalding, M. (1998). Reefs at Risk: a Map Based Indicator of Threats to the World's Coral Reefs. Washington DC, World Resources Institute.

Bryden, H. L., McDonagh, E. L. and King, B. A. (2003). Changes in ocean water mass properties: oscillations or trends? *Science* **300**, 2086–8.

Buckley, L. B. and Roughgarden, J. (2004). Biodiversity conservation: Effects of changes in climate and land use. *Nature* **430**, 34.

Buesseler, K. O., Andrews, J. E., Pike, S. M. and Charette, M. A. (2004). The effects of Iron fertilization on carbon sequestration in the Southern Ocean. *Science* **304**, 414–17.

Bugmann, H. K. M. (1997). Gap models, forest dynamics and the response of vegetation to climate change. In Huntley, B., Cramer, W., Morgan, A. V., Prentice, H. C. and Allen, J. R. M. (eds.). Past and Future Rapid Environmental Changes: The Spatial and Evolutionary Responses of Terrestrial Biota. Berlin, Springer Verlag NATO ASI Series, pp. 441–53.

Bugmann, H. K. M. (2001). A review of forest gap models. *Climate change* **5**, 259–305.

Bugmann, H. K. M. and Pfister, C. (2000). Impacts of interannual climate variability on past and future forest composition. *Regional Environmental Change* **1**, 1–19.

Burn, G. (2003). After the flood. The Guardian Review, 15 November, pp. 4–6.

Burns, S. J., Fleitman, D., Mather, A., Kramer, J. and Al-Subbary, A. A. (2003). Indian Ocean climate and an absolute chronology over Dansgaard-Oeschger Events 9–13. *Science* **301**, 1365–7.

Caillon, N., Jouzel, J., Severinghaus, J. P., Chappellaz, J. and Blunier, T. (2003a). A novel method to study the phase relationship between Antarctic and Greenland climate. *Geophysical Research Letters* **30**, 4-1–4-4.

Caillon, N., Severinghaus, J. P., Jouzel, J. *et al*. (2003b). Timing of atmospheric CO_2 and Antarctic temperature changes across termination III. *Science* **299**, 1728–31.

Calov, R., Ganapolski, A., Petoukhov, V. and Claussen, M. (2002). Large-scale instabilities of the Laurentide ice sheet simulated in a fully coupled climate-system model. *Geophysical Research Letters* **29** (24), 69, 1–4.

Calvo, E., Pejero, C., Logan, C. A. and De Deckker, P. (2004). Dust-induced changes in phytoplankton composition in the Tasman Sea during the last four glacial cycles. *Paleoceanography* **19**, PA2020.

Campbell, B. D., Stafford-Smith, D. M. and the GCTE Pasture and Rangelands Network members. (2000). A synthesis of recent global change research on pasture and rangeland production: reduced uncertainties and their management implication. *Agriculture, Ecosystems and Environment* **82**, 39–55.

Camuffo, D., Secco, C., Brimblecombe, P. and Martin-Vide, J. (2000). Sea storms in the Adriatic and the western Mediterranean during the last millennium. *Climatic Change* **46**, 209–23.

Canadell, J. G. and Pataki, D. (2002). New advances in carbon cycle research. *Trends in Ecology and Evolution* **17**, 156–8.

Carcaillet, C., Almquist, H., Asnong, H. *et al*. (2002). Holocene biomass burning and global dynamics of the carbon cycle. *Chemosphere* **49**, 845–63.

Carrington, D. P., Gallimore, R. G. and Kutzbach, J. E. (2001). Climate sensitivity to wetlands and wetland vegetation in mid-Holocene North Africa. *Climate Dynamics* **17**, 151–7.

Carslaw, K. S., Harrison, R. G. and Kirby, J. (2003). Cosmic rays, clouds, and climate. *Science* **298**, 1732–7.

Cazenave, A. and Nerem, R. S. (2004). Present-day sea-level change: observations and causes. *Review of Geophysics* **42**, RG3001.

CENR (2000). *Integrated Assessment of Hypoxia in the Northern Gulf of Mexico*. Washington DC, National Science and Technology Council Committee on Environment and Natural Resources.

Chameides, W. L., Kasibhatla, P. S., Yienger, J. and Levy, I. H. (1994). Growth of Continetntal-scale metro-agro-plexes, regional ozone pollution, and world Food production. *Science* **264**, 74–7.

Chameides, W. L., Yu, H., Bergin, M. *et al*. (1999). Case study of the effects of atmospheric aerosols and regional haze on agriculture: an opportunity to enhance crop yields in China through emission controls. *Proceedings of the National Academy of Sciences USA* **96**, 13626–33.

Chapin, F. S., III, Zavaleta, E. S., Eviner, V. T. *et al*. (2000a). Consequences of changing biodiversity. *Nature* **405**, 234–42.

Chapin, F. S., III, McGuire, A. D., Randerson, J. *et al*. (2000b). Arctic and boreal ecosystems of western North America as components of the climate system. *Global Change Biology* **6**, 211–23.

Chapman, D. S., Bartlett, M. G. and Harris, R. N. (2004). Comment on 'ground vs. surface air temperature trends: implications for borehole surface temperature reconstructions' by M. E. Mann and G. Schmidt. *Geophysical Research Letters* **31**, L07205.

Chapman, M. R., Shackleton, N. J. and Duplessy, J. -C. (2000). Sea surface temperature variability during the last glacial-interglacial cycle: assessing the magnitude and pattern of climate change in the North Atlantic. *Palaeogeography, Palaeoclimatology, Palaeoecology* **157**, 1–25.

Chappell, J. M. A. and Shackleton, N. J. (1986). Oxygen isotopes and sea-level. *Nature* **324**, 137–8.

Charles, D. F., Binford, M. W., Furlong, E. T. *et al*. (1990). Palaeoecological investigations of recent lake acidification in the Adirondack Mountains, N. Y. *Journal of Paleolimnology*, **3**, 195–241.

Chase, T. N., Pielke, R. A., Kittel, T. G. F. *et al*. (2002). Relative climatic effects of landcover change and elevated carbon dioxide combined with aerosols: a comparison of model results and observations. *Journal of Geophysical Research D. Atmospheres* **106**, 31685–91.

Cheddadi R., Yu, G., Guiot, J., Harrison, S. P. and Prentice, I. C. (1997) The climate of Europe 6000 years ago. *Climate Dynamics* **13**, 1–9.

Chen, D., Cane, M. A., Kaplan, A., Zeblak, S. E. and Huang, D. (2004). Predictability of El Niño over the past 148 years. *Nature* **428**, 733–5.

Chen, F. H., Bloemendal, J., Wang, J. M., Li, J. J. and Oldfield, F. (1997). High-resolution multi-proxy climate records from Chinese loess: evidence for rapid climatic changes over the last 75 kyr. *Palaeogeography, Palaeoclimatology, Palaeoecology* **130**, 323–35.

Chen, F. H., Qiang, M. R., Feng, Z. D., Wang, H. B. and Bloemendal, J. (2003). Stable East Asian monsoon climate during the last interglacial (Eemian) indicated by palaeosol S1 in the western part of the Chinese Loess Plateau. *Global and Planetary Change* **36**, 171–9.

Chepstow-Lusty, A., Bennett, K. D., Fjeldsa, J., Kendall, A., Galliano, W. and Tupayachi-Herrera, A. (1997). Tracing 4000 years of environmental history in the Cuzco area, Peru, from the pollen record. *Mountain Research and Development* **18**, 159–72.

Chock, D. P., Song, Q., Hass, H., Schell, B. and Ackermann, I. (2003). Comment on 'Control of fossil-fuel particulate black carbon and organic matter, possibly the most effective method of slowing global warming' by M. Z. Jacobson. *Journal of Geophysical Research* **108**, 12–1–12–3.

Church, J. A. and Gregory, J. M. (2001). Changes in sea-level. Climate change 2001: the scientific basis. In Houghton, J. T. *et al.* (eds.). Contribution of Working Group 1 to the Third Assessment Report of the Intergovernmental Panel on Climate Change. Cambridge, Cambridge University Press.

Claquin, T., Roelandt, C., Kohfeld, K. E., *et al.* (2002). Radiative forcing of climate by ice-age atmospheric dust. *Climate Dynamics* **20**, 193–202.

Clark, P. U., Marshall, S. J., Clarke, G. K. C., Hostetler, S. W., Licciardi, J. M. and Teller, J. T. (2001). Freshwater forcing of abrupt climate change during the last Glaciation. *Science* **293**, 283–7.

Clark, P. U., McCabe, A. M., Mix, A. and Weaver, A. J. (2004). Rapid rise of sea-level 19 000 years ago and its global implications. *Science* **304**, 1141–4.

Clark, P. U., Pisias, N. G., Stocker, T. F. and Weaver, A. J. (2002). The role of the thermohaline circulation in abrupt climate change. *Nature* **415**, 863–9.

Clark, W. C. and Dickson, N. M. (2003). Sustainability science: the emerging research program. *Proceedings of the National Academy of Sciences* **100**, 8059–61.

Clarke, K. C., Hoppen, S. and Gaydos, L. (1997). A self-modifying cellular automaton model of historical urbanization in the San Francisco Bay area. *Environment and Planning B: Planning and Design* **24**, 247–61.

Clarke, R. T. (2003). Frequencies of extreme events under conditions of changing hydrological regime. *Geophysical Research Letters* **30** (3), 24–1–24–4.

Claussen, M., Brovkin, V., Ganapolski, C. and Petoukhov, V. (2003). Climate change in North Africa: the past is not the Future. *Climatic Change* **57**, 99–118.

Claussen, M., Kubatzki, C., Brovkin, V. *et al.* (1999). Simulation of an abrupt change in Saharan vegetation at the end of the mid-Holocene. *Geophysical Research Letters* **24**, 2037–40.

Clement, A. C., Seager, R. and Cane, M. A., (2000). Suppression of El Niño during the mid-Holocene by changes in the Earth's orbit. *Paleoceanography* **15**, 731–7.

CLIMAP (1981). Seasonal reconstructions of the Earth's surface at the last glacial maximum. *Geological Society of America, Map Chart Series*, MC-36.

Co2science (2003). Atmospheric methane concentration: no longer rising. In *Journal Reviews*. www.co2science.org/journal/2003/v6n43cl.htm.

Coale K. H., *et al.* (1996). A massive phytoplankton bloom induced by ecosystem-scale iron fertilization experiment in the equatorial Pacific Ocean. *Nature* **383**, 495–501.

Cohen, A. S. (2003). The History and Evolution of Lake Systems. Oxford, Oxford University Press.

Cole, J. E. (2003). Holocene coral records: windows on tropical climate variability. In McKay, A., Battarbee, R. W., Birks, H. J. B. and Oldfield, F. (eds.). Global Change in the Holocene. London, Arnold, pp. 168–84.

Colinvaux, P. A., Oliveira, P. E. D. and Bush, M. B, (2000). Amazonian and neotropical plant communities on glacial time-scales: The failure of the aridity and refuge hypotheses. *Quaternary Science Reviews* **19**, 141–69.

Comiso, J. C. (2002). A rapidly declining Arctic perennial ice cover. *Geophysical Research Letters* **29** (20), 1956.

Compton, J. S. (2001). Holocene sea-level fluctuations inferred from the evolution of depositional environments of the southern Langebaan Lagoon salt marsh, South Africa. *The Holocene* **11**, 395–406.

Cook, E. R., Meko, D. M. and Stockton, C. W. (1997). A new assessment of possible solar and lunar forcing of the bidecadal drought rhythm in the western United States. *Journal of Climate* **10**, 1343–56.

Cook, E. R., Meko, D. M., Stahle, D. W. and Cleaveland, M. K. (1999). Drought re constructions for the continental United States. *Journal of Climate* **12**, 1145–62.

Cook, E. R., Palmer, J. G. and D'Arrigo, R. D. (2002). Evidence for a 'Medieval warm Period' in a 1100 year tree-ring reconstruction of past austral summer temperatures in New Zealand. *Geophysical Research Letters* **29** (14), 12–1–12–4.

Cooper, S. R. (1995). Chesapeake Bay watershed historical land use: impact on water quality and diatom communities. *Ecological Applications* **5**, 703–23.

Cooper, S. R. and Brush, G. S. (1991). Long-term history of Chesapeake Bay anoxia. *Science* **254**, 992–6.

(1993). A 2500 year history of anoxia and eutrophication in Chesapeake Bay. *Estuaries* **16**, 617–26.

Costanza, R., d'Arge, R., de Groot, R. *et al.* (1997). The value of the world's ecosystem services and natural capital. *Nature* **387**, 253–60.

Coulthard, T. J. and Macklin, M. G. (2001). How sensitive are river systems to climate and land-use changes? A model-based evaluation. *Journal of Quaternary Science* **16**, 347–51.

Covey, K. (2000). Beware the elegance of the number zero. *Climatic Change* **44**, 409–11.

Cowling, S. A. and Sykes, M. T. (1999). Physiological significance of low atmospheric CO_2 for plant–climate Interactions. *Quaternary Research* **52**, 237–42.

Cowling, S. A., Sykes, M. T. and Bradshaw, R. H. W. (2001). Palaeovegetation-model comparisons, climate change and tree succession in Scandinavia over the past 1500 years. *Journal of Ecology* **89**, 227–36.

Cramer, W., Bondeau, A., Woodward, F. I. *et al.* (2001). Global response of terrestrial ecosystem structure and function to CO_2 and climate change: results from six dynamic global vegetation models. *Global Change Biology* **7**, 357–73.

Crowley, T. J. (2000). Causes of climate change over the past 1000 years. *Science* **289**, 270–7.

Crowley, T. J. and Kim, K. -Y. (1999). Modeling the temperature response to forced climate change over the last six centuries. *Geophysical Research Letters* **26**, 1901– 4.

Crowley, T. J. and Lowery, T. S. (2000). How warm was the medieval warm period? *Ambio* **29**, 51–4.

Crutzen, P. J. (1995). My life with O_3, NO_x and other $YZOO_x$s. Les Prix Nobel (The Nobel Prizes) 1995. Stockholm, Almqvist and Wiksell, pp. 123–57.

Crutzen, P. J. (2003). The ozone hole. In Steffen W., Sanderson, A., Tyson, P. D *et al.* Global Change and the Earth System; A Planet under Pressure. Berlin, Springer Verlag, p. 236.

Crutzen, P. J. and Stoermer, E. (2001). The 'Anthropocene'. *International Geosphere Biosphere Programme Global Change Newsletter* **41**, 12–13.

Cubasch, U., Meehl, G. A., Boer, G. J. *et al*. (2001). Projections of future climate change. In
    Houghton, J. T. *et al*. (eds.). Climate Change 2001. Cambridge, Cambridge University
    Press.

Cuffey, K. M., Clow, G. D., Alley, R. B. *et al*. (1995). Large arctic temperature change at the
    Wisconsin–Holocene glacial transition. *Science* **270**, 455–8.

Cullen, H., Kaplan, A., Arkin, P. A. and deMenocal, P. B. (2002). Impact of the North Atlantic
    Oscillation on Middle Eastern climate and streamflow. *Climatic Change* **55**, 315–38.

Curry, R., Dickson, B. and Yashayaev, I. (2003). A change in the freshwater balance of the
    Atlantic Ocean over the past four decades. *Nature* **426**, 826–9.

D'Arrigo, R., Frank, D., Jacoby, G. and Pederson, N. (2001). Spatial responses to major
    volcanic events in or about AD 536, 934 and 1258: frost rings and other dendrochronolo-
    gical evidence from Mongolia and northern Siberia. Comments on R. B. Stothers,
    'Volcanic dry fogs, climate cooling and plague pandemics in Europe and the Middle East'
    (*Climatic Change*, **42**, 1999). *Climatic Change* **49**, 239–46.

Dahlin, B. H. (2002). Climate change and the end of the classic period in Yucatan: resolving a
    paradox. *Ancient Mesoamerica* **13**, 327–40.

Dahl-Jensen, D., Mosegaard and K., and Gunderstrup, N. *et al*. (1998). Past temperature
    directly from the Greenland ice sheet. *Science* **252**, 268–71.

Dallimore, S. R., Collett, T. S., Weber, M. and Uchida, T. (2002). Drilling program investigates
    permafrost gas hydrates. *EOS*, **83**(18), 193–198.

Dalrymple, G. B. and Lanphere, M. A. (1969) Potassium–Argon Dating: Principles,
    Techniques and Applications to Geochronology. San Francisco, W. H. Freeman.

Dansgaard, W. (1964). Stable isotopes in precipitation. *Tellus* **16**, 436–68.

Darwin, C. [1859] (1964). On the Origin of Species by Means of Natural Selection or the
    Preservation of Favoured Races in the Struggle for Life. Cambridge, MA, Harvard
    University.

De Angelis, H. and Skvarca, P. (2003). Glacier surge after ice sheet collapse. *Science* **299**,
    1559–62.

deMenocal, P. B. (2001) Cultural responses to climate change during the late Holocene.
    Science **292**, 667–73.

deMenocal, P. B., Ortiz, J., Guilderson, T. *et al*. (2000). Abrupt onset and termination of the
    African Humid Period: rapid climate responses to gradual insolation forcing. *Quaternary
    Science Reviews* **19**, 347–61.

De Noblet, N., Prentice, I. C., Joussaume, S. *et al*. (1996). Possible role of atmospheric biosphere
    interactions in triggering the last glaciation. *Geophysical Research Letters* **23**, 3191–94.

De Noblet-Ducoudré, N., Claussen, M. and Prentice, I. C. (2000). Mid-Holocene greening of
    the Sahara: first results of the GAIM 6000 year experiment with two asynchronously
    coupled atmosphere/biome models. *Climate Dynamics* **16**, 643–59.

De Silva, S. L., and Zielinski, G. A. (1998). Global influence of the AD 1600 eruption of
    Huaynaputina, Peru. *Nature* **393**, 455–58.

Dean, W., Anderson, R., Bradbury, J. P. and Anderson, D. (2002). A 1500 year record of
    climatic and environmental change at Elk Lake, Minnesota: varve thickness and gray-scale
    density. *Journal of Palaeolimnology* **27**, 287–99.

DeFries, R. (2004). Determining rates of tropical deforestation. In Steffen W., Sanderson,
    A., Tyson, P. D. *et al*. Global Change and the Earth System; A Planet under Pressure.
    Berlin, Springer Verlag, p. 99.

DeFries, R., Houghton, R. A., Hansen, M. *et al.* (2002). Carbon emissions from tropical deforestation and regrowth based on satellite observations from the 1980's and 90's. *Proceedings of the National Academy of Sciences, USA* **99**, 14256–61.

Del Genio, A. D. (2002). The dust settles on water vapour feedback. *Science* **296**, 665–6.

de Laat, A. T. J., and Maurellis, A. N. (2004). Industrial CO_2 emissions as a proxy for anthropogenic influence on lower tropospheric temperature trends. *Geophysical Research Letters* **31**, LO5204.

Delaygue, G., Stocker, T. F., Joos, F. and Plattner, G. -K. (2003). Simulations of atmospheric radiocarbon during abrupt oceanic circulation changes: trying to reconcile models and reconstructions. *Quaternary Science Reviews* **22**, 1647–58.

Delgrado, C., Rosegrant, M., Steinfield, H., Ehui, S. and Courbois, C. (1999). Livestock to 2020: The Next Food Revolution. Washington DC, IFPRI.

Delmonte, B., Petit, J. R. and Maggi, V. (2002). Glacial to Holocene implications of the new 27 000-year dust record from the EPICA Dome C (East Antarctica) ice core. *Climate Dynamics* **18**, 647–60.

Delmotte, M., Chappellaz, J., Brook, E. *et al.* (2004). Atmospheric methane during the last four glacial–interglacial cycles: rapid changes and their link with Antarctic temperature. *Journal of Geophysical Research* **109**, D12104, 1–13.

Delworth, T. L. and Dixon, K. W. (2000). Implications of the recent trend in the Arctic/North Atlantic oscillation for the North Atlantic thermohaline circulation. *Journal of Climatology* **13**, 3721–7.

Denton, G. H. and Hendy, C. H. (1994). Younger Dryas age advance of Franz Josef glacier in the southern Alps of New Zealand. *Science* **264**, 1434–7.

Diaz, H. F., Eischeid, J. K., Duncan, C. and Bradley, R. S. (2003). Variability of freezing levels, melting season indicators and snow cover for selected high-elevation and continental regions in the last 50 years. *Climatic Change* **59**, 33–52.

Diaz, S. (2001). Complex interactions between plant diversity, succession and elevated CO_2. *Trends in Ecology and Evolution* **16**, 667.

Diaz, S. and Cabindo, M. (2001). Vive la difference: plant functional diversity matters to ecosystem processes. *Trends in Ecology and Evolution* **16**, 646–55.

Diaz, S., Symstad, A. J., Chapin, F. S., III, Wardle, D. A. and Huenneke, L. F. (2002). Functional diversity revealed by removal experiments. *Trends in Ecology and Evolution* **18**, 140–6.

Dickens, G. R. (2004). Hydrocarbon-driven warming. *Nature* **429**, 513–15.

Dickson, R., Yashayaev, I., Meincke, J., Turrill, W., Dye, S. and Holfort, J. (2002). Rapid freshening of the deep North Atlantic Ocean over the past four decades. *Nature* **416**, 832–7.

Dillehay, T. D. (2002). Climate and human migrations. *Science* **298**, 764–5.

Dillehay, T. D., Kolata, A. L. and Pino, M. Q. (2004). Pre-industrial human and environment interactions in northern Peru during the late Holocene. *The Holocene* **14**, 272–81.

Ding, Z., Rutter, N. W. and Liu, T. S. (1994). Towards an orbital timescale for Chinese loess deposits. *Quaternary Science Reviews.* **13**, 39–70

Dlugokencky, E. J., Houweling, S., Bruhwiler, L. *et al.* (2003). Atmospheric methane levels off: temporary pause or a new steady-state? *Geophysical Research Letters* **30** (19), ASC 5–1–5–4.

Dodson, J. R. Taylor, D., Ono, Y. and Wang, P. (eds.). (2004). Climate, human and natural systems of the PEPII transect. *Quaternary International* **118–119**, 1–203.

Doll, P. (2002). Impact of climate change and variability on irrigation requirements: a global perspective. *Climatic Change* **54**, 269–93.

Dore, J. F., Lukas, R., Sadler, D. W. and Karl, D. W. (2003). Climate-driven changes to the atmospheric CO_2 sink in the subtropical North Pacific Ocean. *Nature* **424**, 754–7.

Dorn, W., Dethloff, K., Rinke, A. and Roeckner, E. (2003). Competition of NAO regime changes and increasing greenhouse gases and aerosols with respect to Arctic climate projections. *Climate Dynamics* **21**, 447–58.

Douglass, D. H., Pearson, B. D., Singer, S. F., Knappenberger, P. C. and Michaels, P. D. (2004). Disparity of tropospheric and surface temperature trends: new evidence. *Geophysical Research Letters* **31**, L12307.

Duffy, P. B., Govindaswamy, B., Iorio, J. P. *et al.* (2003). High-resolution simulations of global climate, part 1: present climate. *Climate Dynamics* **21**, 371–90.

Dutta, K. (2002). Coherence of tropospheric $^{14}CO_2$ with El Niño/Southern Oscillation. *Geophysical Research Letters* **29**, 48-1–48-4.

Dyson, T. (1996). Population and Food: Global Trends and Future Prospects. New York, Routledge.

Edwards, R. L., Chen, J. H. and Wasserburg, G. J. (1987). ^{238}U-^{234}U-^{230}Th-^{232}Th systematics and the precise measurement of time over the past 500 000 years. *Earth and Planetary Science Letters* **81**, 175–92.

Edmunds, W. M., Fellman, E. and Baba Goni, I. (1999). Environmental change, lakes and groundwater in the Sahel of northern Nigeria. *Journal of the Geological Society London* **156**, 345–55.

Elliot, M., Labeyrie, L. and Duplessy, J. -C. (2002). Changes in North Atlantic deep-water formation associated with Dansgaard-Oeschger temperature oscillations (10–100 ka). *Quaternary Science Reviews* **21**, 1153–65.

Elton, C. S. (1958). Ecology of Invasions by Animals and Plants. London, Chapman & Hall.

EPICA community members (2004). Eight glacial cycles from an Antarctic ice core. *Nature* **429**, 623–8.

Eschenbach, W. W. (2004). Ecology: climate-change effect on Lake Tanganyika? *Nature* **430**, 207–14.

Esper, J., Cook, E. R. and Schweingruber, F. H. (2002). Low frequency signals in long tree-ring chronologies for reconstructing past temperature variability. *Science* **295**, 2250–3.

Esper, J., Frank, D. C. and Wilson, R. J. S. (2004). Climate reconstructions: low-frequency ambition and high frequency ratification. *EOS* **85** (12), 113–20.

Fagan, B. (2004). The Long Summer: How Climate Changed Civilization. Granta, Basic Books.

Fairbanks, R. G. (1989). A 17 000 year glacio-eustatic sea-level record: influence of glacial melting rates on the Younger Dryas event and deep ocean circulation. *Nature* **342**, 637–42.

Falkowski, P., Scholes, R. J., Boyle, E. *et al.* (2000). The global carbon cycle: a test of our knowledge of Earth as a system. *Science* **290**, 291–6.

Fang, X. M., Ono, Y., Fukusawa, H., Pan, B. T. *et al.* (1999) Asian summer monsoon instability during the past 60 000 years: magnetic susceptibility and pedogenic evidence from the western Chinese Loess Plateau. *Earth and Planetary Science Letters* **168**, 219–32.

FAO (2000). The State of World Fisheries and Aquaculture. Rome, Food and Agriculture Organization of the United Nations.

Feichter, J., Saussen, R., Grassl, H. and Fiebig, M. (2003). Comment on 'Control of fossil-fuel particulate black carbon and organic matter, probably the most effective method of slowing global warming' by M. Z. Jacobson. *Journal of Geophysical Research* **108**, 10-1–10-2.

Finney, B. P., Gregory-Eaves, I., Sweetman, J., Douglas, M. S. V., and Smol, J. P. (2000). Impacts of climatic change and fishing on Pacific salmon abundance over the past 300 years. *Science* **290**, 795–9.

Fiore, A. M., Jacob, D. J., Field, B. D., Streets, D. G. and Fernandes, S. D. (2002). Linking ozone pollution and climate change: the case for controlling methane. *Geophysical Research Letters* **29**, 25-1–25-11.

Fisher D. A. and Koerner, R. M. (2003). Holocene ice core climate history: a multi-variable approach. In Mackay, A. W., Battarbee, R. W., Birks, H. J. B. and Oldfield F. (eds.). Global Change in the Holocene. London, Arnold pp. 281–93.

Fleitmann, D., Burns, S. J., Mudelsee, M. *et al.* (2002). Holocene variability in the Indian Ocean Monsoon: a stalagmite-based high-resolution oxygen isotope record from southern Oman. *PAGES News* **10** (2), 7–8.

Flückiger, J., Dällenbach, A., Blunier, T. *et al.* (1999). Variations in atmospheric N_2O concentration during abrupt climatic changes. *Science* **285**, 227–30.

Flückiger, J., Monnin, E., Stauffer, B. *et al.* (2001). High-resolution Holocene N_2O ice core rcord and its relationship with CH_4 and CO_2. *Global Biogeochemical Cycles* **16**, 10-1–10-7.

Flückiger, J., Monnin, E., Stauffer, B. *et al.* (2004). N_2O and CH_4 variations during the last glacial epoch: insight into global processes. *Global Biogeochemical Cycles* **18**, GB1020.

Foley, J., Kutzbach, J., Coe, M. and Levis, S. (1994). Feedbacks between climate and boreal forests during the Holocene epoch. *Nature* **371**, 52–4.

Foley, J. A., Prentice, I. C., Ramankutty, N. *et al.* (1996). An integrated biosphere model of land surface processes, terrestrial carbon balance and vegetation dynamics. *Global Biogeochemical Cycles* **10**, 603–28.

Forster, P. M. de F. and Collins, M. (2004). Quantifying the water vapour feedback associated with post-Pinatubo global cooling. *Climate Dynamics*, **23**, 207–14.

Foster, D. R. (2002). Insights from historical geography to ecology and conservation: lessons from the New England landscape. *Journal of Biogeography* **29**, 1269–75.

Foster, D. R., Motzkin, G. and Slater, B. (1998). Land-use history as long-term broad-scale disturbance: regional forest dynamics in Central New England. *Ecosystems* **1**, 96–119.

Foster, D. R., Clayden, S., Orwig, D. A., Hall, B. and Barry, S. (2002). Oak, chestnut and fire: climatic and cultural controls of long term forest dynamics in New England, USA. *Journal of Biogeography* **29**, 1359–79.

Foster, D. R., Motzkin, G., Bernardos, D. and Cardoza, J. (2002). Wildlife dynamics in the changing New England landscape. *Journal of Biogeography* **29**, 1337–1358.

Foukal, P. (2003). Can slow variations in solar luminosity provide missing link between the sun and climate? *EOS* **84** (22), 205–8.

Francis, R. C. and Hare, S. R. (1994). Decadal scale regime shifts in the large marine ecosystems of the north-east Pacific: a case for historical science. *Fisheries Oceanography* **3**, 279–91.

Frankus, P., Bradley, R. S., Abbott, M. B., Partridge, W. and Keimig, F. (2002). Paleoclimate studies of minerogenic sediments using annually resolved textural parameters. *Geophysical Research Letters* **29**, 1988–2002.

Free, M. and Robock, A. (1999). Global warming in the context of the Little Ice Age. *Journal of Geophysical Research-Atmospheres* **104**, 19057–70.

Friborg, T., Soegaard, H., Chriistensen, T. R., Lloyd, C. R. and Panikov, N. S. (2003). Siberian wetlands: where a sink is a source. *Geophysical Research Letters* **30**, (21) 2129.

Frumkin, A., Carmi, I., Bopher, A. *et al*. (1999). A Holocene millennial-scale climatic cycle from a speleothem in Nahal Qanah Cave, Israel. *The Holocene* **9**, 677–82.

Fu, C. (2003). Potential impacts of human-induced land cover change on east Asia monsoon. *Global and Planetary Change* **37**, 219–29.

Fu, Q., Johnson, C. M., Warren, S. G. and Seldel, D. J. (2004). Contribution of stratospheric cooling to satellite-inferred tropospheric temperature trends. *Nature* **429**, 55–8.

Funtowicz, S. O. and Ravetz, J. R. (1992). The emergence of Post-Normal Science. In von Schomberg, (ed.), Science, Politics and Morality. Dordrecht, Kluwer.

Funtowicz, S. O. and Ravetz, J. R. (1993). Science for the Post-Normal Age. *Futures* **25**, 739–55.

Gagan, M. K., Ayliffe, L. K., Beck, J. W. *et al*. (2000). New views of tropical paleoclimates from corals. *Quaternary Science Reviews* **19**, 45–64.

Gagan, M. K., Hendy, E. J., Haberle, S. G. and Hantaro, W. S. (2004). Post-glacial evolution of the Indo-Pacific Warm Pool and El Niño Southern oscillation. *Quaternary International* **118–119**, 127–43.

Gaillard, M. -J., Birks, H. J. B., Ihse, M. and Runborg, S. (1998). Pollen/landscape calibration based on modern pollen assemblages from surface-sediment samples and landscape mapping – a pilot study in south Sweden. In Gaillard, M. -J., Berglund, B., Frenzel, B. and Huckriede, U. (eds.). Quantification of land surface cleared of forest during the Holocene. Palaeoklimaforschung/Palaeoclimatic Research, 27, Stuttgart, Gustav Fischer Verlag, pp. 31–55.

Gaillard, M.-J., Dearing, J. A., El-Daoushy, F., Enell, F. and Håkansson, H. (1991). A late Holocene record of land use history, lake trophy and lake-level fluctuations at Lake Baresjø (south Sweden). *Journal of Paleolimnology* **6**, 51–81.

Galloway, J. N. (2004). The global nitrogen cycle: past, present and future. In Steffen W., Sanderson, A., Tyson, P. D. *et al*. Global Change and the Earth System; A Planet Under Pressure. Berlin, Springer Verlag., pp. 122–3.

Galloway, J, N., Aber, J. D., Erisman, J. W. *et al*. (2003). The nitrogen cascade. *Bioscience* **53**, 341–56.

Ganachaud, A. and Wunsch, C. (2000). Improved estimates of global ocean circulation, heat transport and mixing from hydrographic data. *Nature* **408**, 453–7.

Garstang, M., Ellery, W. N., McCarthy, T. S. *et al*. (1998). The contribution of aerosol- and water-borne nutrients to the Okovanga Delta ecosystem, Botswana. *South African Journal of Science* **94**, 223–9.

Gasse, F. (2000). Hydrological changes in the African tropics since the last glacial maximum. *Quaternary Science Reviews* **19**, 189–211.

Gasse, F., Fontes, J. Ch., Plaziat, J. C. *et al*. (1987). Biological remains, geochemistry and stable isotopes for the reconstruction of environmental and hydrological changes in the Holocene lakes from north Sahara. *Palaeogeography, Palaeoclimatology, Palaeoecology* **60**, 1–46.

Gaston, K. J. (2000). Global patterns in biodiversity. *Nature* **405**, 220–7.

Gedalof, Z., Mantua, N. J. and Peterson, D. L. (2002). A multi-century perspective of variability in the Pacific decadal oscillation: new insights from tree rings and coral. *Geophysical Research Letters* **29** (24), 57-1–57-4.

Gehrels, W. R. (2001). Discussion on sea-level changes over the past 1000 years in the Pacific. *Journal of Coastal Research* **17**, 244–5.

Gehrels, W. R., Belknap, D. F., Black, D. F. and Newnham, R. M. (2002). Rapid sea-level rise in the Gulf of Maine, USA, since AD 1800. *The Holocene* **12**, 383–9.

Geist, H. J. and Lambin, E. F. (2002). Proximate causes and underlying driving forces of tropical deforestation. *Bioscience* **52**, 143–50.

Georgia, P. J. (2002a). The IPCC's 'political' scientist. *CEI NewsCenter (www.cei.org)*. April 18.

Georgia, P. J. (2002b). Global-warming nonsense: an economics journal publishes Junk. *CEI NewsCenter (www.cei.org)*. August 2.

Giannini, A., Saravanan, R. and Chang, P. (2003). Oceanic forcing of Sahel rainfall on interannual to interdacadal timescales. *Science* **302**, 1027–30.

Gildor, H. (2003). When Earth's freezer is left ajar. *EOS* **84** (23), 215.

Gillett, N. P., Wehner, W. F., Tett, S. F. B. and Weaver, A. J. (2004). Testing the linearity of the response to combined greenhouse gas and sulphate aerosol forcing. *Geophysical Research Letters* **31**, L14201.

Gitz, V. and Ciais, P. (2003). Amplifying the effects of land-use change on future CO_2 levels. *Global Biogeochemical Cycles* **17** (1), 24-1–24-15.

Glenn, E., Stafford Smith, M. and Squires, V. (1998). On our failure to control desertification: implications for global change issues and a research agenda for the future. *Environmental Science and Policy* **1**, 71–8.

Goldewijk, K. K. (2003). Estimating global land use change over the past 300 years: the HYDE database. *Global Biogeochemical Cycles* **15**, 417–34.

Goldewijk. K. K. and Battjes, J. J. (1997). A Hundred Year Database for Integrated Environmental Assessments. Bilthoven, The Netherlands, National Institute of Public Health and Environment.

Goodwin, I. D. (2003). Unravelling climatic influences on late Holocene sea-level variability. In Mackay, A., Battarbee, R. W., Birks, H. J. B. and Oldfield, F. (eds.). Global Change in the Holocene, London, Arnold, pp. 406–421.

Goolsby, D. A. (2000). Mississippi basin nitrogen flux believed to cause gulf hypoxia. *EOS* **29**, 321–7.

Govindaswamy, B., Duffy, P. B. and Coquard, J. (2003). High-resolution simulations of global climate, part 2: effects of increased greenhouse gases. *Climate Dynamics* **21**, 391–404.

Gregg, W. W., Conkright, M. E., Ginoux, P., O'Reilly, J. E. and Casey, N. W. (2003). Ocean primary production and climate: global decadal changes. *Geophysical Research Letters* **30** (15), 1809.

Gregory, J. M., Huybrechts, P. and Raper, S. C. B. (2004). Threatened loss of the Greenland ice-sheet. *Nature* **428**, 616.

Gregory, P. J. and Ingram, J. S. (2000a). Global change and food and forest production: future scientific challenges. *Agriculture, Ecosystems and Environment* **82**, 3–14.

Gregory, P. J. and Ingram, J. S. (2000b). Food and forestry: global change and global challenges. *Agriculture, Ecosystems and Environment* **82**, 1–2.

Gregory, P. J, Ingram, J. S. I., Anderson, R. *et al.* (2002). Environmental consequences of alternative practices for intensifying crop production. *Agriculture, Ecosystems and Environment* **88**, 279–90.

Grissino-Mayer, H. D. (1996). A 2129-year annual reconstruction of precipitation for Northwestern New Mexico, USA. In Meko, D. M., Swetnam, T. W. and Dean, J. S. (eds.).

Tree-Rings, Environment and Humanity. Tucson, University of Arizona Press, pp. 191–204.

Grissino-Mayer, H. D. and Watson, E. (2000). Tree-ring data document sixteenth century mega-drought over North America. *EOS* **81**, 121–125.

Grossman, E. E., Fletcher, C. P. III and Richmond, B. M. (1998). The Holocene sea-level highstand in the equatorial Pacific: analysis of the insular palaeosea-level database. *Coral Reefs* **17**, 309–27.

Grove, J. M. (1988). The Little Ice Age, London, Methuen.

Grove, J. M. (2001). The initiation of the 'Little Ice Age' in regions round the North Atlantic. *Climatic Change* **48**, 53–82.

Grubb, M., Vrolijk, C. and Brack, D. (1999). The Kyoto Protocol: A Guided Assessment. New York, Brookings.

Grubler, A. and Nakicenovic, M. E. (2001). Identifying dangers in an uncertain climate. *Nature* **412**, 15.

Grudd, H., Briffa, K., Karlen, W. *et al.* (2002). A 7400-year tree-ring chronology in northern Swedish Lapland: natural climatic variability expressed on annual to millennial timescales. *The Holocene* **12**, 657–67.

Gu L., Baldocchi, D. D., Wofsy, S. C. *et al.* (2003). Response of a deciduous forest to the Mount Pinatubo eruption: enhanced photosynthesis. *Science* **299**, 2035–8.

Guiot, J., Pons, A., de Beaulieu, J. -L. and Reille, M. (1989). A 140 000 year climatic reconstruction from two European pollen records. *Nature* **338**, 309–13.

Guo, Z., Biscaye, P., Wei, L. *et al.* (2000). Summer monsoon variations over the last 1.2 Ma from the weathering of loess-soil sequences in China. *Geophysical Research Letters* **27**, 1751–4.

Gyalistras, D., Schar, C., Davies, H. C. and Wanner, H. (1998). Future Alpine climate. In Views from the Alps. Boston, MIT Press, pp. 171–223.

Haan, D. and Raynaud, D. (1998) Ice core record of CO_2 variations during the last two millennia: atmospheric implications and chemical interactions within the Greenland ice. *Tellus Series B–Chemical and Physical Meteorology* **50**, 253–62.

Haberle, S. G. and Lusty, A. C. (2000). Can climate influence cultural development? A view through time. *Environment and History* **6**, 349–69.

Hajdas, I., Bonani, G., Moreno, P. I. and Ariztegui, D. (2003). Precise radiocarbon dating of late-glacial cooling in mid-latitude South America. *Quaternary Research* **59**, 70–8.

Hall, B., Motzkin, G., Foster, D. R., Syfert, M. and Burk, J. (2002). Three hundred years of forest and land-use change in Massachusetts, USA. *Journal of Biogeography* **29**, 1319–35.

Hall, V. A. and Pilcher, J. R. (2002). Late-Quaternary Icelandic tephras in Ireland and Great Britain: detection, characterization and usefulness. *The Holocene* **12**, 223–30.

Hammond, P. (1995). The current magnitude of biodiversity. In Heywood, V. (ed.). Global Biodiversity Assessment. Cambridge, Cambridge University Press.

Hansen, J. E. (2002). A brighter future. *Climatic Change* **52**, 435–40.

Hansen, J. E., Sato, M., Ruedy, R., Lacias, A. and Oinas, V. (2000). Global warming in the twenty-first century: an alternative scenario. *Proceedings of the National Academy of Sciences* **97**, 9875–80.

Harremoes, P. and Turner, R. K. (2001). Methods for integrated assessment. *Regional Environmental Change* **2**, 57–65.

Harris, R. N. and Chapman, D. S. (2001). Mid-Latitude (30° – 60° N) climatic warming inferred from combining borehole temperatures with surface air temperatures. *Journal of Geophysical Research* **30** (21), 2116.

Harrison, S. P. and Digerfeldt, G. (1993). European lakes as palaeohydrological and palaeoclimatic indicators. *Quaternary Science Reviews* **12**, 233–48.

Harrison, S. P., Kohfield, K. E., Roelandt, C. and Claquin, T. (2001). The role of dust in climate changes today, at the last glacial maximum and in the future. *Earth Science Reviews* **54**, 43–80.

Harte, J. (2002). Toward a synthesis of the Newtonian and Darwinian worldviews. *Physics Today*, October 2002, 29–34.

Harte, J., Ostling, A., Green, J. L. and Kinzig, A. (2004). Biodiversity conservation: climate change and extinction risk. *Nature* **430**, in press.

Harvey, L. D. (2000). Upscaling in global change research. *Climatic Change* **44**, 223–63.

Harvey, L. D. (2003). Characterizing and comparing the control run variability of eight coupled AOGCMs and of observations. Part 2: precipitation. *Climate Dynamics* **21**, 647–58.

Hassan, F. A. (1981). Historic Nile floods and their implications for climatic change. *Science* **212**, 1142–5.

Haug, D. and Kaupenjohann, M. (2001). Parameters, prediction, post-normal science and the precautionary principle – a roadmap for modelling for decision-making. *Ecological Modelling* **144**, 45–60.

Haug, G. H., Hughen, K. A., Sigman, D. M., Peterson, L. C. and Roehl, U. (2001). Southward migration of the intertropical convergence zone through the Holocene. *Science* **293**, 1304–8.

Haug, G. H., Gunther, D., Peterson, L. C. *et al.* (2003). Climate and the Maya. *PAGES News* **11**, 28–30.

Hay, S. I., Cox, J., Rogers, D. J. *et al.* (2002). Climate change and the resurgence of malaria in the east African highlands. *Nature* **415**, 905–9.

Hayne, M. and Chappell, J. (2001). Cyclone frequency during the last 5000 years at Curaçao Island, north Queensland. *Palaeogeography, Palaeoclimatology, Palaeoecology* **168**, 207–19.

Hays, J. D., Imbrie, J. and Shackleton, N. J. (1976). Variations in the Earth's orbit: pacemaker of the ice ages. *Science* **194**, 1121–32.

He, Y., Zhang, Z., Theakstone, W. H. *et al.* (2003). Changing features of the climate and glaciers in Chin's monsoonal temperate glacier region. *Journal of Geophysical Research* **108**, D17, 2067.

Hegerl, G. C., Crowley, T. J., Baum, S. K., Kim, K-Y. and Hyde, W. T. (2003). Detection of volcanic and greenhouse gas signals in paleo-reconstructions of northern hemisphere temperature. *Geophysical Research Letters* **30** (5), 46-1–46-4.

Heikkila, M. and Seppä, H. (2003). A 11 000 yr palaeotemperature reconstruction from the southern boreal zone in Finland. *Quaternary Science Reviews* **22**, 541–54.

Heinrich, H. (1988). Origin and consequences of cyclic ice rafting in the northeast Atlantic Ocean during the past 130 000 years. *Quaternary Research* **29**, 142–52.

Heintzenberg, J, Raes, F. and Schwartz, S. E. *et al.* (2003). Tropospheric aerosols. In Brasseur, G. P, Prinn, R. G. and Pszenny, A. P. (eds.). *Atmospheric Chemistry in a Changing World*. Berlin, Springer Verlag.

Hendy, E. J., Gagan, M. K. and Lough, J. M. (2003). Chronological control of coral records using luminescent lines and evidence for non-stationary ENSO teleconnections in northeastern Australia. *The Holocene* **13**, 187–99.

Hendy, I. L., Kennett, J. P., Roark, E. B. and Ingram, B. L. (2002). Apparent synchroneity of sub-millennial scale climate events between Greenland and Santa Barbara Basin, California from 30–10 ka. *Quaternary Science Reviews* **21**, 1167–84.

Hinrichs, K-U., Hmelo, L. R. and Sylva, S. P. (2003). Molecular fossil record of elevated methane levels in late Pleistocene coastal Waters. *Science* **299**, 1214–16.

Hodell, D. A., Curtis, J. H. and Brenner, M. (1995). Possible role of climate in the collapse of Classic Maya civilization. *Nature* **375**, 391–4.

Hoegh-Guldberg, O. (1999). Climate change, coral bleaching and the future of the world's coral reefs. *Marine & Freshwater Research* **50**, 839–66.

Hoelzmann, P., Keding, B., Berke, H., Kroepelin, S. and Kruse, H.-J. (2001). Environmental change and archaeology: lake evolution and human occupation in the eastern Sahara during the Holocene. *Palaeogeography, Palaeoclimatology, Palaeoecology* **169**, 193–217.

Holland, T. (2003). The oracle is always right. The Telegraph, December 20, Books Section, p. 3.

Holzhauser, H. and Zumbuehl, H. J. (2002). Reconstruction of minimum glacier extensions in the Swiss Alps. *PAGES Newsletter* **10** (3), 23–5.

Howarth, R. W., Billen, G., Swaney, D. *et al.* (1996). Regional nitrogen budgets and riverine N and P fluxes for the drainage to the north Atlantic Ocean: natural and human influences. *Biogeochemistry* **35**, 75–9.

Hoyt, D. V. and Schattén, K. H. (1993). A discussion of plausible solar irradiance variations, 1700–1992. *Journal of Geophysical Research* **98**, 18895–906.

Huang, C. C., Zhao, S., Pang, J. *et al.* (2003). Climatic aridity and the relocations of the Zhou culture in the southern Loess Plateau of China. *Climatic Change* **61**, 361–78.

Huang, S. (2004). Merging information from different resources for new insights into climate change in the past and future. *Geophysical Research Letters* **31**, L13205.

Huang, S, Pollack, H. N. and Shen, P. Y. (2000). Temperature trends over the past five centuries from borehole temperatures. *Nature* **403**, 756–8.

Huang, Y., Street-Perrott, F. A., Metcalfe, S. E. *et al.* (2001). Climate change as the dominant control on glacial-Interglacial variations in C_3 and C_4 plant abundance. *Science* **293**, 1648–51.

Hughen, K. A., Eglinon, T. I., Xu, L. and Makou, M. (2004). Abrupt tropical vegetation response to rapid climate changes. *Science* **304**, 1955–8.

Hughen, K. A., Overpeck, J. T., Peterson, L. C. and Trumbore, S. (1996). Rapid climate changes in the tropical Atlantic region during the last deglaciation. *Nature* **380**, 51–4.

Hughen, K. A., Overpeck, J. T., Lehman, S. J. *et al.* (1998). Deglacial changes in ocean circulation from an extended radiocarbon calibration. *Nature* **391**, 65–8.

Hughen, K. A., Lehman, S., Southon, J. *et al.* (2004). [14]C Activity and global carbon cycle changes over the past 50 000 years. *Science* **303**, 202–7.

Hughes, M. K. and Diaz, H. F. (1994). Was there a 'Medieval Warm Period' and if so, where and when? *Climatic Change* **26**, 109–42.

Hughes, M. K. and Funkhouser, G. (1998). Extremes of moisture availability reconstructed from tree rings for recent millennia in the Great Basin of western North America. In Innes, M. and Beniston, J. L. (eds.). The Impacts of Climate Variability on Forests. Berlin, Springer. pp. 99–107.

Hughes, T. P., Baird, A. H., Bellwood, D. R. *et al.* (2003). Climate change, human impacts and the resilience of coral reefs. *Science* **301**, 929–33.

Hulme, M. (2001). Climatic perspectives on Sahelian desiccation: 1973–1998. *Global Environmental Change* **11**, 19–29.

Hulme, M., Barrow, E. M., Arnell, N. W., Harrison, P. A., Johns, T. C. and Downing, T. E. (1999). Relative impacts of human-induced climate change and natural climate variability. *Nature* **397**, 688–91.

Hurrell, J. W., Kushnir, Y., Ottersen, G. and Visbeck, M. (2003). An overview of the North Atlantic Oscillation. *Geophysical Monograph* **134**, 1–35.

IMAGE-team (2001). The IMAGE 2.2 Implementation of the SRES Scenarios. Bilthoven, the Netherlands, National Institute for Public health and Environment (RIVM).

Imbrie, J., and Kipp, N. G. (1971). A new micropaleontological method for paleoclimatology: application to a late Pleistocene Caribbean core. In Turekian K. K. (ed.). The Late Cenozoic Glacial Ages. New Haven, CT, Yale University Press, pp. 71–181.

IPCC (1996). Climate Change 1995: The Science of Climate Change. Cambridge, Cambridge University Press.

IPCC TAR (2001). Climate Change 2001: Synthesis Report. Cambridge, Cambridge University Press.

Irigolen, X., Hulsman, J. and Harris, R. P. (2004). Global biodiversity patterns of marine phytoplankton and zooplankton. *Nature* **429**, 863–7.

Irino, T., Ikehara, K., Katayama, H., Ono, Y. and Tada, R. (2001). East Asian monsoon signals recorded in the Japan Sea sediments. *PAGES News* **9** (2), 7–8.

Isdale, P. J., Stewart, B. J., Tickle, J. S. and Lough, J. M. (1998). Palaeohydrological variation in a tropical river catchment: a reconstruction using fluorescent bands in corals of the Great Barrier Reef, Australia. *The Holocene* **8**, 1–8.

Jackson, J. B. C., Kirby, M. X., Berger, W. H. *et al.* (2001). Historical overfishing and the recent collapse of coastal ecosystems. *Science* **293**, 629–37.

Jacobs, S. S., Giulivi, C. F. and Mele, P. A. (2002). Freshening of the Ross Sea during the late twentieth century. *Science* **297**, 386–9.

Jacobson, M. Z. (2002). Control of fossil-fuel particulate black carbon and organic matter, probably the most effective method of slowing global warming. *Journal of Geophysical Research* **107** (D19), 4410.

Jaenisch, T. and Patz, J. (2002). Assessment of associations between climate and infectious diseases: a comparison of the reports of the intergovernmental panel on climate change (IPCC), the National Research Council (NRC), and United States Global Change Research Program (USGCRP). *Global Change and Human Health* **3**, 67–72.

Jäger J. (2004). Sustainability science. In Steffen W., Sanderson, A., Tyson, P. D. *et al.* (eds.). Global Change and the Earth System; A Planet Under Pressure. Berlin, Springer Verlag, p. 296.

Jahnke, R. A. (2000). The Phosphorus cycle. In Jacobson, M. C., Carlson, R. J., Rodhe, H. and Orians, G. H. (eds.). Earth System Science: From Biogeochemical Cycles to Global Change. London, Academic Press. pp. 360–76.

Jayaraman, A. (1999). Results on direct radiative forcing of aerosols obtained over the tropical Indian Ocean. *Current Science* **76**, 924–30.

Jayaraman, A. and Mitra, A. P. (2004). The Asian Brown Cloud. In Steffen W., Sanderson, A., Tyson, P. D. *et al.* (eds.). Global Change and the Earth System; A Planet Under Pressure. Berlin, Springer Verlag, p. 110.

Jensen, M. N. (2003). Consensus on ecological impacts remains elusive. *Science* **299**, 38.

Jia, G. J., Epstein, H. E. and Walker, D. A. (2003). Greening of arctic Alaska, 1981–2001. *Geophysical Research Letters* **30** (20), 2067.

Johnsen, S. J., Dahl-Jensen, D., Gundestrup, N. *et al.* (2001). Oxygen isotope and paleotemperature records from six Greenland ice-core stations: Camp Century, Dye-3, GRIP, GISP2, Renland and NorthGRIP. *Journal of Quaternary Science* **16**, 299–307.

Jones, G. S., Tett, S. F. B. and Stott, P. A. (2003) Causes of atmospheric temperature change 1960–2000: a combined attribution analysis. *Geophysical Research Letters* **5**, 32-1–32-4.

Jones, P. D. and Mann, M. E. (2004). Climate over past millennia. *Review of Geophysics* **42**, 2003RG000143.

Jones, P. D., Briffa, K. R. and Osborn, T. J. (2003). Changes in the northern hemisphere annual cycle: implications for palaeclimatology? *Journal of Geophysical Research* **108** (D18), 4588.

Jones, P. D., Briffa, K. R., Barnett, T. P. and Tett, S. F. B. (1998). High-resolution palaeoclimatic records for the last millennium: interpretation, integration and comparison with General Circulation Model control run temperatures. *The Holocene* **8**, 455–71.

Jones, P. D., Briffa, K. R., Osborn, T. J., Moberg, A. and Bergström, H. (2002). Relationships between circulation strength and the variability of growing season and cold-season climate in northern and central Europe. *The Holocene* **12**, 643–56.

Jones, R. N. (2001). An environmental risk assessment/management framework for climate change impact assessments. *Natural Hazards* **23**, 197–230.

Jones, R. T., Marshall, J. D., Crowley, S. F. *et al.* (2002). A high resolution, multi-proxy late-glacial record of climate change and intrasystem response in northwest England. *Journal of Quaternary Science* **17**, 329–40.

Joos, F., Gerber, S., Prentice, I. C., Otto-Bliesner, B. L. and Valdes, P. (2004). Transient simulations of Holocene atmospheric carbon dioxide and terrestrial carbon since the Last Glacial Maximum. *Global Biogeochemical Cycles* **18**, GB2002, 1–18.

Joos, F., Plattner, G-K., Stocker, T. F., Koertzinger, A. and Wallace, D. W. R. (2003). *EOS* **84** (21), 197–201.

Jouzel, J., Hoffmann, G., Koster, R. D. and Masson, V. (2000). Water isotopes in precipitation: data/model comparison for present-day and past climates. *Quaternary Science Reviews* **19**, 363–79.

Jouzel, J., Alley, R. B., Cuffey, K. M. *et al.* (1997). Validity of temperature reconstruction from water isotopes in ice cores. *Journal of Geophysical Research* **102**, 26471–87.

Kaiser, D. P. and Qian, Y. (2002). Decreasing trends in sunshine duration over China for 1954–1998: indications of increased haze pollution? *Geophysical Research Letters* **29** (21), 38-1–38-4.

Kalnay, E. and Cai, M. (2003). Impact of urbanization and land-use change on climate. *Nature* **423**, 528–31.

Kaplan, J. O., Bigelow, N. H., Prentice, I. C. *et al.* (2003). Climate change and Arctic ecosystems II: modeling, paleodata-model comparisons, and future projections. *Journal of Geophysical Research* **108** (D19), 8171.

Karlsen A. W., Cronin T. M., Ishman S. E. *et al.* (2004). Historical trends in Chesapeake Bay dissolved oxygen based on benthic Foraminifera from sediment cores. *Estuaries* **23**, 488–508.

Karpuz, N. K. and Jansen, E. (1992). A high-resolution diatom record of the last deglaciation from the SE Norwegian Sea; documentation of rapid climatic changes. *Paleoceanography* **7**, 499–520.

Kaspi. Y., Sayag, R. and Tziperman, E. (2004). A 'triple sea-ice state' mechanism for the abrupt warming and synchronous ice sheet collapses during Heinrich events. *Paleoceanography* **19**, PA3004.

Kassas, M. (1995). Desertification: a general review. *Journal of Arid Environments* **30**, 15.

Kates, R. W. and Parris, T. M. (2003). Long-term trends and a sustainability transition. *Procdings of the National Academy of Sciences* **100**, 8062–7.

Kaufman, D. S., Ager, T. A., Anderson, N. J. *et al.* (2004). Holocene thermal maximum in the western Arctic (0° and 180° W). *Quaternary Science Reviews* **23**, 529–60.

Kaufmann, R. K. and Stock, J. H. (2003). Testing hypotheses about mechanisms for the unknown carbon sink: a time series analysis. *Global Biogeochemical Cycles* **17** (2), 1072.

Keith, D. W. (2001). Geoengineering. *Nature* **409**, 420.

Kennett, J., Cannariato, K. G., Hendy, I. L. and Behl, R. J. (2000). Carbon isotopic evidence for methane hydrate instability during Quaternary interstadials. *Science* **288**, 128–33.

Kennett, J. P., Cannariato, K. G., Hendy, I. L. and Behl, R. J. (2003). Methane Hydrates in Quaternary Climate Changes: the Clathrate Gun Hypothesis. New York, American Geophysical Union.

Kerr, R. A. (2004). Getting warmer, however you measure it. *Science* **304**, 805–7.

Kessler, W. S. (2002). Is ENSO a cycle or a series of events? *Geophysical Research Letters* **29**, 23.

Khodri, M., Leclainche, Y., Ramstein *et al.* (2001). Simulating the amplification of orbital forcing by ocean feedbacks in the last glaciation. *Nature* **410**, 570–4.

Kim, S. J. (2004). The effect of atmospheric CO_2 and ice sheet topography on LGM climate. *Climate Dynamics*, **22**, 639–51.

Kitching, R. (2000). Biodiversity, hotspots and defiance. *Trends in Ecology and Evolution* **15**, 484–5.

Knapp, P. A., Grissino-Mayer, H. D. and Soule, P. T. (2002). Climatic Regionalization and the spatio-temporal occurrence of extreme single-year drought events (1500–1998) in the Interior Pacific Northwest, USA. *Quaternary Research* **58**, 226–33.

Knapp, S. and Mallet, J. (2003). Refuting refugia? *Science* **300**, 71–2.

Knorr, G. and Lohmann, G. (2003). Southern Ocean origin for the resumption of Atlantic themonaline circulation during deglaciation. *Nature* **424**, 532–6.

Knox, J. (2000). Sensitivity of modern and Holocene floods to climate change. *Quaternary Science Reviews* **19**, 439–57.

Knutti, R., Stocker, T. F., Joos, F. and Plattner, G.-K. (2002). Constraints on radiative forcing and future climate change from observations and climate model ensembles. *Nature* **416**, 719–23.

Knutti, R., Stocker, T. F., Joos, F. and Plattner, G.-K. (2003). Probabilistic climate change projections using neural networks. *Climate Dynamics* **21**, 257–72.

Kohfield, K. E. and Harrison, S. P. (2000). How well can we simulate past climates? Evaluating the models using global palaeoenvironmental datasets. *Quaternary Science Reviews* **19**, 321–46.

Koren, I., Kaufman, Y. J., Remer, L. A. and Martins, J. V. (2004). Measurement of the effect of Amazon smoke on inhibition of cloud formation. *Science* **303**, 1342–5.

Körner, C. (2003). Slow in, rapid out – carbon flux studies and Kyoto targets. *Science* **300**, 1242–3.

Krakauer, N. Y. and Randerson, J. T. (2003). Do volcanic eruptions enhance or diminish net primary productivity? Evidence from tree rings. *Global Biogeochemical Cycles* **17** (4), 1118.

Krajick, K. (2004). All downhill from here? *Science* **303**, 1600–2.

Kumar, K. K., Rajagopalan, B., and Cane, M. A. (1999). On the weakeneing relationship between the Indian monsoon and ENSO. *Science* **284**, 2156–9.

Kunkel, K. E., Easterling, D. R., Hubbard, K. and Redmond, K. (2004). Temporal variations in frost-free season in the United States: 1895–2000. *Geophysical Research Letters* **31**, L0321.

Kutzbach, J. E. and Guetter, P. J. (1986). The influence of changing orbital parameters and surface boundary conditions on climate simulations for the past 18 000 years. *Journal of Atmospheric Sciences* **43**, 1726–59.

Kutzbach, J. E. and Liu, Z. (1997). Response of the African monsoon to orbital forcing and ocean feedbacks in the middle Holocene. *Science* **278**, 440–3.

Labeyrie, L., Cole, J., Alverson, K. and Stocker, T. (2003). The history of climate dynamics in the Late Quaternary. In Alverson, K., Bradley, R. S. and Pedersen, T. F. (eds). Paleoclimate, Global Change and the Future. Berlin, Springer Verlag, pp. 33–63.

Lachenbruch, A. H. and Marshall, B. V. (1986). Changing climate: geothermal evidence from permafrost in the Alaskan Arctic. *Science* **234**, 689–96.

Lackner, K. S. (2003). A guide to CO_2 sequestration. *Science* **300**, 1677–8.

Laj, C., Kissel, C., Mazaud, A., Channell, J. E. T. and Beer, J. (2000). North Atlantic paleointensity stack since 75 ka (NAPIS-75) and the duration of the Laschamp event. *Philosophical Transactions Royal Society London* **358**, 1009–25.

Laj, P., Ghermandi, G., Cecchi, R. *et al.* (1997). Distribution of Ca, Fe and S between soluble and insoluble material in the Greenland Ice Core Project ice core. *Journal of Geophysical Research* **102**, 26615–24.

Lamb, H. H. (1965). The early Medieval warm epoch and its sequel. *Palaeogeography, Palaeoclimatology, Palaeoecology* **1**, 13–37.

Lambert, F. H., Stott, P. A., Allen, M. R. and Palmer, M. A. (2004). Detection and attribution of changes in twentieth century land precipitation. *Geophysical Research Letters* **31**, L10203.

Lambin, E. F. and Geist, H. J. (2003). Regional differences in tropical deforestation. *Environment* **45** (6), 22–7.

Lambin, E. F., Geist, H. J. and Lepers, E. (2003). Dynamics of land-use and land-cover change in tropical regions. *Annual Review of Environmental Resources* **28**, 1–14.

Lambin, E. F., Turner, B. L., Geist, H. J. (2001). The causes of land-use and land-cover change: moving beyond the myths. *Global Environmental Change* **11**, 261–69.

Lamy, F., Kaiser, J., Ninnemann, U. *et al.* (2004). Antarctic timing of surface water changes off Chile and Patagonian ice sheet response. *Science* **304**, 1959–62.

Lang, A. (2003). Phases of soil erosion-derived colluviation in the loess hills of South Germany. *Catena* **51**, 209–21.

Lang, A., Hatté. C., Rousseau, D. -D. *et al.* (2003). High-resolution chronologies for loess: comparing AMS ^{14}C and optical dating results. *Quaternary Science Reviews* **22**, 953–9.

Langenfields, R. L., Francey, R. J., Pak, B. C. *et al.* (2002). Interannual growth rate variations in atmospheric CO_2 and its $\delta^{13}C$, H_2, CH_4 and CO between 1992 and 1999 linked to biomass burning. *Global Biogeochemical Cycles* **16**, 21-1–21-7.

Langeweg, F. and Gutierrez-Espeleta, E. E. (2001). Human security and vulnerability in a scenario context: challenges for UNEP's global environmental outlook. *IHDP Update: Newsletter of the International Human Dimensions Programme on Global Environmental Change* **2**, 11–2.

Lapenis, A. G., Lawrence, G. B., Andreev, A. A. *et al.* (2004). Acidification of forest soil in Russia: from 1893 to present. *Global Biogeochemical Cycles* **18**, GB1037, 11–13.

Larson, D. O., Neff, H., Greybill, D. A., Michaelsen, J. and Ambos, E. (1996). Risk, climatic variability and the study of southwestern prehistory: an evolutionary perspective. *American Antiquity* **61**, 217–41.

Latif, M., Roeckner, E. Mikolajewicz, U. and Voss, R. (2000). Tropical stabilization of the thermohaline circulation in the greenhouse warming simulation. *Journal of Climatology* **13**, 1809–13.

Lau, K.-M. and Weng, H. (1999). Interannual, decadal–interdecadal, and global warming signals in sea surface temperature during 1955–97. *Journal of Climate* **12**, 1257–67.

Lauritzen, S. -E. (2003). Reconstructing Holocene climate records from speleothems. In McKay, A., Battarbee, R. W., Birks, H. J. B. and Oldfield, F. (eds.). Global Change in the Holocene. London, Arnold, pp. 242–63.

Lauritzen, S. -E. and Lundberg, J. (1999). Calibration of the speleothem delta function: an absolute temperature record from the Holocene in northern Norway. *The Holocene* **9**, 659–70.

Lawton, J. H. and May, R. M. (1995) Extinction Rates. Oxford, Oxford University Press.

Lean, J., Beer, J. and Bradley, R. S. (1995). Reconstruction of solar irradiance since 1610: implications for climate change. *Geophysical Research Letters* **22**, 3195–8.

Lean, J., Skumanich, A. and White, O. (1992). Estimating the sun's radiative output during the Maunder Minimum. *Geophysical Research Letters* **19**, 1591–4.

Leatherman, S. P., Douglas, B. C. and LeBrecque, J. L. (2003). Sea level and coastal erosion require large-scale monitoring. *EOS* **84** (**2**), 13–16.

Lee, K., Choi, S. -D., Park, G. -H. *et al.* (2003). An updated anthropogenic CO_2 inventory in the Atlantic Ocean. *Global Biogeochemical Cycles* **17** (4), 116.

Leemans, R. (2003). The IMAGE 2 Integrated assessment modelling framework. In Steffen, W., Sanderson, A., Tyson, P. D. *et al.* (eds.). Global Change and the Earth System; A Planet Under Pressure. Berlin, Springer Verlag, p. 206.

LeGrand, P. and Alverson, K. (2001). Variations in atmospheric CO_2 during glacial cycles from an inverse ocean modeling perspective. *Paleoceanography* **16**, 604–16.

Lelieveld, J., Crutzen, P. J., Ramanathan, N. *et al.* (2001). The Indian Ocean Experiment: widespread pollution from south and southeast Asia. *Science* **291**, 1031–6.

Leng, M. J. (2003). Stable isotopes in lakes and lake sediment archives. In McKay, A., Battarbee, R. W., Birks, H. J. B. and Oldfield, F. (eds.). Global Change in the Holocene. London, Arnold, pp. 124–39.

Leng, M. J. and Marshall, J. D. (2004). Paleoclimate interpretation of stable isotope data from lake sediment archives. *Quaternary Science Reviews*, **23**, 811–31.

Lenton, T. M. and Cannell, M. G. R. (2002). Mitigating the extent and rate of global warming. *Climatic Change* **52**, 255–62.

Levitus, S., Antonov, J. I., Boyer, T. P. and Stephens, C. (2000). Warming of the world ocean. *Science* **287**, 2225–9.

Lewis, M. (2003). Common sense. *CEI NewsCenter (www.cei.org)*. June 4.

Lieberman, B. (2003). Ozone depletion's lessons for global Warming . *CEI NewsCenter* (www.cei.org). October 1.

Lindzen, R. S. and Giannitsis, C. (2002). Reconciling observations of global temperature changes. *Geophysical Research Letters* **29** (12), 10.

Lindzen, R. S., Chou, M.-D. and Hou, A. Y. (2001). Does the Earth have an adaptive infrared Iris? *Bulletin of the American Meteorological Society* **82**, 417–32.

Lintner, B. R. (2002). Characterizing the global CO_2 interannual variability with empirical orthogonal function/principal component (EOF/PC) analysis. *Geophysical Research Letters* **29**, 27-1–27-5.

Lipp, J., Trimborn, P., Graf, W., Edwards, T. and Becker, B. (1995). Climate signals in a ^{2}H and ^{13}C chronology (1882–1989) from tree rings of Spruce (*Picea abies L.*), Schussbach Forest, Germany. In Dean, J. S., Meko, D. M. and Swetnam, T. W. (eds.). Tree rings, Environment and Humanity, Radiocarbon, 1996, 603–10.

Liski, J., Korotkov, V., Prins, C. F. L., Karjalainen, T., Victor, D. G. and Kauppi, P. E. (2003). Increased carbon sink in temperate and boreal forests. *Climatic Change* **61**, 89–99.

Liu, Z., Kutzbach, J. E. and Wu, L. (2000). Modeling climate shift of El Niño variability in the Holocene. *Geophysical Research Letters* **27**, 2265–8.

Liu, Z., Harrison, S. P., Kutzbach, J. and Otto-Bliesner, B. (2004). Global monsoons in the mid-Holocene and oceanic feedback. *Climate Dynamics* **22**, 157–82.

Livingstone, D. A. (2003). Global climate change strikes a tropical lake. *Science* **301**, 468–9.

Löffler, H. (2004). Origin of lake basins. In O'Sullivan, P. E. and Reynolds, C. S. (eds.). The Lake Handbook: Limnology and Limnetic Ecology, Oxford, Blackwell, vol. 1.

Lohmann, U. and Lesins, G. (2002). Stronger constraints on the anthropogenic indirect aerosol effect. *Science* **298**, 1012–15.

Lomborg, B. (2001). The Sceptical Environmentalist: Measuring the Real State of the World. Cambridge, Cambridge University Press.

Long, D., Ballantyne, J. and Bertoia, C. (2002). Is the number of Antarctic icebergs really increasing? *EOS* **83** (42), 471–4.

Lorius, C., Jouzel, J., Raynaud, D., Hansen, J. and Le Treut, H. (1990). The ice core record: climate sensitivity and future greenhouse warming. *Nature* **347**, 139–45.

Lotter, A. F. (1999). Late-glacial and Holocene vegetation history and dynamics as shown by pollen and plant macrofossil analyses in annually laminated sediments from Soppensee, central Switzerland. *Vegetation History and Archaebotany* **8**, 165–84.

Lotter, A. F. (2003). Multi-proxy climatic reconstructions. In Mackay, A., Battarbee, R. W., Birks, H. J. B. and Oldfield, F. (eds.). Global Change in the Holocene. London, Arnold, pp. 373–83.

Lowe, J. A., Gregory, J. M. and Flather, R. A. (2001). Changes in the occurrence of storm surges around the United Kingdom under a future climate scenario using a dynamic storm surge model driven by the Hadley Centre climate models. *Climate Dynamics* **18**, 179–88.

Lowe, J. J. and Walker, M. J. C. (1997). Reconstructing Quatternary Environments, 2nd edn. Harlow, Pearson Prentice Hall.

Luterbacher, J., Dietrich, D., Xoplaki, E., Grosjean, M. and Wanner, H. (2004). European seasonal and annual temperature variability, trends, and extremes Since 1500. *Science* **303**, 1499–1503.

Luterbacher, J., Schmutz, C., Gyalistras, D., Xopalski, E. and Wanner, H. (1999). Reconstruction of monthly NAO and EU indices back to AD 1675. *Geophysical Research Letters* **26**, 2745–8.

MacArthur, R. H. (1995). Fluctuatuions of animal populations and a measure of community stability. *Ecology* **36**, 533–6.

Mackay, A., Battarbee, R. W., Birks, H. J. B. and Oldfield, F. (eds.). (2003). Global Change in the Holocene. London, Arnold.

MacAyeal, D. R. (1993) Binge/purge oscillations of the Laurentide ice sheet as a cause of the North Atlantic's Heinrich events. *Paleoceanography* **8**, 775–84.

MacCracken, M. C. (2002). Do the uncertainty ranges in the IPCC and US national assessments account adequately for possible overlooked climatic influences. *Climatic Change* **52**, 11–23.

Mace, G. M., Gittleman, J. L. and Purvis, A. (2003). Preserving the tree of life. *Science* **300**, 1707–9.

Macklin, M. G. (1999). Holocene river environments in prehistoric Britain: human interaction and impact. *Quaternary Proceedings* **7**, 521–30.

Macklin, M. G. and Lewin, J. (2003). River sediments, great floods and centennial-scale Holocene climate change. *Journal of Quaternary Science* **18**, 101–5.

Maggs, R., Wahid, A., Shamsi, S. R. A. and Ashmore, M. R. (1995). Effects of ambient air pollution on wheat and rice yield in Pakistan. *Water, Air and Soil Pollution* **85**, 1311–6.

Magny, M. (1993). Solar influences on Holocene climatic changes. *Quaternary Research* **40**, 1–9.

Maher, B. A. and Dennis, P. F. (2001). Evidence against dust-mediated control of glacial-interglacial changes in atmospheric CO_2. *Nature* **411**, 176–180.

Maher, B. A. and Thompson, R. (1999). Palaeomonsoons I: the magnetic record of Paleoclimate in the terrestrial loess and palaeosol sequences. In Maher, B. A. and Thompson, R. (eds.). Quaternary Climates, Environments and Magnetism. Cambridge, Cambridge University Press, pp. 81–125.

Mahowald, N., Kohfeld, K., and Mansson, M. (1999). Dust sources and deposition during the last glacial maximum and current climate: a comparison of model results with paleodata from ice cores and marine sediments. *Journal of Geophysical Research* **104**, 895–916.

Manabe, S. and Stouffer, R. J. (1995). Simulation of abrupt climate change induced by freshwater input to the North Atlantic Ocean. *Nature* **378**, 165–7.

Manabe, S. and Stouffer, R. J. (2000). Study of abrupt climate change by a coupled ocean–atmosphere model. *Quaternary Science Reviews* **19**, 285–99.

Manley, G. (1974). Central England temperatures: monthly means 1659–1973. *Quarterly Journal of the Royal Meteorological Society* **100**, 389–405.

Mann, M. E. (2002a). The value of multiple proxies. *Science* **297**, 1481–2.

Mann, M. E. (2002b). Large-scale climate variability and connections with the Middle East in past centuries. *Climatic Change* **55**, 287–314.

Mann, M. E. and Jones, P. D. (2003). Global surface temperatures over the past two millennia. *Geophysical Research Letters* **30** (15), 5-1–5-4.

Mann, M. E. and Schmidt, G. (2003). Ground vs. surface air temperature trends: implications for borehole surface temperature reconstructions. *Geophysical Research Letters* **30** (12), 1607.

Mann, M. E., Bradley, R. S. and Hughes, M. K. (1998). Global-scale temperature patterns and climate forcing over the past six centuries. *Nature* **392**, 779–787.

Mann, M. E., Bradley, R. S. and Hughes, M. K. (1999). Northern hemisphere temperatures during the past millennium: inferences, uncertainties and limitations. *Geophysical Research Letters* **26**, 759–762.

Mann, M. E., Bradley, R. S. and Hughes, M. K. (2004). Corrigendum: Global-scale temperature patterns and climate forcing over the past six centuries. *Nature* **430**, 105.

Mann, M. E., Rutherford, R. S., Bradley, R. S., Hughes, M. K. and Keimig, F. T. (2003). Optimal surface temperature reconstructions using terrestrial borehole data. *Journal of Geophysical Research* **109**, D11107.

Mann, M. E., Amman, C., Bradley, R. S. *et al*. (2003). On past temperatures and anomalous late-twentieth century warmth. *EOS* **84** (27), 256–7.

Mann, M. E., Gille, E., Bradley, R. S. *et al*. (2000). Annual temperature patterns in past centuries: an interactive presentation. *Earth Interactions* **4**, 1–29.

Mantua, N. J., Hare, S. R., Zhang, Y., Wallace, J. M. and Francis, R. C. (1997). A Pacific interdecadal climate oscillation with impacts on salmon production. *Bulletin of the American Meteorological Society* **78**, 1069–79.

Marchal, O., Stocker, T. F., Joos, F. *et al*. (1999). Modelling the concentration of atmospheric CO_2 during the Younger Dryas climate event. *Climate Dynamics* **15**, 341–54.

Markgraf, V. (ed.). (2001). Interhemispheric climate linkages. New York, Academic Press.

Marsh, N. D. and Svensmark, H. (2000). Low cloud properties influenced by cosmic rays. *Physics Review Letters* **85**, 5004–7.

Marsh, N. D. and Svensmark, H. (2004). Comment on 'Solar influences on cosmic rays and clous formation: a reassessment' by Bomin Sun and Raymond S. Bradley. *Journal of Geophysical Research* **109**, D1425.

Marshall, J. D., Jones, R. T., Crowley, S. F., Oldfield, F., Nash, S. and Bedford, A. (2002). A high resolution late-glacial isotopic record from Hawes Water, northwest England. Climate oscillations: calibration and comparison of palaetemperature proxies. *Palaeogeography, Palaeoclimatology, Palaeoecology* **185**, 25–40.

Martens, P. (1998). Health and Climate: Modelling the Impacts of Global Warming and Ozone Depletion. London, Earthscan.

Martinson, D. G., Pisias, N. G., Hays, J. D., Imbrie, J., Moore, T. C. and Shackleton, N. J. (1987). Age dating and the orbital theory of the ice ages: development of a high resolution 0–300 000 year chronostratigraphy. *Quaternary Research* **27**, 1–29.

Maslin, M., Pike, J., Stickley, C and Ettwein, V. (2003). Evidence of Holocene climate variability in marine sediments. In Mackay, A. W., Battarbee, R. W., Birks, H. J. B. and Oldfield, F. (eds.) Global Change in the Holocene. London, Arnold, pp. 185–209.

Maslin, M. and Thomas, E. (2003). Balancing the deglacial carbon budget: the hydrate factor. *Quaternary Science Reviews* **22**, 1729–36.

Mason, B. (2004). Climate change: the hot hand of history. *Nature* **427**, 582–3.

Matthews, H. D. Weaver, A. J. Meissner, K. J. Gillett, N. P. and Eby, M. (2004,). Natural and anthropogenic climate change: incorporating historical land cover change, vegetation dynamics and the global carbon cycle. *Climate Dynamics* **22**, 461–79.

May, R. M. (1973). Stability and Complexity in Model Ecosystems. Princeton, NJ, Princeton University Press.

Mayewski, P. A., Meeker, L. D., Twickler, M. S. *et al.* (1997). Major features and forcing of high-latitude northern hemisphere atmospheric circulation using a 110 000-year-long glaciochemical series. *Journal of Geophysical Research* **102**, 26345–66.

Maynard, K., Royer, J.-F. and Chauvin, F. (2002). Impact of greenhouse warming on the west African summer monsoon. *Climate Dynamics* **19**, 499–514.

McCann, K. S. (2000). The diversity-stability debate. *Nature* **405**, 228–33.

McCarroll, D. and Loader, N. J. (2004). Stable isotopes in tree rings. *Quaternary Science Reviews* **23**, 771–801.

McCarroll, D. and Pawelleck, F. (2001). Stable carbon isotope ratios of *Pinus sylvestris* from northern Finland and the potential for extracting a climate signal from long Fennoscandian chronologies. *The Holocene* **11**, 517–26.

McDermott, F. (2004). Paleo-climate reconstructions from stable isotope variations in speleothems: a review. *Quaternary Science Reviews* **23**, 901–18.

McGlone, M. S. (1995). Late glacial landscape and vegetation change during the Younger Dryas climatic oscillation in New Zealand. *Quaternary Science Reviews* **14**, 867–81.

McGregor, H. V. and Gagan, M. K. (2004). Western Pacific δ^{18}O records of anomalous Holocene variability in the El Niño Southern Oscillation. *Geophysical Research Letters* **31** (15), L11204.

McGuffie, K. and Henderson-Sellers, A. (1999). A Climate Modelling Primer, 2nd Edition. New York, Wiley.

McIntyre, S. and McKitrick, R. (2003). Corrections to the Mann *et al.* (1998) proxy data base and northern hemispheric average temperature series. *Energy and Environment* **14**, 751–71.

McManus, J. F., François, R., Gherardi, J.-M., Keigwin, L. D. and Brown-Leger, S. (2004). Collapse and rapid resumption of Atlantic meridional circulation linked to deglacial climate changes. *Nature* **428**, 834–7.

McNeill, B. I., Matear, R. J., Key, R. M., Bullister, J. L. and Sarmiento, J. L. (2003). Anthropogenic CO_2 uptake by the ocean based on the global chlorofluorocarbon data Set. *Science* **299**, 235–8.

Mearns, L. O., Giorgi, F., McDaniel, L. and Shields, C. (2003). Climate scenarios for the southeastern US Based on GCM and regional model simulations. *Climatic Change* **60**, 7–35.

Meehl, G. A. and Arblaster, J. M. (2003). Mechanisms for projected future changes in south Asian monsoon precipitation. *Climate Dynamics* **21**, 659–75.

Meissner, K. J., Weaver, A. J., Matthews, H. D. and Cox, P. M. (2003). The role of land surface dynamics in glacial inception: a study with the Uvic Earth System Model. *Climate Dynamics* **21**, 515–37.

Menon, S., Hansen, J., Nazarenko, L. and Luo, Y. (2002). Climate effects of black carbon aerosols in China and India. *Science* **297**, 250–2.

Menzel, A. (2002). Phenology: its importance to the global change community. *Climate Change* **54**, 379–85.

Menzel, A. and Estrella, N. (2001). Past phenological changes. In Walther, G. R., Burga, C. A. and Edwards, P. J. (eds.). Fingerprints of Climate Change – Adapted Behaviour and Shifting Species Ranges. New York, Kluwer.

Messerli, B., Grosjean, M., Hofer, T., Nuñez, L. and Pfister, C. (2000). From nature-dominated to human-dominated environmental changes. *Quaternary Science Reviews* **19**, 459–79 .

Meybeck, M. and Ragu, A. (1997). Presenting the GEMS-GLORI, a compendium for world river discharges to the oceans. *International Association of Hydrological Sciences* **243**, 3–14.

Meybeck, M. and Vörösmarty, C. (2004). Human-driven changes to continental aquatic systems. In Steffen, W., Sanderson, A., Tyson, P. *et al*. (eds.). Global Change and the Earth System: A Planet Under Pressure. Berlin, Springer Verlag. pp. 112–113.

Mickley, L. J., Jacob, D. J., Field, B. D. and Rind, D. (2004). Climate response to the increase in tropospheric ozone since preindustrial times: a comparison between ozone and equivalent CO_2 forcings. *Journal of Geophysical Research* **109**, D05106.

Midgley, G. F., Hannah, L., Millar, D., Rutherford, M. C. and Powrie, L. W. (2002). Assessing the vulnerability of species richness to anthropogenic climate change in a biodiversity hotspot. *Global Ecology and Biogeography* **11**, 445–51.

Mikolajewicz, U., Crowley, T. J., Schiller, A. and Voss, R. (1997). Modelling teleconnections between the North Atlantic and North Pacific during the Younger Dryas. *Nature* **387**, 384–7.

Milankovitch, M. M. (1941). Canon of insolation and the ice-age problem. Beograd: Koninglich Serbische Akademie. [English translation by the Israel program for Scientific Translations, published by the US Department of Commerce, and the National Science Foundation, Washington DC (1969)].

Miller, L. and Douglas, B. C. (2004). Mass and volume contributions to twentieth-century global sea-level rise. *Nature* **428**, 406–9.

Milly, P. C. D., Wetherald, R. T., Dunne, K. A. and Delworth, T. L. (2002). Increasing risk of floods in a changing climate. *Nature* **415**, 514–7.

Mirza, M., Mirza, Q., Warrick, R. A. and Ericksen, N. J. (2003). The implications of climate change on floods of the Ganges, Brahmaputra and Meghna Rivers in Bangladesh. *Climatic Change* **57**, 287–318.

Mitchell, J. F. B., Karoly, D. J., Hegerl, G. C. *et al*. In IPCC TAR (2001). Detection of Climate Change and Attribution of Causes. Cambridge, Cambridge University Press, pp. 697–738.

Mitrovica, J. X., Tamisea, M. E., Davis, J. L. and Milne, G. A. (2001). Recent mass balance of polar ice sheets inferred from patterns of global sea-level change. *Nature* **409**, 1026–29.

Mitsuguchi, T., Matsumoto, E., Abe, O., Uechida, T. and Isdale, P. J. (1996). Mg/Ca thermo-metry in coral skeletons. *Science* **274**, 961–3.

Mix, A. C., Bard, E. and Schneider, R. (2001). Environmental processes of the ice age: land, oceans, glaciers (EPILOG). *Quaternary Science Reviews* **20**, 627–57.

Mock, C. J. (2002). Documentary records of past climate and tropical cyclones from the southeastern United States. *PAGES Newsletter* **10** (3), 20–21.

Monnin, E., Indermühle, A., Dällenbach, A. *et al*. (2001). Atmospheric CO_2 concentrations over the last glacial termination. *Science* **291**, 112– 114.

Mooney, H. A., Canadell, J., Chapin, F. S. III, *et al*. (1999). Ecosystem physiology responses to global change. In Walker, B., Steffen, W. Canadell, J. and Ingram, J. (eds.). The terrestrial biosphere and global change. Implications for Natural and Managed Ecosystems. Cambridge, Cambridge University Press.

Moore, G. W. K., Holdsworth, G. and Alverson, K. (2001). Extra-tropical responses to ENSO 1736–1985 as expressed in an ice core from the Saint Elias Mountain range in northwestern North America. *Geophysical Research Letters* **28**, 3457–61.

Moore, G. W. K., Holdsworth, G. and Alverson, K. (2002). Climate change in the north Pacific region over the past three centuries. *Nature* **420**, 401–3.

Morgan, V., Delmotte, M., van Ommen, T. *et al.* (2002). Relative timing of deglacial climate events in Antarctica and Greenland. *Science* **297**, 1862–4.

Mörner, N. -A. (2004). Estimating future sea-level changes from past records. *Global and Planetary Change* **40**, 49–54.

Mörner, N. -A., Tooley, M. and Possnert, G. (2004). New perspectives for the future of the Maldives. *Global and Planetary Change* **40**, 177–182.

Morrill, C., Overpeck, J. T. and Cole, J. E. (2003). A synthesis of abrupt changes in the Asian summer monsoon since the last glaciation. *The Holocene* **13**, 465–76.

Mosier, A. R., Bleken, M. A., Chaiwanakupt, P. *et al.* (2002). Policy implications of human-accelerated nitrogen cycling. *Biogeochemistry* **57–58**, 477–516.

Moy, C. M., Seltzer, G. O., Rodbell, D. T. and Anderson, D. M. (2002). Variability of El Niño/southern oscillation activity at millennial timescales during the Holocene epoch. *Nature* **420**, 162–5.

Mueller, D. R., Vincent, W. F. and Jeffries, M. O. (2003). Break-up of the largest Arctic ice shelf and associated loss of an epishelf lake. *Geophysical Research Letters* **30** (20), CRY 1-1–1-4.

Müller P. J., Kirst, G., Ruhland, G., von Storch, I. and Rosell-Mélé, A. (1998). Calibration of alkenone paleotemperature index U^k_{37} based on core tops from the eastern South Atlantic and the global ocean (60° N–60° S). *Geochimica et Cosmochimica Acta* **62**, 1757–71.

Muller, R. A. and MacDonald, G. J. (1997). Glacial cycles and astronomical forcing. *Science* **277**, 215–18.

Munk, W. (2003). Ocean freshening, sea-level rising. *Science* **300**, 2041–3.

Murray, I. (2003a). Hockey stick slapped: climate change's bellisles? *CEI NewsCenter* (www.cei.org). November 3.

Murray, I. (2003b). Tackling junk science. *CEI NewsCenter* (www.cei.org). July 2.

Murray, I. (2003c). Are we all 'damn fools'? *CEI NewsCenter* (www.cei.org). September 22.

Myers, N., Mittermeier, R., Mittermeier, C., da Fonseca, G.-A. and Kent, J. (2000). Biodiversity hotspots for conservation priorities. *Nature* **403**, 853–8.

Nachtergaele, F. (2002). Land degradation assessment in drylands (LADA project). *LUCC Newsletter* **8**, 15.

Nakagawa, T., Kitagawa, H., Yasuda, Y. *et al.* (2003). Asynchronous climate changes in the North Atlantic and Japan during the last termination. *Science* **299**, 688–91.

Nakicenovic, N. and Swart, R. (2001). Special Report on Emission Scenarios. Cambridge, Cambridge University Press.

Neelin, J. D., Chou, C. and Su, H. (2003). Tropical drought regions in global warming and El Niño teleconnections. *Geophysical Research Letters* **30** (24), 5-1–5-4.

Neff, U., Burns, S. J., Mangini, A., Mudalsee, M., Fleitmann, D. and Matter, A. (2001). Strong coherence between solar variability and the monsoon in Oman between 9 and 6 kyr ago. *Nature* **411**, 290–3.

Nemani, R. R., Keeling, C. D., Hashimoto, H. *et al.* (2003). Climate-driven increases in global terrestrial net primary production from 1982 to 1999. *Science* **300**, 1560–62.

Nials, F. L., Gregory, D. A., Graybill, D. A. (1989). Salt river stream flow and Hohokam irrigation systems. In Graybill, D. A., Gregory, D. A., Nials, F. L., Gasser, R., Miksicek, C. and Szuter, C. (eds.). The 1982–1992 excavations at Las Colinas: Environment and

Subsistence 5. Arizona State Museum Archaeological Series. Tuson AZ, University of Arizona, pp. 59–78.

Nicholls, R. J. and Small, C. (2002). Improved estimates of coastal population and exposure to Hazards released. *EOS* **83** (28), 301–5.

Nicholson, S. E. (1982). The Sahel: a Climatic Perspective. Paris, Club du Sahel.

Nicoll, N. (2004). Recent environmental change and prehistoric human activity in Egypt and northern Sudan. *Quaternary Science Reviews*, **23**, 561–80.

Niggermann, S., Mangini, A., Richter, D. K. and Würth, G. (2003). A Paleoclimate record of the last 17 600 years in stalagmites from the B7 cave, Sauerland, Germany. *Quaternary Science Reviews* **22**, 555–67.

Nobre, C. A. (2004). The large-scale biosphere–atmosphere experiment in Amazonia. In Steffen, W., Sanderson, A., Tyson, P. D. *et al.* (eds.). Global Change and the Earth System; a Planet Under Pressure. Berlin, Springer Verlag, p. 278.

Nobre, C. A., Wickland, D. and Kabat, P. I. (2001). The large scale biosphere–atmosphere experiment in Amazonia. *IGBP Global Change Newsletter* **45**, 2–4.

Norby, R. J. (2004). Forest responses to a future CO_2 enriched atmosphere. In Steffen W., Sanderson, A., Tyson, P. D. *et al.* (eds.). Global Change and the Earth System; a Planet Under Pressure. Berlin, Springer Verlag, pp. 158–9.

Noren, A. J., Bierman, P. R., Steig, E. J., Lini, A. and Southron, J. (2002). Millennial-scale storminess variability in the northeastern United States during the Holocene epoch. *Nature* **419**, 821–4.

Nosengo, N. (2003). Fertilized to death. *Nature* **425**, 894–5.

Nott, J. and Hayne, M. (2001). High frequency of 'super-cyclones' along the Great Barrier Reef over the past 5000 years. *Nature* **413**, 508–12.

Nuñez, L., Grosjean, M. and Cartajena, I. (2002). Human occupations and climate change in the Puna de Atacama, Chile. *Science* **298**, 821–4.

Nunn, P. D. (1998). Sea-level changes over the past 1000 years in the Pacific. *Journal of Coastal Research* **14**, 23–30.

Nurse, L. and Sem, G. (2001). Small island states. In McCarthy, J., Canziani, O., Leary, N., Dokken, D. and White, K. (eds.). Climate Change 2001: Impacts, Adaptation and Vulnerability. Cambridge, Cambridge University Press.

Oechel, W. C., Vourlitis, G. L., Hastings, S. J. *et al.* (2000). Oscillation of ecosystem CO_2 exchange in the Alaskan Arctic in response to decadal climate warming. *Nature* **406**, 978–81.

Ogi, M., Yamazaki, K. and Yoshihiro, T. (2003). Solar cycle modulation of the seasonal linkage of the North Atlantic oscillation (NAO). *Geophysical Research Letters* **30** (22), 8-1–8-4.

Okhouchi, N., Eglinton, T. I., Keigwin, L. D. and Hayes, J. M. (2002). Spatial and temporal offsets between proxy records in a sediment drift. *Science* **298**, 1224–6.

Oldfield, F. (1993). Forward to the past: changing approaches to Quaternary palaeoecology. In Chambers, F. M. (ed.) Climate Change and Human Impact on the Landscape. Chapman and Hall, London, pp. 13–22.

Oldfield, F. (in press) Towards developing synergistic linkages between the biophysical and the cultural; a palaeo-environmental perspective. In Hornborg, A., Butzer, K. W., Crumley, C. L., McNeill, J. R. and Martinez-Allier, J. (eds.). World System History and global Environmental change. New York, Columbia University Press.

Oldfield, F. and Alverson, K. (2003). The societal relevance of palaeoenvironmental research. In Alverson, K., Bradley, R. S. and Pedersen, T. F. (eds.). Paleoclimate, Global Change and the Future. Berlin, Springer Verlag, pp. 1–11.

Oldfield, F. and Appleby, P. A. (1984). Empirical testing of ^{210}Pb-dating models for lake sediments. In Haworth, E. Y. and Lund, J. W. G. (eds.). Lake Sediments and Environmental History. Leicester, Leicester University Press, pp. 93–124.

Oldfield. F, Asioli, A, Accorsi, C. A. et al. (2003a). A high resolution late-Holocene palaeo-environmental record from the central Adriatic Sea. Quaternary Science Reviews 22, 319–42.

Oldfield, F. and Dearing, J. A. (2003). The role of human activities in past environmental change. In Alverson, K. D., Bradley, R. S. and Pedersen, T. F. (eds.). Paleoclimate, Global Change and the Future. Berlin, Springer Verlag, pp. 142–62.

Oldfield, F., Richardson N, and Appleby, P. G. (1995). The dating of recent ombrotrophic peat accumulation and evidence for changes in mass balance. The Holocene 5, 141–8.

Oldfield, F., Thompson, R., Crooks, P. R. J. et al. (1997). Radiocarbon dating of a recent high-latitude peat profile: Stor Åmyran, N.Sweden. The Holocene 7, 283–90.

Oldfield, F., Wake, R., Boyle, J. et al. (2003b). The late-Holocene history of Gormire Lake (NE England) and its catchment: a multiproxy reconstruction of past human impact. The Holocene 13, 677–90.

Oppenheimer, M. (1998). Global warming and the stability of the west Antarctic ice sheet. Nature 393, 523–32.

Oppenheimer, M. and Alley, R. B. (2004). The west Antarctic ice sheet and long term climate policy. Climatic Change, 64, 1–10.

Oppo, D. W., McManus, J. F. and Cullen, J. L. (2003). Palaeo-oceanography: deepwater variability in the Holocene epoch. Nature 422, 277.

O'Reilly, C. M., Alin, S. R., Plisnier, P.-D, Cohen, A. S. and McKee, B. A. (2003). Climate-change effect on Lake Tanganyika. Nature 424, 766–8.

Ortlieb, L. (2000). The documentary historical record of El Niño events in Peru: an update of the Quinn record (sixteenth through nineteenth centuries). In Diaz, H. and Markgraf, V. (eds.). El Niño and the Southern Oscillation: Variability, Global and Regional Impacts. Cambridge, Cambridge University Press.

Osborne, T. M., Lawrence, D. M., Slingo, J. M., Challinor, A. J. and Wheeler, T. R. (2004). Influence of vegetation on the local climate and hydrology in the tropics: sensitivity to soil parameters. Climate Dynamics, 23, 45–61.

Osorio, I. G. (2003). The international green agenda. Foundation Watch, November, pp. 1–7.

Overland, J. E., Spillane, M. C. and Soreide, N. N. (2004). Integrated analysis of physical and biological pan-Arctic change. Climatic Change 63, 291–322.

Overpeck, J. T. (1996). Varved sediment records of recent seasonal to millennial scale environmental variability. In Jones, P. D., Bradley, R. S. and Jouzel, J. (eds.). Climate Variations and Forcing Mechanisms of the Last 2000 Years. Berlin, Springer-Verlag.

Overpeck, J., Hughen, K., Hardy, D. et al. (1997). Arctic environmental change of the last four centuries. Science 278, 1251–7.

Overpeck, J., Rind, D., Lacis, A. and Healy, R. (1996). Possible role of dust-induced regional warming in abrupt climate change during the last glacial period. Nature 384, 447–9.

Overpeck, J., Whitlock, C. and Huntley, B. (2003). Terrestrial biosphere dynamics in the climate system: past and Future. In Alverson, K. D., Bradley, R. S. and Pedersen,

T. F. (eds.). Paleoclimate, Global Change and the Future. Berlin, Springer Verlag, pp. 81–103.

Ozanne, C. M. P., Anhuf, D., Boulter, S. L. *et al.* (2003). Biodiversity meets the atmosphere: a global view of forest canopies. *Science* **301**, 183–6.

Pal, J. S., Giorgi, F. and Bi, X. (2004). Consistency of recent European summer precipitation trends and extremes with future regional climate projections. *Geophysical Research Letters* **31**, L13202.

Palmer, T. N. and Raisanen, J. (2002). Quantifying the risk of extreme seasonal precipitation events in a changing climate. *Nature* **415**, 512–4.

Parmesan, C. and Yohe, G. (2003). A globally coherent fingerprint of climate change impacts across natural systems. *Nature* **421**, 37–42.

Parris, T. M. and Kates, R. W. (2003). Characterizing a sustainability transition. *Proceedings of the National Academy of Sciences* **100**, 8068–73.

Parry, M., Arnell, N., McMichael, A., *et al.* (2001). Millions at risk: defining critical climate change threats and targets. *Global Environmental Change* **11**, 181–3.

Paul, F. (2002). Combined technologies allow rapid analysis of glacier changes. *EOS* **83** (23), 253–61.

Pauling, A., Luterbacher, J. and Wanner, H. (2003). Evaluation of proxies for European and North Atlantic temperature field reconstructions. *Geophysical Research Letters* **30** (15), 2-1–2-4.

Paulsen, D. E., Li, H.-C. and Ku, T.-L. (2003). Climate variability in central China over the last 1270 years revealed by high-resolution stalagmite records. *Quaternary Science Reviews* **22**, 691–701.

Pearson, R. G., Dawson, T. P., Berry, P. M. and Harrison, P. A. (2002). SPECIES: a spatial evaluation of climate impacts on the envelope of species. *Ecological Modelling* **154**, 289–300.

Pedersen, T. F., François, R., François, L., Alverson, K. and McManus, J. (2003). The late Quaternary history of biogeochemical cycling of carbon. In Alverson, K., Bradley, R. S. and Pedersen, T. F. (eds). Paleoclimate, Global Change and the Future. Berlin, Springer Verlag, pp. 63–79.

Penner, J. E. (2003). Comment on 'Control of fossil-fuel particulate black carbon and organic matter, probably the most effective method of slowing global warming' by M. Z. Jacobson. *Journal of Geophysical Research* **108**, 14-1–14-5.

Penner, J. E., Dong, X. and Chen, Y. (2004). Observational evidence of a change in radiative forcing due to the indirect aerosol effect. *Nature* **427**, 231–4.

Penner, J. E., Zhang, S. Y. and Chuang, C. C. (2003). Soot and smoke aerosol may not warm climate. *Journal of Geophysical Research* **108** (D24), 4731.

Penuelas, J. and Boada, M. (2003). A global change-induced biome shift in the Montseny mountains (NE Spain). *Global Change Biology* **9**, 131–40.

Peterson, B. J., Holmes, R. M., McClelland, J. W. *et al.* (2002). Increasing river discharge to the Arctic ocean. *Science* **298**, 2171–3.

Peterson, G., Allen, C. R. and Holling, C. S. (1998). Ecological resilience, biodiversity and scale. *Ecosystems* **1**, 6–18.

Petit, J. R., Jouzel, J., Raynaud, D. *et al.* (1999). Climate and atmospheric history of the past 420 000 years from the Vostok ice core, Antarctica. *Nature* **399**, 429–36.

Petit, R. J., Aguinagalde, I., de Beaulieu, J.-L. *et al.* (2003). Glacial refugia: hotspots but not melting pots of genetic diversity. *Science* **300**, 1563–5.

Petit-Maire, N. (1999). Variabilité naturelle des environnements terrestres: les deux extrêmes climatiques (1800 ± 2000 and 8000 ± 1000 yrs BP). *Earth and Planetary Sciences* **328**, 273–9.

Petschel-Held, G. (2001). Actors and their environment – syndromes of land-use change in developing countries. *International Geosphere Biosphere Programme. Global Change NewsLetter* **48**, 27.

Pfister, C. (1992). Monthly temperature and precipitation in central Europe 1525–1979: quantifying documentary evidence on weather and its effects. In: Bradley, R. S. and Jones, P. D. (eds.). Climate Since AD 1500. London, Routledge.

Pfister, C.,Brazdil, R. and Barriendos, M. (2002). Reconstructing past climate and natural disasters in europe using documentary evidence. *PAGES News* **10** (3), 6–8.

Pfister, C. and Wanner, H. (eds.). (2002). *PAGES News* **10** (3), 2, and 8–26.

Philipona, R., Dürr, B., Marty, C., Ohmura, A. and Wild, M. (2004). Radiative forcing – measured at Earth's surface – corroborates the increasing greenhouse effect. *Geophysical Research Letters* **31**, L03202.

Pielke, R. A. (2002). Overlooked issues in the US national climate and IPCC assessments – an Editorial essay. *Climatic Change* **52**, 1–11.

Pielke, R. A. and Chase, T. N. (2004). Comment on 'Contributions of anthropogenic and natural forcing to recent tropopause height changes'. *Science* **303**, 1771.

Pilkey, O. H. and Cooper, J. A. G. (2004). Society and sea-level rise. *Science* **303**, 1781–2.

Pimm, S. L., Russell, G. J., Gittleman, J. L. and Brooks, T. M. (1995). The future of biodiversity. *Science* **269**, 347–50.

Pinot, S., Ramstein, G., Harrison, S. P. *et al.* (1999). Tropical Paleoclimates of the last glacial maximum: comparison of Paleoclimate Modelling Intercomparison project (PMIP): simulations and palaeodata. *Climate Dynamics* **15**, 857–74.

Pittock, A. B. (2002). What we know and don't know about climate change: reflections on the IPCC TAR. *Climatic Change* **53**, 393–411.

Plattner, G.-K., Joos, F. and Stocker, T. F. (2002). Revision of the global carbon budget due to changing air-sea oxygen fluxes. *Global Biogeochemical Cycles* **16**, 43-1–43-8.

Podgorny, I. A., Li, F. and Ramanathan, V. (2003). Large aerosol radiative forcing due to the 1997 Indonesian forest fire. *Geophysical Research Letters* **30**, 28-1–27-4.

Pollack, H. N. and Smerdon, J. E. (2004). Borehole climate reconstructions; spatial structure and hemispheric averages. *Journal of Geophysical Research* **109**, D11106.

Popper, K. R. (1963). Conjectures and refutations. London, Routledge & Keegan Paul.

Prentice, I. C., Farquhar, G. D., Fasham, M. J. R. *et al.* (2001). The carbon cycle and atmospheric CO_2. In Houghton, J., *et al.* (eds). Climate Change 2001: the Scientific Basis. Contribution of Working Group I to the IPCC Third Assessment Report. Cambridge, Cambridge University Press, pp. 183–237.

Prentice, I. C., Jolly, D. and BIOME 6000 members (2000). Mid-Holocene and glacial maximum vegetation geography of the northern continents and Africa. *Journal of Biogeography* **27**, 507–19.

Prentice, I. C., Sykes, M. T., Lautenschlager, M. *et al.* (1993). Modelling the global vegetation patterns and terrestrial carbon storage at the last glacial maximum. *Global Ecology and Biogeography Letters* **3**, 67–76.

Pretty, J. N., Morrison, J. L. L. and Hine, R. E. (2002). Reducing food poverty by increasing agricultural sustainability in developing countries. *Agriculture, Ecosystems and Environment* **88**, 1–18.

Procopio, A. S., Artaxo, P., Kaufman, Y. J., Remer, L. A., Schafer, J. S. and Holben, B. N. (2004). Multiyear analysis of Amazonian biomass burning smoke radiative forcing of climate. *Geophysical Research Letters* **31**, L03108.

Purvis, A. and Hector, A. (2000). Getting the measure of biodiversity. *Nature* **405**, 212–9.

Ramrath, A., Sadori, L. and Negendank, J. F. W. (2000). Sediments from Lago di Mezzano, central Italy: a record of late glacial/Holocene climatic variations and anthropogenic impact. *The Holocene* **10**, 87–95.

Rahmstorf, S. (2002). Ocean circulation and climate during the past 120 000 years. *Nature* **419**, 207–14.

Rahmstorf, S. (2003). Timing of abrupt climate change: a precise clock. *Geophysical Research Letters* **30** (10), 17-1–17-4.

Rahmstorf, S. and Alley, R. B. (2002). Stochastic resonance in glacial climates. *EOS* **83**, 129–135.

Rahmstorf, S. and Ganapolski, A. (1999). Long term global warming scenarios computed with an efficient coupled climate model. *Climatic Change* **43**, 353–67.

Raisanen, J., Hansson, U., Ullerstig, A. *et al.* (2003). European climate in the late twenty-first century: regional simulations with two driving global models and two forcing scenarios. *Climate Dynamics* **21**, 13–31.

Ramanathan, V., Crutzen, P. J., Lelieveld, J. *et al.* (2001). The Indian Ocean Experiment: an integrated assessment of climate forcing and effects of the great Indo-Asian haze. *Journal of Geophysical Research* **106**, 28371–98.

Ramankutty, N. and Foley, J. A. (1999). Estimating historical changes in global land cover: croplands from 1700 to 1992. *Global Biogeochemical Cycles* **13**, 997–1027.

Raymo, M. (1992). Global climate change: a three million year perspective. In Kukla, G. J. and Went, E. (eds.). Start of a Glacial. Berlin, Springer Verlag.

Raymond, C. F. (2002). Ice sheets on the move. *Science* **298**, 2147–8.

Raynaud, D., Blunier, T., Ono, Y., and Delmas, R. J. (2003). The late Quaternary history of atmospheric trace gases and aerosols: interactions between climate and biogeochemical cycles. In Alverson, K. D., Bradley, R. S. and Pedersen, T. F. (eds.). Paleoclimate, Global Change and the Future. Berlin, Springer-Verlag, pp. 13–31.

Reddy, M. S. and Venkataram, C. (2002). Inventory of aerosol and sulphur dioxide emissions from India, Parts I and II. *Atmospheric Environment* **36**, 677–712.

Redman, C. L. (1999) Human Impact on Ancient Environments. Tuscon AZ, The University of Arizona Press.

Reeburgh, W. S. (1997). Figures summarizing the global cycles of biogeochemically important elements. *Bulletin of the Ecological Society of America* **78**, 260–7.

Reichenau, T. G. and Esser, G. (2003). Is interannual fluctuation of atmospheric CO_2 dominated by combined effects of ENSO and volcanic aerosols? *Global Biogeochemical Cycles* **17** (4), 1094.

Reichert, B. K., Schnurr, R. and Bengtsson, L. (2002). Global ocean warming tied to anthropogenic forcing. *Geophysical Research Letters*, **29**, 20-1–20-4.

Ren, G, and Zhang, L. (1998). A preliminary mapped summary of Holocene pollen data for northeast China. *Quaternary Science Reviews* **17**, 669– 88.

Renberg, I. (1990). A 126 000 year perspective of the acidification of Lille Oresjon, southwest Sweden. *Philosophical Transactions of the Royal Society of London Series B* **327**, 357–61.

Renssen, H., Goosse, H. and Fichefet, T. (2003). On the non-linear response of the ocean thermohaline circulation to global deforestation. *Geophysical Research Letters* **30** (2), 10.

Revenga, C., Brunner, J., Henninger, N., Kassem, K. and Payne, R. (2000). Pilot Analysis of Global Ecosystems: Freshwater Systems. Washington DC, World Resources Institute.

Rex, M., Salawitch, von der Gathen, P. *et al.* (2004). Arctic ozone loss and climate change. *Geophysical Research Letters* **31**, L04416.

Ribbe, J. (2004). Oceanography: the southern supplier. *Nature* **427**, 23–4.

Ridgwell, A. J. (2003). Implication of the glacial CO_2 'iron hypothesis' for Quaternary climate change. *Geochemistry, Geophysics, Geosystems – Research Letters* **4** (9), 1–10.

Ridgwell, A. J. and Watson, A. J. (2002). Feedback between Aeolian dust, climate, and atmospheric CO_2 in glacial time. *Paleoceanography* **17** (4). 11-1–11-7.

Rignot, E. and Thomas, R. H. (2002). Mass balance of polar ice sheets. *Science* **297**, 1502–6.

Riley, J. (2001a). Indicator quality for assessment of impact of multidisciplinary systems. *Agriculture, Ecosystems and Environment* **8**, 121–8.

  (2001b). Multidisciplinary indicators of impact and change: key issues for identification and summary. *Agriculture, Ecosystems and Environment* **8**, 245–59.

Rind, D. (2000). Relating paleoclimate data and past temperature gradients: some suggestive rules. *Quaternary Science Reviews* **19**, 381–90.

Rind, D. (2003). The Sun's role in climate variations. *Science* **296**, 673–7.

Rind, D., Lean, J. and Healy, R. (1999). Simulated time-dependent climate response to solar radiative forcing since 1600. *Journal of Geophysical Research-Atmospheres* **104**, 1973–90.

Roberts, H. M., Wintle, A. G., Maher, B. A. and Hu, M. (2001). Holocene sediment-accumulation rates in the western Loess Plateau, China, and a 2500-year record of agricultural activity, revealed by OSL dating. *The Holocene* **11**, 477–83.

Roderick, M. L. and Farquhar, G. D. (2004). The pan Evaporation paradox. In Steffen W., Sanderson, A., Tyson, P. D. *et al.* (eds.). Global Change and the Earth System; a Planet Under Pressure. Berlin, Springer Verlag, p. 167.

Roderick, M. L., Farquhar, G. D., Berry, S. L. and Noble, I. R. (2001). On the direct effect of clouds and atmospheric particles on the productivity and structure of vegetation. *Oecologia* **128**, 21–30.

Rodhe, H., Dentner, F. and Schulz, M. (2002). The global distribution of acidifying wet deposition. *Environmental Science and Technology* **36**, 4382–8.

Romanovsky, V., Burgess, M., Smith, S., Yoshikawa, K. and Brown, J. (2002). Permafrost temperature records: indicators of climate change. *EOS* **83** (50), 589–94.

Root, T. L., Price, J. T., Hall, K. R. *et al.* (2003). Fingerprints of global warming on wild animals and plants. *Nature* **421**, 57–60.

Rosen, A. M. (1995). The social response to environmental change in early Bronze age Canaan. *Journal of Anthropological Archaeology* **14**, 26–44.

Rosen, A. M. and Rosen, S. A. (2001). Determinist or not determinist? Climate, environment and archaeological explanation in the Levant. In Wolff, S. (ed.) Studies in the Archaeology of Israel and Neighbouring Lands. Chicago, University of Chicago Press.

Rothlisberger, R., Mulvaney, R., Wolff, E. W. *et al.* (2002). Dust and sea salt variability in central east Antarctica (Dome C) over the last 45kyrs and its implications for southern high latitude climate. *Geophysical Research Letters* **29** (20), 24-1–24-4. (Correction published in 2003, **30** (5), 10.).

Rowe, D., Guilderson, T. P., Dunbar, R. B. *et al.* (2003). Late Quaternary lake-level changes constrained by radiocarbon and stable isotope studies on sediment cores from Lake Titicaca, South America. *Global and Planetary Change* **38**, 273–90.

Ruddiman, W. F. (2003a). Orbital insolation, ice volume and greenhouse gases. *Quaternary Science Reviews* **22**, 1597–1629.

(2003b). The anthropogenic greenhouse era began thousands of years ago. *Climatic Change* **61**, 261–93.

(2004). The role of greenhouse gases in orbital scale climate changes. *EOS* **85** (1). 1–7.

Ruddiman, W. F. and Raymo, M. E. (2003). A methane-based timescale for Vostok ice. *Quaternary Science Reviews* **22**, 141–55.

Ruddiman, W. F. and Thompson, J. S. (2001). The case for human causes of increased atmospheric CH_4 over the last 5000 years. *Quaternary Science Reviews* **20**, 1769–77.

Rundgren, M. and Beerling, D. (1999). A Holocene CO_2 record from the stomatal index of subfossil *Salix herbacea L.* leaves from northern Sweden. *The Holocene* **9**, 509–13.

Rundgren, M. and Björk, S. (2003). Late-glacial and early Holocene variations in atmospheric CO_2 concentration indicated by high-resolution stomatal index data. *Earth and Planetary Science Letters* **213**, 191–204.

Rutherford, S. and Mann, M. E. (2004). Correction to 'Optimal surface temperature reconstructions using terrestrial borehole data'. *Journal of Geophysical Research* **109**, D11107.

Saarinen, T. (1999). Paleomagnetic dating of late Holocene sediments in Fennoscandia. *Quaternary Science Reviews* **18**, 889–97.

Saaroni, H., Ziv, B., Edelson, J. and Alpert, P. (2003). Long-term variations in summer temperatures over the eastern Mediterranean. *Geophysical Research Letters* **30** (18), 1946.

Sabine, C. L., Felly, R. A., Gruber, N. *et al.* (2004). The ocean sink for anthropogenic CO_2. *Science* **305**, 367–71.

Sahagian, D. (2000). Global physical effects of anthropogenic hydrological alterations: sea-level and water redistribution. *Global and Planetary Change* **25**, 39–48.

Saloranta, T. M. (2001). Post-normal science and the global climate issue. *Climate Change* **50**, 395–404.

Saltzman, B. (1985). Paleoclimatic modeling. In Hecht, A. D. (ed.). Paleoclimate Analysis and Modeling. Wiley, Chichester. pp. 341–96.

Sanchez, P. A. (2000). Linking climate change research with food security and poverty reduction in the tropics. *Agriculture, Ecosystems and Environment* **82**, 371–83.

Sanderson, A. (2004). The Gulf of Mexico dead zone. In Steffen, W., Sanderson, A., Tyson, P. *et al.* (eds.). Global Change and the Earth System: a Planet Under Pressure. Berlin, Springer Verlag, p. 184.

Santer, B. D., Berger, A., Eddy *et al.* (1993). How can palaeodata be used to evaluate forcing mechanisms responsible for past climate changes? In Eddy, J. A. and Oeschger, H., (eds.). Global Changes in the Perspective of the Past. Chichester, Wiley, pp. 343–67.

Santer, B. D., Wehner, M. F., Wigley, T. M. L. *et al.* (2003). Contributions of anthropogenic and natural forcing to recent tropopause height changes. *Science* **301**, 479–83.

Santer, B. D., Wehner, M. F., Wigley, T. M. L. *et al.* (2004). Response to Comment on 'Contributions of anthropogenic and natural forcing to recent tropopause height changes.' *Science* **303**, 1771.

Sarmiento, J. L. and Gruber, N. (2002). Sinks for anthropogenic carbon. *Physics Today* **56** (5), 30–6.

Sarmiento, J. L., Gruber, N., Brezinski, M. A. and Dunne, J. P. (2004). High-latitude controls of thermocline nutrients and low latitude biological productivity. *Nature* **427**, 56–60.

Sarnthein, M., Kennett, J. P., Allen, J. R. M. *et al.* (2002). Decadal-to-millennial-scale climate variability – chronology and mechanisms: summary and recommendations. *Quaternary Science Reviews* **21**, 1121–8.

Saunders, M. A. and Quian, B. (2002). Seasonal predictability of the winter NAO from North Atlantic sea surface temperatures. *Geophysical Research Letters* **29** (22), 2049.

Schaeffer, M., Selten, F. M. and Opsteegh, J. D. (2002). Intrinsic limits to predictability of abrupt regional climate change in IPCC SRES scenarios. *Geophysical Research Letters* **29** (16), 14-1–14-4.

Schafer, J. S., Eck, T. F., Holben, B. N. *et al.* (2002). Observed reductions of total solar irradiance by biomass-burning aerosols in the Brazilian Amazon and Zambian Savanna. *Geophysical Research Letters* **29**, 4-1–27-8.

Schär, C., Vidale, P. L., Luthi, D. *et al.* (2004). The role of increasing temperature variability in European summer heatwaves. *Nature* **427**, 332–6.

Schellnhuber, H. J. (1999). 'Earth system' analysis and the second Copernican revolution. *Nature* **402**, C19–23.

Schellnhuber, H. J. (2002). Coping with Earth system complexity and irregularity. In Steffen, W., Jäger, J., Carson, D. and Bradshaw, C. (eds.). Challenges of a Changing Earth; Proceedings of the Global Change Open Science Conference. IGBP Global Change Series. Berlin, Springer Verlag, pp. 151–6.

Schiermeier, Q. (2003). Alpine thaw breaks ice over permafrost's role. *Nature* **424**, 712–3.
(2004a). Modellers deplore 'short-termism' on climate. *Nature* **428**, 593.
(2004b). Global warming anomaly may succomb to microwave study. *Nature* **429**, 7.
(2004c). A rising tide. *Nature* **428**, 114.

Schilman, B., Bar-Matthews, M., Almogi-Labin, A. and Luz, B. (2001). Global climate instability reflected by eastern Mediterranean marine records during the late Holocene. *Palaeogeography, Palaeoclimatology, Palaeoecology* **176**, 157–76.

Schimel, D. and Baker, D. (2002). Carbon cycle: the wildfire factor. *Nature* **420**, 29–30.

Schimel, D. S., House, J. I., Hubbarde, K. A. *et al.* (2001). Recent patterns and mechanisms of carbon exchange by terrestrial ecosystems. *Nature* **414**, 169–72.

Schimmelmann, A., Lange, C. B. and Meggers, B. J. (2003). Palaeoclimatic and archaeological evidence for a ~200 recurrence of floods and droughts linking California, Mesoamerica and South America over the last 2000 years. *The Holocene* **13**, 763–78.

Schindler, D. W. (2001). The cumulative effects of climate warming and other human stresses on Canadian freshwaters in the new millennium. *Canadian Journal of Fisheries and Aquatic Sciences* **59**, 18–29.

Schindler, D. W. and Curtis, P. J. (1997). The role of DOC in protecting freshwater subjected to climatic warming and acidification from UV exposure. *Biogeochemistry* **36**, 1–8.

Schmittner, A., Appenzeller, C. and Stocker, T. F. (2000). Enhanced Atlantic freshwater exported during El Niño. *Geophysical Research Letters* **27**, 1163–6.

Schmittner, A., Saenko, O. A. and Weaver, A. J. (2003). Coupling of the hemispheres in observations and simulations of glacial climate change. *Quaternary Science Reviews* **22**, 659–671.

Schmittner, A., Yoshimori, M. and Weaver, A. J. (2002). Instability of glacial climate in a model of the ocean-atmosphere-cryosphere system. *Science* **295**, 145–149.

Schmutz, C., Luterbach, J., Gyalistras, D., Xopalski, E. and Wanner, H. (2000). Can we trust proxy-based NAO reconstructions? *Geophysical Research Letters* **27**, 1135–8.

Schneider, S. H. (2001). Earth systems engineering and management. *Nature* **409**, 417–21.
    (2002). Can we estimate the likelihood of climatic changes at 2100? *Climatic Change* **52**, 441–51.

Scholes, M. C., Matrai, P. A., Andreae, M. O., Smith, K. A. and Manning, M. R. (2003). Biosphere–Atmosphere interactions. In Brasseur, G. P., Prinn, R. G. and Pszenny, A. P. (eds.). Atmospheric Chemistry in a Changing World. Berlin, Springer Verlag, pp. 19–71.

Scholes, R. (2002) The past, present and future of carbon on land. In Steffen, W., Jäger, J., Carson, D. and Bradshaw, C. (eds.). Challenges of a Changing Earth; Proceedings of the Global Change Open Science Conference. IGBP Global Change Series. Berlin, Springer Verlag, pp. 81–5.

Schubert, S. D., Suarez, M. J., Pegion, P. J., Koster, R. D. and Bacmeister, J. T. (2004). On the cause of the 1930's dust bowl. *Science* **303**, 1855–9.

Schulze, E. D. and Mooney, H. A. (1993). Biodiversity and Ecosystem Function. Berlin, Springer Verlag.

Schwander, J., Eicher, U. and Ammann, B. (2000). Oxygen isotopes of lake marl at Gerzensee and Leysin (Switzerland), covering the Younger Dryas and two minor oscillations and their correlation to the GRIP ice core. *Palaeogeography, Palaeoclimatology, Palaeoecology* **159**, 213–14.

Schwartz, P. and Randall, D. (2004). Abrupt Climate Change. Report prepared by Global Business Network (GBN) for the Department of Defense. At www.gbn.org/ ArticleDisplayServlet.srv?aid=26231.

Seki, O., Ishiwatari, R. and Matsumoto, K. (2002). Millennial scale oscillations in NE Pacific surface waters over the last 82kyr: new evidence from alkenones. *Geophysical Research Letters* **29** (23), 2144.

Serreze, M. C., Walsh, J. E., Chapin, F. S. III *et al*. (2000). Observational evidence of recent change in the northern high-latitude environment. *Climatic Change* **46**, 159–207.

Severinghaus, J. P. and Brook, E. J. (1999). Abrupt climate change at the end of the last glacial period inferred from trapped air in polar ice. *Science* **286**, 930–4.

Severinghaus, J. P., Jouzel, J., Caillon, N. *et al*. (2004). Comment on 'Greenland–Antarctica phase relations and millennial time-scale climate fluctuations in the Greenland ice-cores' by C. Wunsch. *Quaternary Science Reviews*, **23**, 2053–4.

Severinghaus, J. P., Sowers, T., Brook, E. J., Alley, R. B. and Bender, M. L. (1998). Timing of abrupt climate change at the end of the Younger Dryas interval from thermally fractionated gases in polar ice. *Nature* **391**, 141–6.

Shackleton, N. J., Fairbanks, R. G., Chiu, T. and Parrenin, F. (2004). Absolute calibration of the Greenland timescale: implications for Antarctic timescales and for δ ^{14}C. *Quaternary Science Reviews*, **23**, 1513–22.

Shackleton, N. J., Hall, M. A. and Vincent, E. (2000). Phase relationships between millenial-scale events 64 000–24 000 years ago. *Paleoceanography* **15**, 565–9.

Shackleton, N. J. and Opdyke, N. D. (1973). Oxygen isotope and paleomagnetic stratigraphy of Pacific core V28–238: oxygen isotope temperatures and ice volumes on a 10^5 year and 10^6 year scale. *Quaternary Research* **3**, 39–55.

Shackleton, N. J., Sanchez-Goñi, Pailler, D. and Lancelot, Y. (2003). Marine isotope substage 5e and the Eemian interglacial. *Global and Planetary Change* **36**, 151–5.

Shah, M. (2002). Food in the twenty-first century: global climate of disparities. In Steffen, W., Jäger, J., Carson, D. and Bradshaw, C. (eds.). Challenges of a Changing Earth; Proceedings of the Global Change Open Science Conference. IGBP Global Change Series. Berlin, Springer Verlag, pp. 31–8.

Shemesh, A., Rosqvist, G., Rietti-Shati, M. *et al.* (2001). Holocene climate change in Swedish Lapland inferred from an oxygen isotope record of lacustrine biogenic silica. *The Holocene* **11**, 447–54.

Shennan, I. (1989). Holocene crustal movements and sea-level changes in Great Britain. *Journal of Quaternary Science* **4**, 77–89.

Shennan, I. and Horton, B. (2002). Holocene land- and sea-level changes in Great Britain, *Journal of Quaternary Science* **17**, 511–26.

Shennan, S. (2003). Holocene climate and human populations: an archaeological approach. In Mackay, A. W., Battarbee, R. W., Birks, H. J. B. and Oldfield, F. (eds.). Global Change in the Holocene. London, Arnold, pp. 36–48.

Shindell, D. T. (2003). Whither Arctic climate? *Science* **299**, 215–6.

Shindell, D. T., Schmidt, G. A., Mann, M. E. and Faluvegi, G. (2004). Dynamic winter climate response to large tropical volcanic eruptions since 1600. *Journal of Geophysical Research* **109**, D05104.

Shindell, D. T., Schmidt, G. A., Mann, M. E., Rind, D. and Waple, A. (2001). Solar forcing of regional climate change during the Maunder Minimum. *Science* **294**, 2149–52.

Shotyk, W., Cheburkin, A. K., Appleby, P. G., Frankhauser, A. and Kramers, J. D. (1996). Two thousand years of atmospheric arsenic, antimony, and lead deposition recorded in an ombrotrophic peat bog profile, Jura Mountains, Switzerland. *Earth Planetary Science Letters* **145**, E1–7.

Shotyk, W., Weiss, D., Appleby, P. G. *et al.* (1998). History of atmospheric lead deposition since 12 370 ^{14}C yr BP from a peat bog, Jura Mountains, Switzerland. *Science* **281**, 1635–40.

Showstack, R. (2003). Montreal protocol benefits cited. *EOS* **84** (39), 395.

Siddall, M., Rohling, E. J., Almogi-Labin, A. *et al.* (2003). Sea-level fluctuation during the last glacial cycle. *Nature* **423**, 853–8.

Sigman, D. M. and Boyle, E. A. (2000). Glacial/Interglacial variations in atmospheric carbon dioxide. *Nature* **407**, 859–69.

Singh, G., Wasson, R. J. and Agrawal, D. P. (1990). Vegetational and seasonal climatic changes since the last full glacial in the Thar Desert, northwestern India. *Review of Palaebotany and Palynology* **64**, 351–8.

Smith, L. C., Sheng, Y., Forster, R. R. *et al.* (2003). Melting of small Arctic ice caps observed from ERS scatterometer time series. *Geophysical Research Letters* **30** (20), 2034.

Smith, S. V., Renwick, W. H., Bartley, J. D. and Buddemeier, R. W. (2002). Distribution and significance of small artificial water bodies across the United States landscape. *The Science of the Total Environment* **299**, 21–36.

Smol, J. P., Cumming, B. F., Dixit, A. S. and Dixit, S. S. (1998). Tracking recovery in acidified lakes: a palaeolimnological perspective. *Restoration Ecology* **6**, 318–26.

Snowball, I. F. and Sandgren, P. (2002). Geomagnetic field variations in northern Sweden during the Holocene quantified from varved lake sediments and their implications for cosmogenic nuclide production rates. *The Holocene* **15**, 517–30.

Soden, B. J., Wetherald, R. T., Stenchikov, G. L. and Robock, A. (2002). Global cooling after the eruption of Mount Pinatubo: a test of climate feedback by water vapour. *Science*, **296**, 727–30.

Solomon, S. (2004). The hole truth. *Nature* **427**, 289–91.

Somoza, L., Gardner, J. M., Diaz-del-Rio, V. *et al*. (2002). Numerous methane gas-related sea floor structures identified in Gulf of Cadiz. *EOS* **83** (47), 541–9.

Soon, W. and Baliunas, S. (2003). Lessons and Limits of Climate History: Was the Twentieth Century Climate Unusual? Washington DC, The George C. Marshall Institute.

Soon, W., Baliunas, S., Idso, C., Idso, S. and Legates, D. R. (2003). Reconstructing climatic and environmental changes of the past 1000 years: a reappraisal. *Energy and Environment* **14**, 233– 96.

Solanki, S. K. and Krivova, N. A. (2002). Can solar variability explain global warming since 1970? *Journal of Geophysical Research* **108**, 7-1–7-8.

Sorvari, S., Korhola, A. and Thompson, R. (2002). Lake diatom responses to recent Arctic warming in Finnish Lapland. *Global Change Biology* **8**, 171–81.

Sowers, T. (2001). The N_2O record spanning the penultimate deglaciation from the Vostok ice core. *Journal of Geophysical Research-Atmospheres* **106**, 31903–14.

Sowers, T., Alley, R. and Jubenville, J. (2003). Ice core records of atmospheric N_2O covering the last 106 000 years. *Science* **301**, 945–8.

Spahni, R., Schwander, J., Flückinger, J., Stauffer, B., Chappellaz, J. and Raynaud, D. (2003). The attenuation of fast atmospheric CH_4 variations recorded in polar ice cores. *Geophysical Research Letters* **30** (11), 1571.

Stager, J. C., Mayewski, P. A. and Meeker, L. D. (2002). Cooling cycles, Heinrich Event 1 and the desiccation of Lake Victoria. *Palaeogeography, Palaeoclimatology, Palaeoecology* **183**, 169–78.

Stahle, D. W., Cook, E. R., Cleaveland, M. K. *et al*. (2000) Tree-ring data document sixteenth century megadrought over North America. *EOS* **81**, 121–5.

Staubwasser, M., Sirocko, F., Grootes, P. M. and Segl, M. (2003). Climate change at 4.2 ka BP termination of the Indus valley civilization and Holocene south Asian monsoon variability. *Geophysical Research Letters*, **30** (8), 1425.

Steffen, W. and Crutzen, P. J. (2003). How long have we been in the Anthropocene era? *Climatic Change* **61**, 251–7.

Steffen, W., Sanderson, A., Tyson, P. *et al*. (2004). Global Change and the Earth System: a Planet Under Pressure. Berlin, Springer Verlag.

Stenni, B., Masson-Delmotte,V Johnsen, S. *et al*. (2001) An oceanic cold reversal during the last deglaciation. *Science* **293**, 2074–7.

Stevens, C. J., Dise, N. B., Mountford, J. O. and Gowing, D. J. (2004). Impact of nitrogen deposition on the species richness of grasslands. *Science* **303**, 1876–7.

Stine, S. (1994). Extreme and persistent drought in California and Patagonia during medieval time. *Nature* **369**, 546–9.

Stive, M. J. F. (2003). How important is global waming for coastal erosion? *Climatic Change* **61**, 1–13.

Stocker, T. F. (1998). The seesaw effect. *Science* **282**, 61–2.

  (2003). Global change: south dials north. *Nature* **424**, 496–9.

Stocker, T. F. and Johnsen, S. J. (2003). A minimum thermodynamic model for the bipolar seesaw. *Paleoceanography* **18** (4), 11-1–11-9.

Stocker, T. F. and Schmittner, A. (1997). Influence of CO_2 emission rates on the stability of the thermohaline circulation. *Nature* **388**, 862–5.

Stone, J. O., Balco, G. A., Sugden, D. E. *et al.* (2003). Holocene deglaciation of Marie Byrd land, west Antarctica. *Science* **299**, 99–102.

Stone, R. S., Dutton, E. G., Harris, J. M. and Longenecker, D. (2002). Earlier spring snowmelt in northern Alaska as an indicator of climate change. *Journal of Geophysical Research D: Atmospheres* **107**, 10-1–10-15.

St-Onge, G., Stoner, J. S. and Hillaire-Marcel, C. (2003). Holocene paleomagnetic records from the St. Lawrence Estuary, eastern Canada: centennial- to millennial-scale geomagnetic modulation of cosmogenic isotopes. *Earth and Planetary Science Letters* **209**, 113–30.

Stott, P. A. and Kettleborough, J. A. (2002). Origins and estimates of uncertainty in predictions of twenty-first century temperature rise. *Nature* **416**, 723–6.

Stott, P. A., Allen, M. R. and Jones, G. S. (2003). Estimating signal amplitudes in optimal fingerprinting. Part II: application to general circulation models. *Climate Dynamics* **21**, 493–500.

Stott, L., Poulsen, C., Lund, S. and Thunell, R. (2002). Super ENSO and global climate oscillations at millennial timescales. *Science* **297**, 222–6.

Stott, P. A., Tett, S. F. B., Jones, G. S. *et al.* (2001). Attribution of twentieth century temperature change to natural and anthropogenic causes. *Climate Dynamics* **17**, 1–21.

Streets, D. G., Jiang, K., Hu, X. *et al.* (2001). Recent Reductions in China's greenhouse gas emissions. *Science* **284**, 1835–7.

Stuiver, M., Braziunas, T. F., Becker, B. and Cromer, B. (1991). Climatic, solar, oceanic and geomagnetic influences on late glacial and $^{14}C/^{12}C$ change. *Quaternary Research* **35**, 1–24.

Stuiver, M. and Reimer, P. J. (1993). Extended ^{14}C data base and revised Calib 3.0 ^{14}C age calibration program. *Radiocarbon* **35**, 215–30.

Stuiver, M., Reimer, P. J., Bard, E. *et al.* (1998). INTCAL98 radiocarbon age calibration, 24 000 – 0 cal BP. *Radiocarbon* **40**, 1041–83.

Stute, M., Forster, M., Frischkorn, H. *et al.* (1995). Cooling of tropical Brazil (5 °C) during the last glacial maximum. *Science* **269**, 379–83.

Stute, M. and Talma, S. (1998). Glacial temperatures and moisture transport regimes reconstructed from noble gas and $\delta^{18}O$, Stampriert aquifer, Namibia. In Isotope Techniques in the Study of Past and Current Environmental Changes in the Hydrosphere and the Atmosphere. Vienna, IAEA, Proceedings of Vienna Symposium, pp. 307–28.

Sugita, S., Gaillard, M.-J. and Broström, A. (1999). Landscape openness and pollen records: a simulation approach. *The Holocene* **9**, 409–21.

Sun, B. and Bradley, R. S. (2002). Solar influences on cosmic rays and cloud formation: a reassessment. *Journal of Geophysical Research* **107**, D14, 4211, AAC 5–1.

Sun, B. and Bradley, R. S. (2004). Reply to comment on 'Solar influences on cosmic rays and cloud formation: A reassessment'. *Journal of Geophysical Research* **109**, D14206.

Svensmark, H. and Friis-Christensen, E. (1996). Variation of cosmic ray flux and global cloud coverage – a missing link in solar–climate relationships. *Journal of Atmospheric and Solar-Terrestrial Physics* **59**, 1225–32.

Svensen, H., Planke, S., Malthe-Sorenssen, A. *et al.* (2004). Release of methane from a volcanic basin as a mechanism for initial Eocene global warming. *Natue* **429**, 542–5.

Swetnam, T. W. (1993). Fire history and climate change in Giant Sequoia groves. *Science* **262**, 885–8.

Swiss Biodiverity Forum (2003). Visions in Biodiversity Research. Towards a New, Integrative Biodiversity Science. Bern, Swiss Academy of Sciences.

Sykes, M., Prentice, I. C., Smith, B., Cramer, W. and Venevsky, S. (2001). An introduction to the European terrestrial ecosystem modelling activity. *Global Ecology and Biogeography* **10**, 581–94.

Szeicz, J. M., Haberle, S. G. and Bennett, K. D. (2003). Dynamics of north Patagonian rainforests from fine-resolution pollen, charcoal and tree-ring analysis, Chonos Archipelago, southern Chile. *Austral Ecology* **28**, 413–26.

Taberlet, P. and Cheddadi, R. (2002). Quaternary refugia and persistence of biodiversity. *Science* **297**, 2009–10.

Tada, R., Irino, T. and Koizumi, I. (1999) Land–ocean linkage over orbital and millennial timescales recorded in the late Quaternary sediments of the Japan Sea. *Paleoceanography* **14**, 236–47.

Talbot, M. (1990). A review of the palaeohydrological interpretation of carbon and oxygen isotopic ratios in primary lacustrine carbonates. *Chemical Geology (Isotope Geosciences Section)* **80**, 261–79.

Taylor, K. C., Hammer, C. U., Alley, R. B. *et al.* (1993). Electrical conductivity measurements from the GISP2 and GRIP Greenland ice cores. *Nature* **366**, 549–52.

Taylor, K. C., White, J. W. C., Severinghaus, J. P. *et al.* (2003). Abrupt climate change around 22ka on the Siple coast of Antarctica. *Quaternary Science Reviews* **23**, 7–15.

Teller, J. T., Leverington, D. W. and Mann, J. W. (2002). Freshwater outbursts to the oceans from glacial Lake Agassiz and their role in climate change during the last deglaciation. *Quaternary Science Reviews* **21**, 879–87.

TEMPO Members (1996). Potential role of vegetation feedback in the climate sensitivity of high-latitude regions: a case study at 6000 years BP *Global Biogeochemical Cycles* **10**, 727–37.

Tett. S. F. B., Jones, G. S., Stott, P. A. *et al.* (2002). Estimation of natural and anthropogenic contributions to twentieth century temperature change. *Journal of Geophysical Research* **107**, ACL 10-1–10-24.

Tett, S. F. B., Stott, P. A., Allen, M. R., Ingram, W. J. and Mitchell, J. F. B. (1999). Causes of twentieth-century temperature change near the Earth's surface. *Nature* **399**, 569–72.

Thomas, C. D., Cameron, A., Green, R. E. *et al.* (2004). Extinction risk from climate change. *Nature* **427**, 145–8.

Thomas, J. A., Telfer, M. G., Roy, D. B. *et al.* (2004). Comparative losses of British butterflies, birds and plants and the global extinction crisis. *Science* **303**, 1879–81.

Thompson, L. G. (1995). Late glacial stage and Holocene tropical Ice core records from Huascaran, Peru. *Science* **269**, 46–50.

Thompson, L. G., Mosley-Thompson, E., Davis, M. E., Henderson, K. A. and Lin, P. N. (2000). The tropical ice core record of ENSO. In Diaz, H. F. and Markgraf, V. (eds.). El Niño and the Southern Oscillation: Multiscale Variability and Global and Regional Impacts. Cambridge, Cambridge University Press, pp. 325–56.

Thompson, L. G., Mosley-Thompson, E., Davis, M. E *et al*. (2002). Kilimanjaro ice core record: evidence of Holocene climate change in tropical Africa. *Science* **298**, 598–2.

Thorne, P. W., Jones, P. D., Tett, S. F. B. *et al*. (2003). Probable causes of late twentieth century tropospheric temperature trends. *Climate Dynamics* **21**, 573–91.

Thuiller, W., Araujo, M. B., Pearson, R. G. *et al*. (2004). Uncertainty in predictions of extinction risk. *Nature* **430**, 34.

Tilman, D. (2000). Causes, consequences and ethics of biodiversity. *Nature* **405**, 208–11.

Timmerman, A., Oberhuber, J., Bacher, A. *et al*. (1999). Increased El Niño frequency in a climate model forced by future greenhouse warming. *Nature* **398**, 694–7.

Trenberth, K. E. (2000). Conceptual framework for changes of rainfall and extremes of the hydrological cycle with climate change. *PAGES Newsletter* **8** (1), 13; *CLIVAR Exchanges* **5** (1), 12–3.

Trenberth, K. E. (2004). Rural land-use change and climate. *Nature* **427**, 213.

Trenberth, K. E., Overpeck, J. T. and Solomon, S. (2004). Exploring drought and its implications for the future. *EOS* **85** (3) 27.

Tricot, C. and Berger, A. (1988). Sensitivity of present-day climate to astronomical forcing. In Wanner, H. and Siegenthaler, U. (eds.). Long and Short-Term Variability of Climate. New York, Springer Verlag.

Tsuda, A., Takeda, S., Saito, H. *et al*. (2003). A mesoscale iron enrichment in the western subarctic Pacific induces a large centric diatom bloom. *Science* **300**, 958–61.

Tudhope, A.,W., Chilcott, C. P., McCulloch, M. T. *et al*. (2001) Variability in the El Niño-southern oscillation through a glacial–interglacial cycle. *Science* **291**, 1511–17.

Turner, B. L. II, Kasperson, R. E., Matson, P. A. *et al*. (2003). A framework of vulnerability analysis in sustainability science. *Proceedings of the National Academy of Science* **100**, 8074–9.

Turner, B. L. II, Kasperson, R. E., Meyer, W. B. *et al*. (1990). Two types of global environmental change. *Global Environmental Change* **4**, 15–22.

Tyson, P. D., Lee-Thorp, J., Holmgren, K. and Thackeray, J. F. (2002). Changing gradients of climate change in Southern Africa during the past millennium: implications for population movements. *Climatic Change* **52**, 129–35.

Tzedakis, C. (2003). Timing and duration of last interglacial conditions in Europe: a chronicle of changing chronology. *Quaternary Science Reviews* **22**, 763–8.

UKCIP (2002). Climate Change Scenarios for the United Kingdom. The UKCIP02 Briefing Report. At: www.ukcip.org.uk/ukcip.html.

Urban, F. E., Cole, J. E. and Overpeck, J. T. (2000). Influence of mean climate change on climate variability from a 155 year tropical Pacific coral record. *Nature* **407**, 989–93.

Valdes, P. J. (2003). Holocene climate modelling. In McKay, A., Battarbee, R. W., Birks, H. J. B. and Oldfield, F. (eds.). Global Change in the Holocene. London, Arnold, pp. 20–35.

Van Asselt, M. B.A and Rotmans, J. (2002). Uncertainty in integrated assessment modelling: from positivism to pluralism. *Climatic Change* **54**, 75–105.

van de Plassche, O., (2000). North Atlantic climate–ocean variations and sea-level in Long Island Sound, Connecticut, since 500 cal. yr. AD *Quaternary Research* **53**, 89–97.

van de Plassche, O., van der Borge, K. and de Jong, A. F. M. (1998). Sea level-climate correlation during the past 1400 yr. *Geology* **26**, 319–22.

van de Plassche, O., van der Schrier, G., Weber, S. L., Gehrels, W. R. and Wright, A. J. (2003). Sea-level variability in the northwest Atlantic during the past 1500 years: a delayed response to solar forcing. *Geophysical Research Letters* **30** (18). 1-1–1-4.

van Geel, B., Buurman, J. and Waterbolk, H. T. (1996). Archaeological and palaeoecological indications of an abrupt climate change in the Netherlands and evidence for climatological teleconnections around 2650 BP. *Journal of Quaternary Science* **11**, 451–60.

van Geel, B., van der Plicht J. and Renssen, J. (2002). Major $\Delta^{14}C$ excursions during the late glacial and early Holocene: changes in ocean ventilation or solar forcing of climate change? *Quaternary International* **105**, 71–76.

Vecsei, A. and Berger, W. H. (2004). Increase of atmospheric CO_2 during deglaciation: constraints on the coral reef hypothesis from patterns of deposition. *Global Biogeochemical Cycles* **18**, GB1035.

Vellinga, M. and Wood, R. A. (2002). Global climatic impacts of a collapse of the Atlantic Thermohaline circulation. *Climatic Change* **54**, 251–67.

Vellinga, M. and Wood, R. A. (2004). Timely detection of anthropogenic change in the Atlantic meridional overturning circulation. *Geophysical Research Letters* **31**, L14203.

Verburg, P., Hecky, R. E. and Kling, H. (2003). Ecological consequences of a century of warming in Lake Tanganyika. *Science* **301**, 505–7.

Verschuren, D. (2003). Global change: the heat on Lake Tanganyika. *Nature* **424**, 731–2.

Verschuren, D., Laird., K. R. and Cumming, B. F. (2000) Rainfall and drought in equatorial east Africa during the past 1100 years. *Nature* **403**, 410–14.

Vinnikov, K. and Grody, N. C. (2003). Global warming trend of mean tropospheric temperature observed by satellites. *Science* **302**, 269–272.

Vinther, B. M., Andersen, K. K. and Hansen, A. W. (2003). Improving the Gribraltar/Reykjavik NAO index. *Geophysical Research Letters* **30** (23), 8-1–8-4.

Vitousek, P. M., Mooney, H. A., Lubchenco, J. and Melillo, J. (1997). Human domination of the Earth's ecosystems. *Science* **277**, 494–9.

Voelker, A. H. L. and workshop participants. (2002). Global distribution of centennial-scale records for marine isotope stage (MIS) 3: a database. *Quaternary Science Reviews* **21**, 1185–1214.

Voldoire, A. and Royer, J. F. (2004). Tropical deforestation and climate variability. *Climate Dynamics* **22**, 857–74.

Von Grafenstein, U., Eicher, U., Erlenkauser, Ruch, P., Schwander, J. and Ammann, B. (2000). Isotope signature of the Younger Dryas and two minor oscillations at Gerzensee (Switzerland): palaeoclimatic and palaeolimnological interpretation based on bulk and biogenic carbonates. *Palaeogeography, Palaeoclimatology, Palaeoecology* **159**, 215–29.

Von Grafenstein, U., Erlenkauser, H., Brauer, A., Jouzel, J. and Johnson, S. (1999). A mid-European decadal isotope-climate record from 15 500 to 5000 years BP. *Science* **284**, 1654–7.

Von Grafenstein, U., Erlenkeuser, H.,Muller, J., Jouzel, J. and Johnsen, S. (1998). The cold event 8200 years ago documented in oxygen isotope records of precipitation in Europe and Greenland. *Climate Dynamics* **14**, 73–81.

Von Storch, H. and Stehr, N. (2000). Climate change in perspective. *Nature* **405**, 615.

Vörösmarty, C. J. and Sahagian, D. (2000). Anthropogenic disturbance of the terrestrial water cycle. *BioScience* **50**, 753–65.

Vörösmarty, C. J., Green, P., Salisbury, J. and Lammers, R. B. (2000). Global water resources: vulnerability from climate change and population growth. *Science* **289**, 284–8.

Vörösmarty, C. J., Meybeck, M., Fekete, B. *et al.* (2003). Anthropogenic sediment retention: major global impact from registered river impoundments. *Global and Planetary Change*, **39**, 169–90.

Vörösmarty, C. J., Sharma, K., Fekete, B. *et al.* (1997). The storage and aging of continental run-off in large reservoir systems. *Ambio* **26**, 210–9.

Vose, R. S., Karl, T. R., Easterling, D. R., Williams, C. N. and Menne, M. J. (2004). Impact of land-use change on climate. *Nature* **427**, 213–4.

Waelbroeck. C., Duplessy, J.-C., Michel, E., Labeyrie, L., Paillard, D. and Duprat, J. (2001). The timing of the last deglaciation in North Atlantic climate records. *Nature* **412**, 724–7.

Wagner, G., Beer, J., Laj, C. *et al.*(2000a) Chlorine-36 evidence for the Mono Lake event in the Summit GRIP ice core. *Earth and Planetary Science Letters* **181**, 1–6.

Wagner, G., Masarik, J., Beer, J. *et al.* (2000b). Reconstruction of the geomagnetic field between 20 and 60 kyr BP from cosmogenic radionuclides in the GRIP ice core. *Nuclear Instruments and Methods in Physics Research B* **172**, 597–604.

Waldhardt, R. (2003). Biodiversity and landscape – summary, conclusions and perspectives. *Agriculture, Ecosystems and Environment* **98**, 305–9.

Walker, K. J., Pywell, R. F., Warman, E. A., Fowbert, J. A., Bhogal, A. and Chambers, B. J. (2003). The importance of former land use in defining successful re-creation of lowland heath in southern England. *Biological Conservation*, **116**, 289–303.

Walliser, D. E., Jin, K., Kang, I. -S. *et al.* (2003). AGCM simulations of intraseasonal variability associated with the Asian summer monsoon. *Climate Dynamics* **21**, 391–404.

Walsh, K. J. E., Nguyen, K. -C. and McGregor, J. L. (2003). Fine-resolution regional climate model simulations of the impact of climate change on tropical cyclones near Australia. *Climate Dynamics* **21**, 47–56.

Walther, G. -R., Post, E., Convey, P. *et al.* (2002). Ecological responses to recent climate changes. *Nature* **416**, 389–95.

Wang, C. (2004). A modelling study on the climate impacts of black carbon aerosols. *Journal of Geophysical Research* **109** (D3), 10.

Wang, N., Thompson, L. G., Davis, M. E., Mosley-Thompson, E., Yao. T. and Pu, J. (2003). Influence of variations in NAO and SO on air temperature over the northern Tibetan Plateau as recorded by $\delta^{18}O$ in the Malan ice core. *Geophysical Research Letters* **30** (22),.

Wang, X. and Key J. R. (2003). Recent trends in Arctic surface, cloud and radiation properties from space. *Science* **299**, 1725–7.

Wang, Y. J., Cheng, H., Edwards, R. L. *et al.*, (2001). A high-resolution absolute-dated late Pleistocene monsoon record from Hulu Cave, China. *Science* **294**, 2345–8.

Wang, Z. and Mysak, L. A. (2002). Simulation of the last glacial inception and rapid ice sheet growth in the McGill Paleoclimate Model. *Geophysical Research Letters* **29**, (23), 2102.

Wanner, H. and Luterbacher, J. (2002). The LOTRED approach – a first step towards a 'Palaeoreanalysis' for Europe. *PAGES Newsletter* **10**(3), 9–11.

Wardle, D. A., Walker, L. R. and Bardgett, R. D. (2004). Ecosystem properties and forest decline in contrasting long-term chronosequences. *Science* **305**, 509–13.

Wasson, R. J. and Claussen, M. (2002). Earth system models: a test using the mid-Holocene in the southern hemisphere. *Quaternary Science Reviews* **21**, 819–24.

Watson, A. J., Bakker, D. C. E., Ridgwell, A. J., Boyd, P. W. and Law, C. S., (2000). Effect of iron supply on Southern Ocean CO_2 uptake and implications for glacial atmospheric CO_2. *Nature* **407**, 730–3.

Watson, R. T. and the core writing team (eds.) (2001). Climate Change 2001: Synthesis Report: Contribution of Working Groups I, II and III to the Third Assessment Report of the Intergovernmental Panel on Climate Change. Cambridge, Cambridge University Press.

Watts, W. A., Allen, J. R. M., Huntley, B. and Fritz, S. C. (1996). Vegetation history and climate of the last 15 000 years at Laghi di Monticchio, southern Italy. *Quaternary Science Reviews* **15**, 113–32.

Weaver, A. J., Saenko, O. A., Clark, P. U. and Mitrovica, J. X. (2003). Meltwater pulse 1A from Antarctica as a trigger of the Bølling–Allerød warm interval. *Science* **299**, 1709–12.

Weiss, H. (1997). Late third millennium abrupt climate change and social collapse in west Asia and Egypt. In Dalfes, H. N., Kukla, G. and Weiss, H. (eds.) Third Millennium BC Climate Change and Old World Collapse. NATO ASI Series, pp. 711–22.

Weiss, H. and Bradley, R. S. (2001). What drives societal collapse? *Science* **291**, 609–10.

Weiss, H., Courtney, M. -A., Wetterstrom, W. *et al*. (1993). The genesis and collapse of third millennium north Mesopotamian civilization. *Science* **261**, 995–1004.

WGBU (1996). World in Transition: Ways Towards Global Environmental Solutions. Annual Report 1995, German Advisory Council on Global Change. Berlin, Springer Verlag.

White, R. and Engelen, G. (1997). Cellular automata as the basis for integrated dynamic regional modelling. *Environment and Planning B: Planning and Design* **24**, 235–46.

Whittaker, R. J., Willis, K. J. and Field, R. (2001). Scale and species richness: towards a general, hierarchical theory of species diversity. *Journal of Biogeography* **28**, 453–70.

Wigley, T. M. L. and Raper, S. C. B. (2001). Interpretation of high projections for global-mean warming. *Science* **293**, 451–4.

Wilkinson, C. R. (ed.) (2000). *Status of Coral Reefs of the World 2000*. Townsville, Australia, Global Coral Reef Monitoring Network, Australian Institute of Marine Science.

Willard, D. A., Cronin, T. M. and Verardo, S. (2003). Late-Holocene climate and ecosystem history from Chesapeake Bay sediment cores, USA. *The Holocene* **13**, 201–14.

Wilson, E. O. (1988). Biodiversity. Washington DC, National Academy of Science.

Wilson, K. (2000). Global warming and the spread of disease: the debate heats up. *Trends in Ecology and Evolution* **15**, 488.

Wintle, A. G. (1993). Luminescence dating of Aeolian sands: an overview. In Pye, K. (ed.). The Dynamics and Environmental Context of Aeolian Sedimentary Systems. London, Geological Society, Special Publication No. 72. pp. 49–58.

Wintle, A. G., Clarke, M. L., Musson, F. M., Orford, J. D. and Devoy, R. J. N. (1998). Luminescence dating of recent dunes on Inch Spit, Dingle Bay, southwest Ireland. *The Holocene* **8**, 331–9.

Wohlfart, J., Harrison, S. P. and Braconnot, P. (2004). Synergistic feedbacks between open ocean and vegetation on mid- and high-latitude climates during the mid-Holocene. *Climate Dynamics* **22**, 223–38.

Wolff, E. W., Moore, J. C., Clausen, H. B. and Hammer, C. U. (1997). Climatic implications of background acidity and other chemistry derived from electrical studies of the Greenland Ice Core Project ice core. *Journal of Geophysical Research* **102**, 26325–32.

Wolfram, S. (2002). A new kind of science. Champaign, IL, Wolfram Media.

Woodroffe, C. D., Beech, M. R. and Gagan, M. K. (2003). Mid–Late Holocene El Niño variability in the equatorial Pacific from coral microatolls. *Geophysical Research Letters* **30** (7), 101–4.

Woods, J. J., Schloss, J. A., Mosteller, J. *et al*. (2000). Water level decline in the Ogallala Aquifer. A report on KWO-KGS contract 99–132, Kansas Geological Survey open-file report 2000–29B (v2.0). At /www.kgs.ukans/HighPlains/2000–29B/Decdir.htm.

World Resources Institute. (2000). World Resources: People and Ecosystems: The Fraying Web of Life. Oxford, Oxford University Press.

Wright, H. E. and Thorpe, J. (2003). Climatic change and the origin of agriculture in the Near East. In Mackay, A., Battarbee, R. W., Birks, H. J. B. and Oldfield, F. (eds.). Global Change in the Holocene. London, Arnold, pp. 49–62.

Wu, H., Guo, Z. and Peng, C. (2003). Land use induced changes of organic carbon storage in soils of China. *Global Change Biology* **9**, 305–15.

Wuebbles, D. J. and Hayhoe, K. (2002). Atmospheric methane and global change. *Earth Science Reviews* **57**, 177–210.

Wuebbles, D. J., Jain, A., Edmonds, J., Harvey, D. and Hayhoe, K. (1999). Global change: state of the science. *Environmental Pollution* **100**, 57–86.

Wunsch, C. (2004). Quantitative estimate of the Milankovitch-forced contribution to observed Quaternary climate change. *Quaternary Science Reviews*, **23**, 1001–12.

Xiao, J. L., Porter, S. C., An, Z. S., Kumai, H. and Yoshikawa, S. (1995). Grain size of quartz as an indicator of winter monsoon strength on the loess plateau of central China during the last 130 000 yr. *Quaternary Research* **43**, 22–9.

Xu, L. (1999). From GCMs to river flow: a review of downscaling methods and hydrological modelling approaches. *Progress in Physical Geography* **23**, 229–39.

Xu, Q. (2001). Abrupt change of mid-summer climate in central east China by the influence of atmospheric pollution. *Atmospheric Environment* **35**, 5029–40.

Yasuda, Y. (ed.) (2002). The origins of pottery and agriculture. New Delhi, Lustre Press.

Ye. D., Dong, W. and Jiang, Y. (2003). The northward shift of climatic belts in China during the last 50 years. *IGBP Global Change Newsletter* **53**, 7–9.

Yevich, R. and Logan, J. A. (2003). An assessment of biofuel use and burning of agricultural waste in the developing world. *Global Biogeochemical* **17** (4), 6-1–6-21.

Yokohama, Y., Deckker, P. D., Lambeck, K., Johnston, P. and Fifield, L. K. (2001). Sea-level at the last glacial maximum: evidence from northwestern Australia to constrain ice volumes for oxygen isotope stage 2. *Palaeogeography, Palaeoclimatology, Palaeoecology* **165**, 281–97.

Yu, Z. and Ito, E. (1999). Possible solar forcing of century-scale drought frequency in the northern Great Plains. *Geology* **27**, 263–6.

Zhang, K., Douglas, B. C. and Leatherman, S. P. (2003). Global warming and coastal erosion. *Climatic Change* **64**, 41–58.

Zielinski, G. A. (2000). Use of paleo-records in determining variability within the volcanism-climate system. *Quaternary Science Reviews* **19**, 417–38.

Zielinski, G. A., Mayewski, P. A., Meeker, L. D. *et al*. (1994). Record of explosive volcanism since 70 000 BC from the GISP 2 Greenland ice core and its implications for the volcano-climate system. *Science* **267**, 256–8.

Zillen, L. (2003). Century-scale Holocene geomagnetic field variations and apparent polar wander paths reconstructed from varved lake sediments in Sweden. In: Zillen, L. Setting the Holocene Clock Using Varved Lake Sediments in Sweden. Lundqua Thesis 50. Lund University.

Zillen, L. M., Wastegard, S. and Snowball, I. F. (2002). Calendar year ages of three mid-Holocene tephra layers identified in varved lake sediments in west central Sweden. *Quaternary Science Reviews* **21**, 1583–91.

Zolitschka, B. (1998). A 14 000 year sediment yield record from western Germany based on annually laminated sediments. *Geomorphology* **22**, 1–17.

Zolitschka, B. (2003). Dating based on freshwater- and marine-Laminated sediments. In McKay, A., Battarbee, R. W., Birks, H. J. B. and Oldfield, F. (eds.). Global Change in the Holocene. London, Arnold, pp. 92–106.

Zong, Y. and Chen, X. (2000). The 1998 flood on the Yangtze, China. *Natural Hazards* **22**, 165–84.

Selected websites with information on global change

Carbon Dioxide Information Center www.cdiac.esd.ornl.gov/trends/trends.htm

Department of Environment, Food and Rural Affairs, UK www.defra.gov.uk/environment/climatechange/

Global Change Data and Information System/US Global Change Research Program Gateway to Global Change Data and Information www.globalchange.gov/

Global Change Master Directory www.gcmd.gsfc.nasa.gov/

The National Academies – Global Change Website www.dels.nas.edu/ccgc/

UK Rivers www.ukrivers.net/climate.html

US Environmental Protection Agency www.epa.gov/globalwarming

US Global Change Research Information Office www.gcrio.org

US National Science Foundation – Global Change Research Programs www.geo.nsf.gov/egch/

WGBU German Advisory Council of Global Change www.wbgu.de/wbgu_home_engl.html

UK Climate Impacts Programme www.ukcip.org.uk/

Selected 'sceptical' websites

Capital Research Institute www.capitalresearch.org

Center for the Study of Carbon Dioxide and Global Change www.co2science.org

Competitive Enterprise Institute www.cei.org

Cooler Heads Coalition www.globalwarming.org/

George Marshall Institute www.marshall.org

Global Climate Coalition www.globalclimate.org/index.htm

Global Warming Information Center www.nationalcenter.org/Kyoto.html

John Daly www.johndaly.com/

Lavoisier Group www.lavoisier.com.au

Philip Stott www.probiotech.fsnet.co.uk/

Science and Environmental Policy Project (SEPP) www.sepp.org

Skepticism.net (with many links) www.skepticism.net/global_warming/

Skepticism.Net www.skepticism.net/faq/environment/global_warming/

Index

Page numbers in *italics* refer to figures.